To Orbit and Back Again
How the Space Shuttle Flew in Space

Davide Sivolella

To Orbit and Back Again

How the Space Shuttle Flew in Space

Published in association with
Praxis Publishing
Chichester, UK

Davide Sivolella
Aerospace Engineer
Thomson Airways
London Luton Airport
United Kingdom

SPRINGER–PRAXIS BOOKS IN SPACE EXPLORATION

ISBN 978-1-4614-0982-3   ISBN 978-1-4614-0983-0 (eBook)
DOI 10.1007/978-1-4614-0983-0

Springer Dordrecht Heidelberg London New York

Library of Congress Control Number: 2012945502

© Springer Science+Business Media New York 2014

All rights reserved. This work may not be translated or copied in whole or in part without the written permission of the publisher (Springer Science+Business Media, LLC, 233 Spring Street, New York, NY 10013, USA) except for brief excerpts in connection with reviews or scholarly analysis. Use in connection with any form of information storage and retrieval, electronic adaptation, computer software, or by similar or dissimilar methodology now known or hereafter developed is forbidden.
The use in this publication of trade names, trademarks, service marks, and similar terms, even if they are not identified as such, is not to be taken as an expression of opinion as to whether or not they are subject to proprietary rights.

Cover design: Jim Wilkie
Project copy editor: David M. Harland
Typesetting: BookEns, Royston, Herts., UK

Printed on acid-free paper

Springer is a part of Springer Science + Business Media (www.springer.com)

Contents

Foreword ix
Author's preface xi
Acknowledgments xv
Acronyms xvii

1. **A brain and mind for the Orbiter: the avionics system** 1
   Shuttle data processing system: familiarization 1
   Redundancy 13
   Backup flight software 23
   Crew operations 26
   C/W and on board system management 28

2. **A skeleton for the Orbiter: structure and mechanisms** 35
   Designing the Orbiter structure 35
   Shuttle Orbiter structure: fuselage 39
   Orbiter active vent system 48
   Payload bay doors 51
   Orbiter wing 61
   Body flap 66
   Vertical tail 66
   Robotic manipulator arm 67
   Orbiter structural maintenance 79

3. **Power to orbit: the main engines** 81
   Introduction 81
   Main propulsion system 81
   SSME propellant flow 83
   SSME main components 91
   SSME development 107
   SSME evolution 110
   SSME operations 114

vi Contents

4. Power to orbit: solid rocket boosters 125
   The rocket equation 125
   The search for a booster 127
   SRB structure 129
   Thermal protection 137
   Recovery system 138
   Propellant 140
   SRB thrust vector control 144
   SRB recovery and refurbishment 147

5. Shuttle propulsion: the external tank 157
   Introduction 157
   Ground processing 167
   Launch pad operations 171
   Thermal protection system 172

6. Maneuvering in space: the orbital maneuvering system and reaction control system 179
   Introduction 179
   Development 180
   OMS and RCS configuration 183
   Propellant storage and gauging 191
   Orbital maneuvering system engine 196
   Reaction control system thrusters 201

7. Heart and lung of the Orbiter: the environmental control life support system and electrical control system 207
   Environmental control life support system 207
   Waster collector system 221
   Fire detection and suppression system 225
   Electrical power system 226
   Extended duration Orbiter kit 234

8. The Orbiter's skin: the thermal protection system 237
   Introduction 237
   Reusable surface insulation tiles 239
   Flexible reusable insulation 249
   Reinforced carbon-carbon 250
   Orbiter TPS configurations 252
   The flight experience 263
   On-orbit TPS repair techniques 273
   TPS ground maintenance 288
   TPS flight testing 290

Contents vii

9. Auxiliary power unit and hydraulic system 291
Introduction 291
APU system description 292
Hydraulic system 297

10. Fundamentals of the Shuttle GNC 299
Guidance, Navigation & control 299
Coordinate systems 300
Shuttle navigation hardware 303
Shuttle flight control system hardware 313
State vector propagation 314
The Orbiter flight deck 320
Digital autopilot 332

11. The art of reaching orbit 339
Nominal ascent: first stage 339
Nominal ascent: second stage 352
Nominal ascent: displays 356
Control stick steering 359
Intact ascent abort modes 360
Contingency aborts 382

12. Orbital dancing 389
Maneuvering the Shuttle 389
Orbital rendezvous maneuvers: development 397
Orbital rendezvous maneuvers: operations 417
Orbital flight rules 445

13. Returning home 447
Falling from orbit 447
Atmospheric entry 454
Terminal area energy management 465
Approach and landing 473

Glossary 495
Bibliography 497
Index 499

Foreword

Davide Sivolella has created an extraordinary exposition on the Space Shuttle System; it is everything that he set out achieve and more. I got into it and couldn't get out of it. Despite having assisted in the development of the Shuttle, having taken it into space six times, and all in all worked on the program for twenty years, I read on with curious excitement to discover how and why I flew it the way I did, and how and why the vehicles responded the way they did. This book took me beyond my knowledge of the system; it could have served very nicely as my textbook for flying the Shuttle.

It is massively detailed and massively technical, so isn't for the faint of heart, but that is okay because it is as I have referred to it, an exposition; it is fully explanative all the way through. It is always the *what*, *how*, and *why* of the technology, the *engineering* and the *operations*. It is not a history book, but the history it contains is wonderful; the history is focused on *why* and *how* the Shuttle got to be, *why* it is what it is, and *why* it is flown the way it is. It is a terrific marriage of engineering design, flying characteristics, and operations.

Davide has produced a treatise that could serve as a university textbook on the engineering and operation of spacecraft and space systems. But the real magic of this textbook is that it would illustrate all of its principles in real world terms, providing a case study of a machine which really flew. It would be a classic in scenario-based knowledge and learning, why it was done this way and how it turned out. Davide has chosen the perfect vehicle because the Shuttle is so complex and demands such state-of-the-art and perfect engineering that all the lessons and principles are illustrated, and again the illustrations are from real world experience and data.

The book is details, engineering, technology and operations, but Davide narrates the story with his heart as well as his mind. He does not just tell, he teaches; he is a marvelous mentor who is led by his reader to explain and to illustrate every concept. The book is filled with timely pictures, illustrations, schematics, and diagrams. It is complemented with the ever-present list of acronyms – how could we have a space program without that special language!? It has an extensive and relevant glossary, bibliography, and index. It creates wonderful parallels between the Shuttle systems and biological systems to clarify and enrich one's comprehension of highly technical concepts. Davide is a true mentor, and he has made the huge effort to see that his reader can 'get it'. The book is encyclopedic in detail but is done with a respect and tender-loving-care for its reader.

In conclusion, Davide Sivolella has produced a unique masterpiece. The history is not just chronological, it is the how and why the system ended up the way it did. It is a wonderful marriage of design, requirements, engineering, operations and outcomes. And most importantly, though filled with detailed engineering and technology, it is readable and comprehensible because Davide is at heart a teacher with great empathy who produced a great narrative exposition on the Space Shuttle System.

Story Musgrave
May 15, 2013

Author's preface

Having been born on 31 July 1981, just short of four months after the maiden flight of *Columbia*, I consider myself to be a child of the Space Shuttle. Despite being too small to remember any news about the program during its early years of operations, I do recall reading over and over again an astronomy book that contained a chapter on human space exploration. While I spent hours looking at pictures of Mercury, Gemini and Apollo capsules, and of the Skylab space station, there was one picture that etched an everlasting mark in my mind and heart. It was a schematic of a white winged space vehicle with two huge doors open to expose a cylindrical module (I would later learn that this was the European Spacelab) and a battery of instruments, telescopes, antennas, and so on. What was so captivating about that schematic, was that it showed the inside of the vehicle, its structure, engines, tanks, crew quarters, etc. I spent hours contemplating that colorful schematic, pondering the purpose of everything that it illustrated. That was my first encounter with that magnificent bird that was the Space Shuttle.

As I was growing up, my interest in astronautics became ever more passionate. Back then, I did not know a word of English; I did not have any computer, let alone Internet. However, whenever I could, I watched news about Space Shuttle missions on television, recording as much as I could on VHS so that I could play it again and again. Meanwhile, I began to develop an engineering mind which, in combination with my passion for human space exploration, made me eager to explore the "how" of astronautics. In particular, I wanted to learn in detail about the Space Shuttle; its structure, on board systems, flight procedures, and so on. But still there was not much material available.

The illumination on my way to Damascus, so to speak, came when I started to study aerospace engineering with a view to becoming a rocket scientist. I was then nineteen, was finally studying English, and had an Internet broadband connection. I was now able to read and understand a myriad of workbooks, handbooks, manuals, and technical papers about the things that NASA had achieved; Space Shuttle first and foremost. Recalling my earlier frustration at not having had anything to explain the Space Shuttle in a captivating fashion and with technical correctness, I decided that one day I would write my own book about the Space Shuttle.

Author's preface

Many books have been written about the Space Shuttle, so it is fair to ask why I should add another one. To answer this question, we must look at what is currently available on the subject. Browsing Google readily shows that most of the books are focused on the history of the program, its origin, development, and missions flown. Some books have been written by astronauts, and provide personal insight into the program and the thrill of riding this wonderful bird. But very few pages are devoted to explaining the technology of the Space Shuttle. How did it fly in space? How did it return to Earth? How was electrical power produced on board? Often a very basic description of the systems is provided, together with a quick description of the main components and phases of flight. But this barely scratches the surface! A few books do provide a more detailed explanation, but it is dry and boring, essentially only a system component list or, worse, a mere copy-and-paste from technical manuals or workbooks. A machine as complex as the Space Shuttle really cannot be described in a few pages! To try would be to do an injustice to the thousands of engineers that designed it. On the other hand, unless you have an engineering-oriented mind and a real fondness for the subject, you would very quickly be put off by the first pages of a paper describing the main engine ignition sequence or a rendezvous checklist.

I therefore offer this book about the technology of the Space Shuttle as a bridge to span the vast gap between the books that are too basic and the technical literature that is too detailed. Actually, it is the book that my younger self would have eagerly devoured when yearning to find out everything about the Space Shuttle.

In this regard, the book is split into two logical sections. The first nine chapters explain the systems of the Space Shuttle. The remaining chapters explain how the vehicle reached orbit and accomplished complex tasks such as rendezvous with the Hubble Space Telescope or the International Space Station, prior to returning home. Despite maintaining the necessary technical correctness, the text is written with a narrative style, and draws upon remarks and explanations by individuals who either designed or flew the Space Shuttle. In particular, the text has been arranged in such a manner as to obviate the need to consult a later paragraph. In the rare instances in which this has not been possible, footnotes provide an appropriate road sign.

Bear in mind that this book is *not* an account of the development of the Space Shuttle, and nor is it a log of each mission flown. Other books serve these roles. My objective was to explain in the simplest manner and with technical correctness, how it was possible to fly the Space Shuttle and how it worked. Where I do discuss the development of the Space Shuttle, it is to explain the rationale for choosing a given system configuration.

Although the Space Shuttle was, strictly speaking, an assembly consisting of a winged Orbiter whose belly was linked to a huge external tank that was flanked by two pencil-like solid rocket boosters, in common parlance the Orbiter is referred to as the Space Shuttle. In this book, these two terms are used in an interchangeable manner.

One important feature of this book is the great number of pictures. These are not present simply to fill the space available. On the contrary, in most cases the pictures are instrumental to understanding the adjacent text. So if you find a topic difficult to understand, do study the nearby pictures. I am confident that your queries will be answered immediately.

Author's preface xiii

This book is based on material contained in manuals, workbooks and checklists used by the astronauts themselves during their training and in flight. Unlike them, I did not have the benefit of an instructor to provide answers to my queries. I spent hundreds of hours studying everything until I was able to put all the pieces together and provide an accurate account. But if you spot a mistake, please write to me via the publisher so that, together, we can make the appropriate correction in a future edition.

That said, let me wish you an enjoyable read in the hope that you will find the material informative.

Davide Sivolella
January 2013

Acknowledgments

As every astronaut will readily admit, without the work of thousands of often sadly forgotten people behind the scenes, their missions could not even leave the drawing board.

Similarly, although there is only my name on the cover of this book, it was only with the help of several behind-the-scenes people that I was able to accomplish my mission. Several truly deserve to be mentioned due to the paramount role that they played. First and foremost is space historian David M. Harland, whose tips, advice, and suggestions early on were instrumental in understanding how a book should be written. His assistance continued with the editing of the manuscript, for which I will always be very grateful. I am particularly grateful to Clive Horwood of Praxis, who, even though I did not have previous writing experience, trusted me sufficiently to support this project right from the signing of the contract through to its completion. Again, I will be always very grateful to him for having allowed this project to come into existence. It was author Paolo Ulivi who suggested that I contact Praxis. I must give credit to W. David Woods, whose book *How Apollo Flew to the Moon* was a great inspiration. I fondly remember an evening spent with him and David Harland discussing my wish to write an "engineering book" about the Space Shuttle. And of course thanks to Story Musgrave for his Foreword.

Most of the pictures are NASA copyright, but I must also thank Chris Bergin, administrator of the nasaspaceflight.com website whose L2 forum has given me access to invaluable material, such as manuals and workbooks on the inner working of the Space Shuttle. I am also grateful to him for permission to use some pictures.

My thanks also go to Andrew L. Klausman for patiently answering my questions about the Space Shuttle avionics system, and to Ben Evans, Jim Kirkpatrick and Robert Adamick for their positive reviews of my project during the contracting process. Other people have offered continuous support and encouragement. First of all, my biggest supporter, Monica, the woman who is now my wife, has always been very supportive. My parents Pasquale and Maria have supported my passion for space exploration since I was little, and encouraged me to become an aerospace engineer. Finally, there are friends and colleagues. I will name Simon Delaney but there were many more whose kind words of support and interest in this project will never be forgotten.

Acronyms

AA: Accelerometer Assembly
ACLS: Augmented Contingency Landing Site
ADC: Analog-to-Digital Converter
ADI: Attitude Director Indicator
ADTA: Air Data Transducer Assembly
AFRSI: Advanced Flexible Reusable Surface Insulation
AGL: Above Ground Level
AHMS: Advanced Health Management System
AHSS: Aft Heat Shield Seal
ALT: Approach and Landing Test
AOA: Abort Once Around
APDS: Androgynous Peripheral Docking System
APU: Auxiliary Power Unit
ARS: Atmosphere Revitalization System
ASA: Aerosurface Servoamplifier
ASI: Augmented Spark Igniter
ATCO: Ambient Temperature Catalytic Oxidizer
ATCS: Active Thermal Control System
ATO: Abort To Orbit
ATVS: Ascent Thrust Vector Control
AVS: Active Vent System
BCE: Bus Control Element
BDS: Caribbean island of Bermuda
BFS: Backup Flight Software
BITE: Built-In Test Equipment
BLT: Boundary Layer Transition
BTU: Bus Terminal Unit
CA: Contingency Abort
CAIB: Columbia Accident Investigation Board
CAM: Crew Annunciator Matrix
CCTV: Closed-Circuit Television

CIPAA: Cure-In-Place Ablative Applicator
COAS: Crew Optical Alignment Sight
CPM: Cell Performance Monitor
CPU: Central Processing Unit
CRT: Cathode Ray Tube
CS: Common Set
CSS: Control Stick Steering
CTV: Crew Transport Vehicle
C/W: Caution and Warning
DAP: Digital Autopilot
D&C: Display and Controls
DI: Direct Insertion
DOLILU: Day-Of-Launch I-Load Update
DOP: Diver Operated Plug
DPS: Data Processing System
DK: Display Keyboard
EAS: Equivalent Airspeed
ECAL: East Coast of North America
ECLSS: Environmental Control Life Support System
EDO: Extended Duration Orbiter
EDW: Edwards Air Force Base (California, USA)
EI: Entry Interface
EMI: Electromagnetic Interference
ET: External Tank
EVA: Extra Vehicular Activity
FC: Flight Critical bus
FCMS: Fuel Cell Monitoring System
FCOS: Flight Computer Operating System
FDA: Fault Detection Annunciator
FEM: Finite Element Method
FES: Flash Evaporator System
FOD: Foreign Object Debris
FRCI: Fibrous Refractory Composite Insulation
FRSI: Felt Reusable Surface Insulation
GGVM: Gas Generator Valve Module
GNC: Guidance, Navigation and Control
GPC: General Purpose Computer
GPS: Global Positioning System
GRTLS: Glided Return To Launch Site
GSE: Ground Support Equipment
GUCA: Ground Umbilical Carrier Assembly
HA: orbit apogee
HAC: Heading Alignment Cone
HEPA: High Efficiency Particulate Atmosphere
HHL: Hand-Held LiDAR

HLL: High Level Language
HMC: Health Management Computer
HMF: Hypergolic Maintenance Facility
HP: orbit perigee
HPFTP: High Pressure Fuel Turbopump
HPOTP: High Pressure Oxidizer Turbopump
HPU: Hydraulic Power Unit
HRSI: High temperature Reusable Surface Insulation
HSI: Horizontal Situation Indicator
HST: Hubble Space Telescope
HWT: Heavy Weight Tank
HUD: Head-up Display
ICC: Intercomputer Communication bus
IDS: Image Dissector Tube
IDP: Integrated Display Processor
IGS: Inner Glide Scope
IMU: Inertial Measurement Unit
INRTL: Inertial
IOP: Input/Output Processor
ISS: International Space Station
IV&V: Independent Verification and Validation
ITVC: Intensified Television Camera
KSC: Kennedy Space Center (Florida, USA)
LCD: Liquid Cristal Display
LCS: Laser Camera System
LDEF: Long Duration Exposure Facility
LDO: Loads and DOLILU Officer
LDR: Laser Dynamic Range Imager
LiDAR: Light Detection and Ranging
LPFTP: Low Pressure Fuel Turbopump
LPOTP: Low Pressure Oxidizer Turbopump
LRSI: Low temperature Reusable Surface Insulation
LVLH: Local Vertical Local Horizontal
LWT: Light Weight Tank
LMP: Large Main Parachute
MA: Master Alarm
MAF: Michoud Assembly Facility
MC: Memory Configuration
MCC: Mission Control Center
MDM: Multiplexer/Demultiplexer
MDU: Multifunctional Display Unit
MECO: Main Engine Cut-Off
MEDS: Multifunctional Electronic Display System
MEP: Minimum Entry Point
MFB: Major Function Base

MIA: Multiplexer Interface Adapter
MLS: Microwave Landing System
MM: Major Mode
MMH: Monomethyl Hydrazine
MMU: Mass Memory Unit
MPS: Main Propulsion System
NASA: National Aeronautics and Space Administration
NAV: navigation
NC: phasing maneuver
NCC: corrective combination maneuver
NEP: Nominal Entry Point
NH: height adjustment maneuver
NOAX: Non-Oxide Adhesive eXperimental
NOR: Northup Flight Strip (New Mexico, USA)
NPC: plane change maneuver
NRS: coelliptic maneuver
NSI: NASA Standard Igniter
NSP: Network Signal Processor
NSW: Nose Wheel Steering
NTO: Nitrogen Tetroxide
OBSS: Orbiter Boom Sensor System
OEX: Orbiter Experiment
OGS: Outer Glide Scope
OMS: Orbital Maneuvering System
OPF: Orbiter Processing Facility
OPS: Operational Sequence
ORBT: Optimized R-Bar Targeted
PAD: Propellant Acquisition Device; Preliminary Advisory Data
PAL: Protuberance Air Load
PAPI: Precision Approach Path Indicator
PASS: Primary Avionics Software System
PCMMU: Pulse-Code Modulation Master Unit
PCS: Pressure Control System
PCT: Post Contact Thrust
PEG: Powered Explicit Guidance
PET: Phase Elapsed Time
PFD: Primary Flight Display
PL: payload
POHS: Position Orientation Hold Sub-mode
POR: Point of Resolution
PPA: Powered Pitch Around
PPD: Powered Pitch Down
PPO2: Partial Pressure Oxygen
PRSD: Power Reactants Storage and Distribution
PRTLS: Powered Return To Launch Site

PTU: Pan-Tilt Unit
RA: Radar Altimeter
RCC: Reinforced Carbon-Carbon
RCRS: Regenerable Carbon dioxide Removal System
RCS: Reaction Control System
REF: Reference
RGA: Rate Gyro Assemblies
RPM: R-Bar Pitch Maneuver
RPOP: Rendezvous and Proximity Operation Program
RPTA: Rudder Pedal Transducer Assembly
RHC: Rotational Hand Controller
RS: Redundant Set
RSI: Reusable Surface Insulation
RTLS: Return To Launch Site abort
RTV: Room Temperature Vulcanizing
S&A: Safe and Arm
SAS: Space Adaptation Syndrome
SBTC: Speed Brake and Thrust Controller
SI: Standard Insertion
SIP: Strain Isolation Pad
SILTS: Shuttle Infrared Leeside Temperature Sensor
SLI: Structural Load Indicator
SLWT: Super Light Weight Tank
SM: System Management
SMP: Small Main Parachute
SRB: Solid Rocket Booster
SRMS: Shuttle Remote Manipulator System
SSME: Space Shuttle Main Engine
SSST: Solid State Star Tracker
SSTO: Single Stage To Orbit
STA: Shuttle Tile Ablator; Shuttle Trainer Aircraft
STS: Space Transportation System
SWT: Standard Weight Tank
TACAN: Tactical Air Navigation
TAL: Transoceanic Abort Landing.
TAEM: Terminal Area Energy Management
TAOS: Thrust Assisted Orbiter Shuttle
TCS: Trajectory Control System
TDRS: Tracking and Data Relay Satellite
Ti: Target intercept
TIG: Time of Ignition
THC: Translational Hand Controller
TMR: Triple Modular Redundant
TPI: Terminal Phase Initiation
TPS: Thermal Protection System

T-RAD: Tile Repair Ablator Dispenser
TRS: Teleoperator Retrieval System
TVC: Thrust Vector Control
TZM: Titanium Zirconium Molybdenum
UCD: Urine Collector Device
UPP: User Parameter Processing
USAF: United States Air Force
WCS: Waste Collector System
WIX: Wait For Index
WONG: Weight On Nose Gear
WOW: Weight On Wheel
WP: Waypoint
WSB: Water Spray Boiler
VAB: Vehicle Assembly Building.

1

A brain and mind for the Orbiter: the avionics system

SHUTTLE DATA PROCESSING SYSTEM: FAMILIARIZATION

A "Swiss-knife" computer for the Shuttle

As the human body cannot live and function without the pumping action of the heart, the data processing system (DPS) formed the active heart of the Space Shuttle, for without it the Orbiter simply could not fly. Events such as external tank separation, jet firings, main engine cutoff, communications, and miscellaneous other functions, were so complex and time-critical that only by using computers were they feasible. Even manual control of ascent and re-entry would have been impracticable without computers, since the manual inputs provided by the pilots needed to be elaborated by the computers to produce the desired effect. At a higher level, the DPS performed tasks essential to flying the vehicle (guidance, navigation and control, or GNC), monitoring on board systems (system management, or SM), and both transmitting telemetry to Mission Control and enabling Mission Control to command on board systems. Owing to this "Swiss-knife" character of the Shuttle computers, they were normally called general purpose computers or GPCs.

For reasons that will be explained in the following paragraphs, five GPCs formed the brain of the DPS. In an era in which we are used to knowing which are the most common computer brands and manufactures, the computers used on the Shuttle are hardly known to the general public. In order to lower research and development costs of the Shuttle program, NASA wanted an off-the-shelf computer system. If "space rating" a system involved stricter requirements than a military standard, starting with a military-rated computer would make the next step in certification a lot easier and cheaper. Therefore, in the early 1970s, only two computers for aircraft avionics under development were potentially suitable for the new spaceship: the IBM AP-101B (a derivative of the technology that was already in use by various military and NASA flight programs) and the Singer-Kearfott SKC-2000 (which at that time was under consideration for the B-1 "stealth bomber" program). But both would clearly require extensive modification for use in space. The IBM machine was

2 A brain and mind for the Orbiter: the avionics system

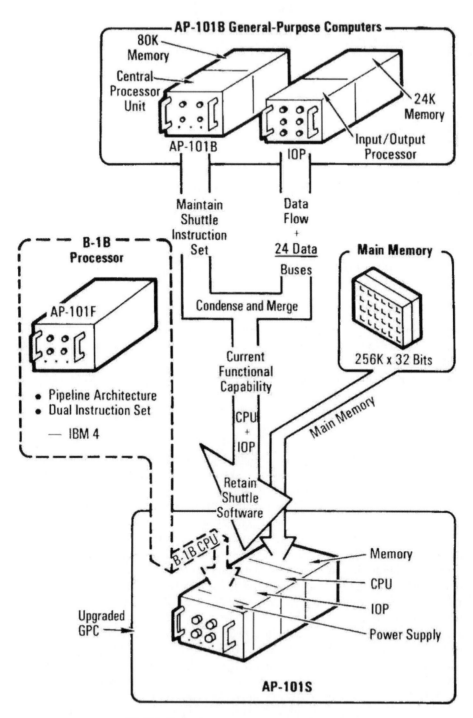

AP-101 General purpose computer schematics.

selected because of the company's success with developing the computers for the Saturn V moonrocket and the Skylab space station, whose systems bore a slight similarity to the avionics configuration planned for the Shuttle. In modern terms, the processing power of the GPC was ridiculously inferior to even the least powerful desktop computer that one can buy, but compared to what was available for space applications back then, they were cutting-edge technology.

Each computer consisted of a central processing unit (CPU), an input/output processor (IOP), one megabyte of memory and various other components housed in an electromagnetic interference (EMI) hardened case. While the CPU performed the instructions to control on board systems and manipulate data, the IOP formatted and transmitted commands to the systems, received and validated response data from the systems, and maintained the status of the interfaces between the CPU and the other computers. In other words, while the CPU was the "number cruncher" the IOP did all the interfacing with the rest of the computers and vehicle systems. The computers were able to perform their functions by control logic embedded in a combination of software and microprogrammed hardware.

Within a few years of initiating the AP-101B design in January 1972, it became evident that an improved GPC would be required. Studies for upgrading the existing AP-101B started in January 1984, and they culminated in the mid-1990s with the introduction into service of the AP-101S. From a configuration point of view, the big difference was that the new computers incorporated the CPU and the IOP in a single avionics box, halving the weight and size, and also reducing the power requirements. From a performance point of view, this upgrade provided 2.5 times the memory capacity and up to three times the processor speed with minimum impact on flight software. For instance, while the old GPCs were capable of 400,000 operations per seconds, the new ones could perform up to 1,000,000 operations per second.

The Shuttle nerves

As in the human body, in which the brain communicates its commands and receives information by means of an extensive network of nerves, the Shuttle computers could communicate with all the on board systems and payloads via discrete signal lines and serial digital data buses. While the discrete signal lines transmitted signals indicating a binary condition such as the position of a given switch or circuit breaker, the data buses transmitted bulk information and data regarding the status of all the systems.

The choice to use data buses was taken early in the program, in a period in which the aviation industry was already starting to implement this configuration in the latest jets. Because sensors, control effectors and associated devices would be distributed all over the Orbiter, the weight of the individual wires required to carry all the signals and commands needed for operating all of its elements would have been prohibitive. In response, the use of multiplexed digital data buses was investigated and baselined. Generally speaking, a data bus physically consists of a pair of insulated wires twisted together and then electrically shielded, and it permits data transmission from a large number of sources on a time-sharing basis to a single or perhaps multiple receivers.

4 A brain and mind for the Orbiter: the avionics system

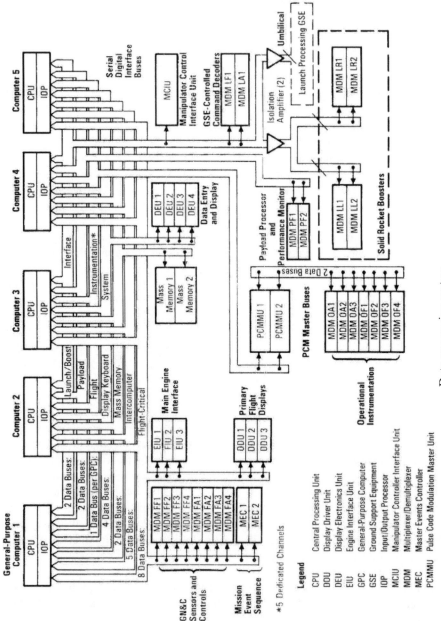

Data processing system.

The Orbiter's data bus network comprised 28 data buses allocated by functional use, criticality, and traffic load, into seven different categories.

Eight data buses belonged to the category of flight-critical buses (FC) since they carried all the data and command traffic associated with guidance, navigation, flight control, mission sequence, and management of critical non-avionics functions.[1] An important part of a space flight is enabling Mission Control to analyze the status of the vehicle and its payload. In the Orbiter, two pulse-code modulation master units (PCMMU) received data from the on board instrumentation and payload as well as from the five GPCs via individual instrumentation/PCMMU data buses. Once in the PCMMU, the data were formatted in operational downlink, which was sent to one of two network signal processor (NSP). In the NSP the operational downlink was combined with the on board recorded voice for transmission to the ground by either the S-band or Ku-band communications systems. The GPCs were linked to the displays and keyboards on the flight deck by four display/keyboard (DK) data buses. And two launch data buses that were used mainly for ground checkout and launch phase activities served as an interface for data gathered from the solid rocket boosters, and once in space they provided an interface with the controller for the remote manipulator system. These buses differed from the others in that they required isolation amplifiers to accommodate the long wire runs to the launch processing system and to isolate the buses when disconnected at liftoff and at solid booster separation. Two payload data buses (PL) provided an interface for payload support operations, system management functions, payload bay door control, and communications antenna switching.

Five intercomputer communication (ICC) data buses connected each GPC to its four counterparts to enable them to communicate among themselves. Interestingly, these buses operated in a slightly different way to all the others. In general, each GPC would "request" data using a specific data bus to prompt the appropriate hardware device to provide that data to the requesting GPC over the same data bus. On the ICC buses, each GPC transmitted to its counterparts without receiving a request for data. In this way, each computer was able to continually know what the others were doing, and this enabled them to remain synchronized.

Finally, the mass memory (MM) data buses allowed each GPC to retrieve flight software from one of the two mass memory units (MMU). In the initial version, each MMU was a coaxially mounted reel-to-reel digital magnetic tape storage device for GPC software and Orbiter systems' data and it could be written to or read from. The tape was 602 feet long, 0.5 inch wide and had nine tracks, eight of which were data tracks and the ninth was a control track. Each track was also divided into files and subfiles for finding particular locations. Later in the program, the tape recorders were replaced by modular memory units that were faster, had greater capacity, and used a solid-state mass memory and a solid-state recorder for serial recording and dumping of digital voice.

---

[1] The reason for allocating eight data buses for flight-critical data was a combination of the requirement for fault tolerance and the need to spread the traffic load.

6 A brain and mind for the Orbiter: the avionics system

Just as the synapses act as ports for exchanging information between neurons in the brain, the GPCs had dedicated connections to enable them to communicate with the digital data buses and hence with the outside world. In fact, each computer had 24 so-called BCE/MIA which acted as input/output ports. How many ports a computer ought to have was the subject of much discussion in the early design phase. At that time the total system bus traffic density was known only to a first approximation, and the catastrophic effects on the system of exceeding the 1 Mbit/sec bus limit provided the motivation to build in a significant margin. The uncertainty in this area, and the desire for functional isolation, resulted in the greatest number that could reasonably be accommodated in the computer input/output processor: twenty-four.

Inside each computer, data were transmitted in parallel along 18-bit buses but on the buses data were transferred in serial form at a 1 MHz rate, so it was necessary to have a "translator" between these different forms of data. The multiplexer interface adapter (MIA) converted serial data into parallel data for the CPU of the computer and vice versa. The other part of the port, the bus control element (BCE), was a microprogrammed processor which could transfer data back and forth between a GPC's memory and the MIA.

The GPCs sent and received commands and data to and from hardware known by the generic name of bus terminal units (BTU), and a multiplexer/demultiplexer (MDM) was one example. Each MDM was connected to a given number of on board sensors, from which it received data to transmit to the GPCs via one of the two data buses to which it was connected. At the same time, an MDM transmitted to a specific sensor the data and commands provided by the GPCs commanding it. In other words, the GPCs were not connected to each individual sensor or system distributed across the Orbiter, as that would have been impractical in terms of wiring and data handling. That is why the serial digital data bus network was implemented. It can be visualized as a sort of tree, where the nourishment is brought from the roots to each single leaf by a network of vessels which branch ever more diversely. In the same manner, the GPCs (the root of the tree) sent commands to the MDMs via the digital data buses (a main branch), from which the information was sent to the specified sensor (a leaf). In the Orbiter though, this was a two-way path.

Each MDM converted and formatted serial digital GPC commands into separate parallel discrete digital and analog commands for various vehicle hardware systems. This was demultiplexing. The opposite process of multiplexing was converting and formatting the discrete digital and analog data from vehicle systems into serial digital data for transmission to the GPCs.

Each MDM included two redundant MIAs, which worked in the same manner as the GPC's ports. Each MIA was part of a redundant channel inside the MDM, which included a sequence control unit (SCU) and an analog-to-digital (A/D) converter. The SCU split the commands provided by the GPCs and directed them to the proper input/output modules (IOM), usually referred to as "cards", for forwarding to one of the subsystems to which a card was connected. The cards available in an MDM were specific to the hardware components accessed by that type of MDM. For this reason, a flight-critical MDM and a solid rocket booster

MDM (for example) were not interchangeable. In fact, flight-critical MDMs could only be swapped amongst themselves. This could be performed in flight, if doing so would restore access to a critically needed piece of Orbiter hardware. The SCU also assembled all the inputs from the various IOM cards into a single bit stream to be sent to the GPCs. The A/D converted any analog input data to digital form prior to being multiplexed by the SCU. The use of MDMs made the system very flexible, in that sensor devices could be added with only minor changes to the MDMs and the flight software.

Thirteen of the 20 MDMs on the Orbiter were incorporated into the DPS. They were connected directly to the GPCs, and were named and numbered by reference to their location in the vehicle and hardware interface. The other seven were part of the vehicle instrumentation system and sent instrumentation data to the PCMMUs to be included in the telemetry transmitted to the ground. The DPS MDMs consisted of flight-critical forward (FF) MDMs 1 through 4, payload (PL) MDMs 1 and 2, and GSE/LPS[2] launch forward (LF1), launch mid (LM1), and launch aft (LA1). One or two flex-MDMs (FMDMs) could also be connected to the PL data buses, depending on the payload needs of a particular flight. Of the seven operational instrumentation MDMs, four were located forward (OF1 to OF4) and three on the aft fuselage (OA1 to OA3).

A mind for the Shuttle: the primary avionics software system

If the five GPCs, 24 data buses and 20 MDMs represented the brain and nervous system of the Orbiter, its mind was the primary avionics software system (PASS). Simply put, PASS contained all the programming needed in order to fly the vehicle through all flight phases and to manage all vehicle and payload systems. Due to the vast number of functions that PASS was required to perform, its code was divided into two major groups: system software and application software.

System software was analogous to an operating system running on each GPC. As such, some of the main functions that it performed were to control GPC input/output, load in new memory configurations, keep track of time, assign computers in the role of commanders and listeners on specific data buses, and exercise the logic involved in transmitting commands over these buses at specific rates. The system software comprised three modules. In particular, the flight computer operating system (FCOS) controlled the processors, monitored vital system parameters, allocated computer resources, provided orderly program interruptions for higher priority activities, and updated computer memory. The system control program initialized each GPC and also arranged for multiple GPC operation during flight-critical phases. Flight crew commands or requests were processed by the user interface program.

While the system software served as a housekeeper, the application software was

---

[2] Ground support equipment/launch processing sequencer.

8 A brain and mind for the Orbiter: the avionics system

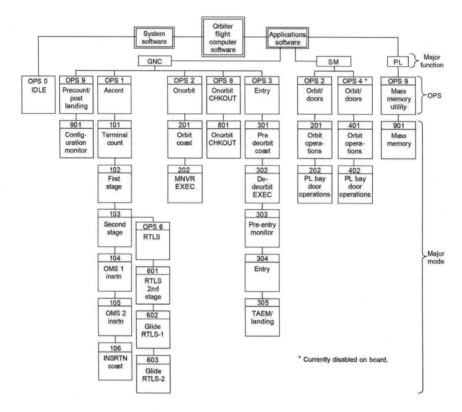

Orbiter flight computer software.

that part of PASS which performed the actual duties required to fly and operate the Orbiter. At this point, it is important to note that the development of the GPCs and PASS were pursued in parallel, and it was soon realized that the memory available in each computer would be insufficient to store all of the flight software that was being developed. This had two important consequences: firstly, the addition to the DPS of mass memory units, and secondly the need to divide PASS into a number of small software modules, some of which would be used only during specific flight phases. This division was organized in three levels.

On the first level, PASS was divided into three so-called major functions, defined as follows:

1. Guidance, Navigation and Control (GNC): This major function had all the software necessary to perform flight-critical functions such as navigation sensor management, control of aerosurfaces for maneuvers, and trajectory calculations.
2. System Management (SM): All non-avionics systems were managed via this major function, including the electrical, environmental and communication systems. It also contained payload-related software.

3. Payloads (PL): Despite its name, this software did not support operations with the payload during flight. It was only used when preparing the vehicle at KSC, to load content into the MMUs. This major function was said to be unsupported, which meant that at any given time it was not being processed by any of the GPCs.

On the second fragmentation level, each major function was then split into a given number of submodules, referred to as operational sequences (OPS), each containing the instructions to control a particular phase of the flight. In its turn, on the third level of fragmentation, each operational sequence comprised a series of submodules called major modes (MM) which provided the instructions for a specific portion of a given mission phase.

Owing to the need for the flight crew to interact with the software on a daily basis for checking software configuration, monitoring on board system status, providing instructions for undertaking mission-related maneuvers, and so on, PASS offered a series of additional software blocks for each OPS to generate data and information to show on the displays on board. The highest priority blocks were linked to the major modes and generated the so-called major mode displays, or base pages, that provided information on the current portion of the mission phase. At the same time,

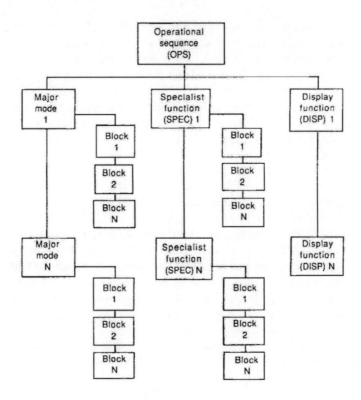

OPS substructure.

the crew could interact with this data via keyboard entries. Sequencing from base page to base page could be initiated manually by keyboard entry from the crew or, in some cases, automatically in response to a specific event or condition detected by the software.

The second-highest priority software blocks were called specialist functions, and generated SPEC display pages that enabled the crew to monitor the operations of the Orbiter. As with the base pages, the SPEC displays could be altered by the crew via keyboard entry, but unlike the base pages they could be recalled only with keyboard entry. It is important to remember that the difference between base pages and SPEC pages was that the former allowed monitoring and alteration of the primary functions within an OPS, but the latter applied only to secondary or background functions.

The lowest priority software blocks carried out so-called display functions, that is to say they generated display pages, known as DISP, on which data and information were shown but could not be modified by the crew. The DPS Dictionary, a document of several hundred pages, provided a detailed description of every possible display (base pages, SPECs and DISPs) that a crew member might require to access.

Once again, it is worth remembering that this seemingly complicated subdivision of the primary software derived from the scarce memory capability of the processors when the Shuttle was designed.

When a new flight phase was initiated, the appropriate software had to be loaded into the GPCs in a process called OPS transition, where all major modes of an OPS were loaded into GPC memory. Once an OPS was loaded, the crew could manually initiate the new phase of the mission. The operational sequence and major mode transitions were generally performed by the flight crew, but during ascent and the final part of re-entry all major mode transitions were carried out automatically by software because these phases of the flight were very critical and the workload on the crew was already intense. Irrespective of how they were made, all transitions had to be "legal", meaning that several preconditions had to be satisfied. Of these, the most interesting was that the transition had to be logical: for example there would be no need to transition the GPCs back to the terminal countdown software from any post-launch major mode. The system would refuse to perform such a transition.

It is important to note that at any given moment, only one OPS could be present in a GPC's memory. However, this was not true for ascent, where the GPCs running PASS had both the OPS for a nominal ascent and that for the Return To Launch Site (RTLS) launch abort mode. Because of the need to maintain the Orbiter stable and controllable during the transition from nominal ascent to RTLS, there would not have been enough time for loading the PASS software for the abort. For this reason it was already loaded and ready to be activated immediately if the need arose.

The memory configuration (MC) comprised the combination of system software and application software for a specific operational sequence. When new application software was to be loaded, the crew had to recall from the memory units one of the eight memory configurations available. In order to improve the redundancy provided by the two memory units, a copy of that portion of software concerning re-entry (OPS3) was also stored in the upper memory of each GPC running PASS, in

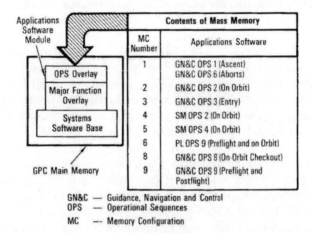

Application software memory configuration.

the so-called G3 archive. In this way the software was always available, ready for an emergency re-entry. In addition, this archive allowed a quick loading of re-entry software in the GPC lower memory for executing a Transoceanic Abort Landing (TAL) and Abort Once Around (AOA) during ascent without wasting time loading it from the memory units.

In an effort to continuously optimize the utilization of the scarce memory space available, and also to preserve vital data during software transitions, the application software was further split in two components named major function base (MFB) and OPS overlay. The MFB was the application software that was common to all major modes of a given major function. For example, each operative sequence in the GNC major function used a different scheme for calculating the Orbiter's state vector,[3] but in transitioning from one OPS to the next it was essential to maintain the information about the current state vector. This data would be contained in the MFB for GNC. An MFB would also contain portions of flight software that were common to all of the OPS of a given major function. During an OPS transition, only that part of the code that was not common to the other OPS of the major mode would be loaded, which is why it was called OPS overlay.

HAL

In Stanley Kubrick's movie *2001: A Space Odyssey* the crew of *Discovery One* on course to Jupiter are attacked by a sentient computer named HAL that is unwilling to be deactivated.

The Orbiter of the Shuttle had its own HAL, but there was no danger of the crew

---

[3] The state vector is a vectorial representation of the position and velocity of a spacecraft along its orbit. See Chapter 10 for details.

being overpowered. HAL was the name of the programming language used to write the application software for the Shuttle. Prior to the development of the Shuttle, flight software for space applications was written in assembly language, or something close to that level. Generally speaking, assembly language is very powerful since it allows a strict control of the processor memory and registers, making the code optimized for the specific computer on which it is to run. The downside, however, is the lengthy, expensive and difficult development due to the nature of the language, which uses a complex syntax and grammar that is susceptible to errors. Furthermore, software in assembly language is specific to the machine, it is not portable. To overcome this, in the 1960s high level languages (HLL) were invented with a syntax and grammar that was closer to human programmers than machines. This characteristic makes a HLL a powerful tool, because it facilitates faster development and the code is much easier to read and modify. Of course, the computer still needs to receive instructions in its own binary code language, a task carried out by a translator known as a compiler. In this way, software written in a HLL is created independently of the computers on which it will run, making it portable and flexible. The downside of a HLL language is that it does not permit a programmer to directly manipulate the processor's memory and registers, slightly penalizing the performance.

Having written the flight software for Apollo in assembly language, and realizing that the flight software for the Shuttle would be considerably more elaborate, NASA opted for a HLL. However, because none of the languages available at that time were optimized for real-time computing, it was decided to develop a new one specifically for this kind of task. HAL not only supports vector arithmetic, it can also schedule tasks according to programmer-defined priority levels. Because NASA directed the development of the language from the beginning, it strongly influenced its final form and specifically the way in which it could handle real-time processing. In developing HAL, NASA adopted a syntax and grammar similar to that which programmers were already accustomed to, and provided a variety of tools that could be used for creating real-time programs.

It seems that the name HAL had nothing to do with the heuristically programmed algorithmic computer of *2001: A Space Odyssey*. Some have suggested that it is an acronym for Higher Avionics Language, but it may simply derive from an engineer named Hal who was involved in the early development.

The development by NASA of HAL was criticized by managers in a community used to assembly language systems. They felt that it would have been better to write optimized code in assembly language rather than produce less efficient software by a high level language. To settle the controversy, NASA told two teams to race against each other to produce some test software with one team using assembly language and the other using HAL. The running times of the software written in HAL was only 10 to 15 per cent longer than its counterpart in assembly language. It was therefore decided that the system software would be written in assembly language, since it would be modified only very rarely, and the application software and the redundancy management code would be written in HAL.

It is worth mentioning that unfortunately HAL did not succeed in NASA's initial intention of making it the primary programming language for space applications. Its only other programs were for the Jupiter-bounded Galileo mission and some ground applications of the Deep Space Network. It was abandoned in favor of Ada, another language optimized for real-time programming, developed in the mid-1970s by the Department of Defense for military applications and which is still in widespread use.

REDUNDANCY

When *Columbia* arose in the skies above Florida on 12 April 1981, it marked the first time that a new spaceship was test flown on its maiden launch with a human crew on board. Even if, with insight, this seems to have been a gamble, it must be admitted that it was possible thanks to the incredible reliability of the avionics system.

For the Apollo program, the issue of providing a reliable data processing system was addressed by building a special-purpose computer with an incredible high level of expensive quality control. For the Shuttle NASA, facing budget cuts, decided to employ off-the-shelf hardware as much as possible. To compensate for the reduced reliability that this would mean, a new architecture was devised for incorporating into the hardware network the redundancy and reliability needed to enable a crew to safely fly the inaugural flight.

The concept of redundancy was not new to NASA – the guidance system of the Saturn V used triple modular redundant (TMR) circuits, meaning that there was one computer with redundant components. This was also the philosophy implemented for the Apollo spacecraft, but the near-fatal Apollo 13 mission showed that extensive damage elsewhere in the vehicle could disable its computer. One of the many lessons learned from that mission was that by spreading redundancy among several simplex circuit computers distributed around the spacecraft the effects of such catastrophic failures could be minimized. For the Skylab space station, along with TMRs, it was decided to install two identical computers, each of which was capable of performing all the functions of the mission. Only one of these computers was active at any time, the other was switched off but available for immediate use in the event of a problem involving the first. The disadvantage of this system was that the computer which stepped in would have to find out where its partner had left off by referring to the contents of a 64-bit transfer register in the common section built with TMR circuits. This would have required some time, but it would not have been a problem since for Skylab the computers were not responsible for navigation or high frequency flight control functions. In a failure, it would have been permissible for the attitude of the vehicle to drift temporarily without causing a serious problem. For the Shuttle, things were much more complex. In this regard, it is worth examining in depth the mission and vehicle design drivers that dictated the overall system architecture.

Mission-derived requirements

The significant differences between the Shuttle and previous spacecraft included the requirement for much more complex and extensive on-orbit operations in support of a much wider variety of payloads, and the requirement to make precisely controlled unpowered runway landings. Along with the longstanding NASA rule that a mission must be aborted unless at least two means of safely returning to Earth were available, these requirements profoundly affected the design approach. To illustrate, previously the concept of safe return could be reduced to a relatively simple backup process, like a second set of pyrotechnics for extracting the parachutes or using more parachutes than were necessary for a nominal splashdown. So relatively simple backup systems were developed which, although less effective than the primary operational system, would nevertheless comply with the mission rule of assuring a safe re-entry. For the Shuttle this approach was inadequate. Atmospheric entry through final approach and landing imposed a performance requirement on the systems of the Orbiter as severe as any mission phase, meaning that a backup system with reduced performance was not feasible. Simply put, because the complexity of the re-entry required much more maneuvering than previous capsules, the backup systems of the Orbiter had to be capable of performing the same operations as the primary systems in order to ensure a safe return, albeit probably with reduced precision.

Also, the economic impact of frequently aborting missions on a user-intensive program such as the Shuttle meant that to abort after suffering a single failed system would be unacceptable. Therefore, a comprehensive fail operational/fail safe (FO/FS) philosophy was applied to all systems. For the avionics, this requirement meant that it had to remain fully capable of performing the operational mission after any single failure (fail operational) and capable of returning safely to a runway landing after any two failures (fail safe).

Another constraint derived from experience on previous programs concerned the use of built-in test equipment (BITE) as a means of detecting component failure. This requirement was justified by the many recorded cases of BITE circuitry failures that had led to false doubts of the operability of a unit. Again this was unacceptable for the Shuttle due to the high annual rate of missions which it was intended to fly. The much preferred method of fault detection, which was the one chosen, was to compare actual operational data produced by one device or subsystem with that produced by devices or subsystems operating in parallel and performing the same function.

Vehicle-derived requirements

The Orbiter was an unstable airframe that could not have been flown manually even for the brief ascent/re-entry aerodynamic phases without full-time control stability augmentation. Although considered early in the program for post-entry aerodynamic flight control, cable/hydraulic boost systems were eliminated owing to their weight and mechanization difficulties, and instead an augmented fly-by-wire approach was baselined.

Digital flight control systems were successfully used in the Apollo program and NASA was well aware of their advantages, so digital flight control was baselined for the Shuttle. However, the full-time augmentation requirement placed the digital flight control computation system in the safety-critical path, which in turn dictated a high degree of redundancy.

The control authority necessary to achieve all the Shuttle vehicle requirements, particularly during ascent and re-entry, created a situation in which a control actuator hard-over command, issued erroneously, could cause structural failure and the loss of the vehicle if the command were to be allowed to remain in effect for as little as 10 to 400 milliseconds depending on the mission phase. This situation affected the design in at least two important ways. First, it imposed a requirement for actuator hard-over prevention irrespective of the failure condition. Second, because of the reaction time required, it eliminated any reliance on direct manual intervention from consideration in the reaction to a failure, in turn requiring a fully automatic redundancy fault-down approach. The concept adopted to prevent hard-overs was to use hydraulic actuators with multiple command inputs to a "secondary actuator". These secondary actuator inputs were hydraulically force-summed, and the resultant command was sent to the so-called "primary or power actuator", which was nothing less than a massive steel rod connected to the aerosurface. If one of the inputs diverged from the rest, as in the event of an erroneous hard-over command, the effect of its secondary stage output would be overpowered by the other secondary stage outputs and the control effector operated correctly. To make such a system work, multiple independently computed commands to the secondary actuator inputs had to be provided.

These mission and vehicle requirements led to an avionics system that relied on coupled parallel multi-strings, tight synchronization, and redundancy management to accommodate any failure that could jeopardize a safe re-entry and divergence of the commands.

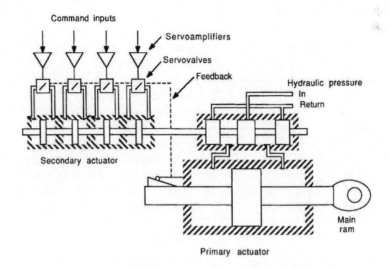

Typical actuator scheme.

16 A brain and mind for the Orbiter: the avionics system

Looking for the right network architecture

As very often happens in any branch of engineering, the problem of finding the best avionics architecture to satisfy the requirements of the Shuttle avionics gave rise to several different redundancy management schemes; in this case three.

The first scheme was to run a number of totally independent sensors, computers and actuator strings.[4] But this approach had a fatal flaw. Consider a scheme that had two independent strings at a critical point of the mission: two equal but opposite commands could be issued. In landing for example, one string might issue the elevon pitch-down command and the other one might issue the elevon pitch-up command, potentially resulting in loss of control of the vehicle. Another flaw in this design was that an analysis had to be made of what values were reasonable for every sensor, and how an average should be defined. It was also hard to set a tolerance level that would reject bad data whilst not losing good data near that limit. A multi-independent-string system, furthermore, would not be very fault tolerant. If the computer commanding a given string were to be lost, then all the sensors and effectors connected to that string would be irretrievably lost.

Attention switched to the so-called master/slave scheme, in which one computer would be in charge of reading all the sensors and the other computers would be in a listening mode, gathering information. The problem with this scheme was the time it would require for a backup computer to take over if the master computer failed. For some very critical flight phases, there could be as little as four-tenths of a second of reaction time from when the master failed until the control of the vehicle began to be lost. A particularly critical phase was the final flare immediately prior to touchdown, when it was necessary to command an extremely rapid excursion of the elevons so as to touch down at the proper rate of descent. If the master failed just at that moment, it would have been impossible for the crew to switch to the backup computer in time. Another critical point was about 60 seconds into the ascent, where the aerosurfaces had to be moved to relieve the aerodynamic pressure on the wings. If a failure of the master occurred just at this moment, again the inability of the flight crew to switch to the backup computer sufficiently rapidly would have caused the loss of the vehicle. One way to overcome the slow reaction time of the flight crew would have been to arrange for an automatic switchover, but this raised the prospect of a faulty computer erroneously jumping into the automatic switchover code and seizing command of the vehicle. This was not considered a likely failure, but there was sufficient concern to rule out the master/slave scheme.

The scheme that was adopted was a distributed command approach in which all of the computers process the same information simultaneously, yet remain closely synchronized in order to implement a rapid switchover. The first question that had to be addressed in pursuing this approach was, very simply, how many computers? This had to be considered along with the level of the redundancy required to satisfy

---

[4] Generally speaking, a string comprises a GPC and all the units it directly commands over the flight-critical data buses.

the mission-driven requirement of a fail operational/fail safe system. In order to comply with these requirements it would be necessary to incorporate quadruple redundancy involving four computers, each with an independent string for the same information. In this way, a minimum of three strings would guarantee identification of a diverging or disabled unit based on the comparison of actual data produced by devices running in parallel carrying out the same functions, thereby satisfying the "fail operational" requirement. A fourth string would allow for a second failure to occur without losing control of the vehicle, thereby satisfying the "fail safe" requirement.

Nevertheless, the final system had not four computers but *five*. In the beginning, to achieve the desired level of safety, the requirement was set of guaranteeing a "fail operational/fail operational/fail safe" approach. This would have meant incorporating five computers. However, reliability projections for fly-by-wire aircraft had shown that triple computer system failures were expected to cause loss of an aircraft three times in a million flights, whereas quadruple computer system failures would do so only four times in one thousand million flights! The cost considerations in terms of equipment and time led NASA to lower its requirement to fail operational/ fail safe, which allowed the number of computer to be reduced to four. Since five computers were already procured and designed into the system, the fifth machine was kept with the initial intention of being loaded with the system management software. Then it was decided to add the system management functions to the primary software, releasing the fifth computer to serve as a repository for the backup flight software which was at that time under development.

Strangely enough, for the orbital flight tests there were *six* computers! In fact, as the first Shuttle flights loomed, Arnold Aldrich, in charge of the Shuttle Office at the Johnson Space Center, wrote a memo arguing for a sixth computer to be carried as a spare. He pointed out that because 90 per cent of avionics component failures were expected to be computer failures, and a minimum of three computers and the backup should exist for a nominal re-entry, aborts would then have to take place after one failure. By carrying a spare computer preloaded with re-entry software, the primary system could be brought back to full strength. And indeed the sixth computer was dubbed "re-entry in a suitcase" and carried on the early flights.

Synchronization and redundancy management

In essence, redundancy in the Orbiter's avionics relied on the fact that each computer could perform all the functions necessary for a particular mission phase. But for true redundancy it was required that each computer be able to listen to all of the other computers on all of the buses (even though each computer commanded only a few of the buses) so that they could be aware of all the data generated in the current phase. Furthermore, all of the computers had also to be able to process data at the same time as the others. To preserve redundancy, a failed computer had to be able to "drop out" without causing any functional degradation. To achieve all this, it was necessary to devise a means of synchronizing all of the computers.

In the beginning, the Shuttle's designers thought it would be possible to run the redundant computers separately and then just compare answers periodically to make sure the data and computation matched. This turned out to be a poor solution, as even small differences in the oscillators that acted as clocks within the computers could soon cause the computers to get out of step. The first step towards a solution was the proposal to synchronize the computers at their input and output points. This concept was later expanded to include synchronization at points of process changes, when the system transitioned from one software module to another.

In practice what happened was that all of the computers running some part of the PASS were in the so-called common set (CS), all communicating with one another, sharing the basic status information that they needed to know about each other over the ICC data buses 6.25 times per second. Typical information exchanged by this CS synchronization process were input/output errors, fault messages, GPC failure status, keyboard entries, memory configuration tables, system level display information, etc. Synchronization within the CS enabled all of its computers to perform regular checks to verify that they were all correctly executing the flight software.

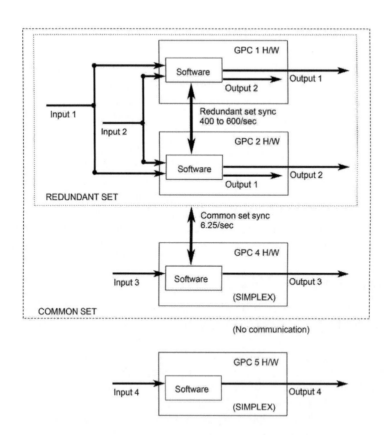

GPC synchronization. (Courtesy of www.nasaspaceflight.com)

Due to the criticality involved in flying the vehicle, the GNC major function was designed to run simultaneously in multiple GPCs, in what was called the redundant set (RS). These computers had to simultaneously execute the same part of the flight software with the same inputs. In this case the synchronization check to verify that each computer was at the same place in the flight software was carried out 400 times per second.

It is important to understand that the difference between common and redundant sets was the kind of information received. In a redundant set, all of the computers had to receive the same input and (owing to the tight synchronization) produce the same output. If one computer failed, then the others would have been able to process the software without interruption or degradation in their performance. Guidance, navigation and control would not be affected at all by a computer dropping out of the redundant set.

Synchronization of the redundant set worked liked this: when the software being processed by the GPCs received an input, delivered an output, or branched to a new process, it sent a 3-bit discrete signal and waited 4 milliseconds to receive similar discretes from the other computers. The discretes were coded to provide information beyond just saying "here I am". For example, 010 meant an I/O operation had been achieved without error, while 011 meant the opposite. If a computer either sent the wrong synchronization code or was late, the other computers detecting either of these conditions concluded that the computer had failed and thereafter refused to listen to it or acknowledge its presence, thereby removing it from the redundant set. Each GPC could vote itself out of the redundant set in the event of noting an error in the input received over a bus to which it was connected. This last function was very important, since in order to maintain consistently identical inputs for the GPCs in the redundant set, input transactions involving these computers had to be protected. This meant that if two or more GPCs in the redundant set failed to receive and process the input data, they would all ignore the data.[5] The protected transaction capability was maintained through the use of sync codes and certain I/O error processing techniques whereby if any member of the redundant set noted an I/O error then this information would be exchanged by all of the GPCs in the set. If more than one GPC detected an error, all the GPCs would stop listening to that unit or element. Special logic was incorporated to ensure that a single faulty GPC could not prevent the other computers in the set from receiving the data that they needed. If a single GPC was the only member of the set to detect the same I/O error two consecutive times, it would force itself to fail-to-sync and drop out the redundant set and possibly also drop out of the common set.

Computers in the common set could receive different input, run different software and issue different output, and by means of a less tight synchronization could verify that they were all in good health and able to properly process software. If a computer

---

[5] Typically, after the first detection of an input error, the transaction was tried again in the next data cycle.

20 A brain and mind for the Orbiter: the avionics system

failed a sync point, it was dropped out the common set. It is worth recalling that although protected transactions were used primarily in the redundant set, this philosophy also applied to transactions over the ICC buses in the common set. Again, if more than one GPC in the common set detected an I/O error on an ICC transaction, this information was shared with the other GPCs in order that they could reject the data from that ICC bus. Also, the same logic applied as in the other protected transactions. If a single GPC detected two successive I/O errors on an ICC transaction whilst the other GPCs did not, that GPC would fail-to-sync and vote itself out the set.

Fortunately, the software needed for redundancy management required only 5 to 6 per cent of a GPC's central processor resources. One reason why the redundancy management software was able to be kept so light was that NASA decided to move voting to the actuators, rather than before commands were sent over the data buses. To understand how this worked, it is necessary to remember that each actuator was quadruple redundant. Also, if a failed GPC issued an incorrect command, the commands of the good GPCs would prevail, physically voting down the erratic GPC and giving the crew time to remove it from the redundant set. The only serious possibility was that three computers would fail simultaneously, negating the effects of the voting. If the crew received proper warning, they could engage the backup flight system. It is important to remember that only the crew could physically remove a GPC from the redundant set. The option of having the software undertake this task was rejected in order to ensure that an unknown error in the software could not erroneously remove a computer that was not actually faulty.

Since only GNC could be run simultaneously on more than one computer, PASS SM and PL were run on only one computer in a configuration called "simplex". A computer running in simplex configuration could be in the common set (running for example SM) or not be part of the set at all, such as when it was in a "frozen" state or if it had been loaded with the backup flight software.[6] During ascent and re-entry, four computers formed the common and redundant set, since they were all loaded with the PASS GNC software. In noncritical mission phases such as on-orbit, the computers were reconfigured. Two were left in the redundant set to handle guidance and navigation functions, such as maintaining the state vector. A third machine was loaded with the system management software to control life support, power, and the payload. A fourth machine would be loaded with the descent software and powered down (in the jargon "freeze dried") for an emergency descent and to protect against a failure of the two memory units. Finally, the fifth computer held the BFS software.

---

[6] A computer that was loaded with the application software of a particular memory configuration was said to be "frozen" or "freeze-dried" if it did not process the software. The procedure was to place a computer in sleep mode after having been frozen, meaning it was in a condition where it drew only the power necessary to maintain the contents of its volatile memory. Flight procedures called for freeze-drying one computer (usually GPC 3) soon after reaching orbit and storing in its memory the re-entry software in order to have an additional source along with the memory units.

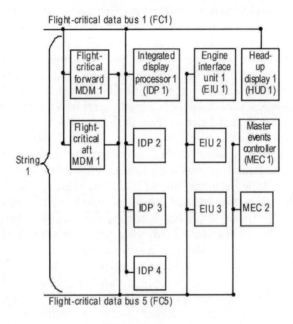

Components of string 1.

Management of the redundant set also involved the manner in which the GPCs communicated with the Orbiter's hardware. The digital data bus network had eight flight-critical buses. FC 1 to 4 connected the GPCs to the four flight-critical forward (FF) MDMs, the four flight-critical aft (FA) MDMs, the four integrated display processors, and the two head-up displays. FC 5 to 8 connected the GPCs to the same four FF MDMs and FA MDMs, plus the two master event controllers and the three main engine interface units. These FC buses were grouped to create four strings, each one composed of one FC data bus from each of the two FC-bus groups, along with a fixed distribution of hardware connected to them. If we consider a redundant set of four GPCs, each string was assigned to a computer and that GPC would act as master and commander over the flight-critical hardware elements of that particular string. That GPC would also passively listen to the commands and data transmitted over the other three strings, each commanded by one of the other GPCs of the set. In other words, when a GPC sent a request for data to the hardware on its string, the other GPCs would hear this and receive the same data returned on that string, but without having any power of ruling over that string. This transaction would occur in parallel with the other three strings, so that all GPCs in the set would get a copy of all of the data from all four strings.

For example, the DPS could be configured to have each GPC commanding one of the four aft right-firing reaction control system jets. Because of the redundancy set synchronization, identical inputs going through identical processing had to yield identical outputs. This meant that as the four GPCs independently executed the GNC algorithms from the same set of inputs, they would issue the same control command. If they decided that a small +Yaw correction was required, they would all issue the command to fire an aft right jet. Supposing that the jet with the highest priority was the –R2R[7] one and it was controlled by string 2, all four GPCs would issue the same command to fire the –R2R jet, but this would occur only in response to the command given by GPC 2, the computer that was commanding string 2. Although at first sight this might appear to be a complicated way of organizing and managing the hardware, it gave the DPS incredible power and flexibility not only by guaranteeing nominal mission operations in the event of one string being lost due to a GPC failure but also by allowing a safe return to Earth if a second string were lost. To better illustrate this, consider a failure occurring during ascent or re-entry. For both phases, the redundant set was composed of four GPCs running PASS GNC, each of them controlling only one-quarter of all the flight-critical hardware. If we suppose that GPC 1 fails, then we must consider string 1 to have been lost and with it that portion of the hardware that it was commanding. But the three good GPCs are still commanding the flight-critical hardware and the mission is able to continue smoothly and without any degradation in performance.

In the event of a GPC failure during ascent or re-entry, the flight rules allowed the crew to attempt a manual change of configuration, or restringing, in which one of the good GPCs was permitted to take command of the string of the failed GPC in order to restore full capability of the flight hardware. Although rehearsed during training, it is debatable whether this could have been done for real, owing to the already high workload on the crew in a very dynamic phase of the mission where their attention was devoted to verifying that the vehicle was flying the proper trajectory. However, restringing was performed without any problem on-orbit, when the redundant set was shrunk to only two GPCs. In this configuration, the two GPCs had to command two strings each.

If a GPC failed-to-sync and dropped out the redundant set then the remaining computers performed so-called "bus masking" to terminate the command/listening mode of the string commanded by the failed GPC. For example, for a nominal string assignment on ascent of GPC 1/string 1, GPC 2/string 2, etc., if GPC 3 failed-to-sync with GPCs 1, 2 and 4, then GPCs 1, 2 and 4 would mask FC buses 3 and 7 (string 3) commanded by GPC 3. Similarly, if GPC 3 was still processing software it would mask FC 1 and 5, 2 and 6, and 4 and 8 (strings 1, 2 and 4) commanded respectively by GPCs 1, 2 and 4. Bus masks of this type were reset as appropriate after actions such as an OPS transition or string reassignment. Thus, continuing the example, if string 3 was reassigned to GPC 4, GPCs 1, 2, and 4 would remove their bus masks on FC 3 and 7. Bus masking was also performed in nominal situations when a GPC was

---

[7] The reaction control system consisted of a complex arrangement of small jet thrusters located in the forward and aft fuselage sections of the Orbiter, and they were used for small maneuvers and attitude changes. When a given maneuver had to be performed, the fact that more than one jet could be used meant the flight software looked up tables of priority rankings that specified which jet would be used first in order to execute a given maneuver. See Chapter 6 for details of the reaction control system.

required neither to transmit nor to receive data on a particular data bus. For example, the GPC set running PASS GNC would perform a bus masking on the PL data buses on-orbit, when the PL buses were assigned to the SM GPC. In turn, the SM GPC would do the same for the FCs.

On 28 November 1983, *Columbia* lifted off launch pad 39A to begin STS-9, the first flight of the European Spacelab, a laboratory carried inside the payload bay that greatly enhanced the capability of the Orbiter to perform scientific experiments. The mission went very smoothly, and on 8 December the crew of six was about to bring home the rich crop harvested in the orbital fields. However, a surprise was in store. As Brewster H. Shaw, the pilot, recalls, "About the time that we were reconfiguring the computers, we had a couple of thruster firings ... and we got the big X-pole file on the CRT, meaning that the computer had failed. This is the first computer failure we had on the program ... So I get out the emergency procedures checklist ... We started going through the steps and everything. And in just a couple of minutes we had another one fail the same way, a firing of the jets and the computer failed." Due to the severity of the situation, Mission Control decided to waive off the first de-orbit opportunity in order to try to determine what was going on with the computers. The crew eventually managed to recover one of the failed GPCs, but surprisingly, "When the nose gear slapped down, one of the GPCs that had recovered failed again," Shaw remembers. An investigation was immediately started. This found that both GPCs failed for a trivial reason. As Shaw continues, "It turns out there were little, itty bitty slivers of solder that were loose in those two computers, and when those jets fired and the solder was floating in there, it made the solder sit down across two memory locations, changing the state of a memory location. The computer, which is always doing a self-test, sees this memory location change value and it says, 'Something's wrong. I'm outta there.' And it self-failed. And the same thing happened to two of those computers." No time was lost in putting into practice the lessons learned. Shaw continues, "So we went up to Oswego, New York, where IBM had a plant that built these computers ... and watched them do particle impact noise detection tests where they put microphones on the GPC box, then put it on a shaker and listened for loose particles inside. It became a standard screening criteria after that time."

BACKUP FLIGHT SOFTWARE

When the design of the Shuttle began, NASA had proven its mettle by successfully landing astronauts on our nearest celestial neighbor, the Moon. It was an engineering triumph. Often called the "fourth crew member", the Apollo spacecraft had a digital computer to perform all guidance and navigation tasks. For redundancy, there was an analog flight control system with both automatic and manual modes. In addition, a direct mode enabled the crew themselves to operate the maneuvering jets.

The Shuttle was a much more complex spaceship than Apollo, and for this reason it could be controlled only by means of digital computers. There was no possibility of an analog backup. Owing to its total reliance on computers, the synchronization and redundant management schemes were developed to provide a strong resilience to the hardware failures that might occur at any time in a flight. However, this strategy did not offer protection from errors embedded in the PASS software which, despite the redundancy scheme, might prove fatal. Software will never be 100 per cent free from latent bugs.

This concern arose when the Approach and Landing Test (ALT) program was being devised to study the aerodynamic and handling qualities of an Orbiter during gliding approaches using the first Orbiter built, *Enterprise*. In this case, it was highly desirable to have secondary flight software available for use if the primary software failed. Its function would be to stabilize the Orbiter long enough for the crew to bail out. The question was how to achieve this redundancy. The answer came in the form of three alternatives: (1) increasing the internal PASS redundancy, (2) duplicating the PASS in a version programmed by a different set of programmers isolated from the primary programmers, or (3) implementing a reduced-capability backup system by a semi-isolated set of programmers. The third option was chosen, since it achieved the additional measure of protection achievable within cost and schedule constraints. The first option was not pursued because it was felt that every practical internal measure had been taken by the PASS programmers. The second option was considered too costly and fraught with duplication of functions not required for a secondary system. The new software was named backup flight software (BFS) and was baselined as a simplified version of the primary software. To assure complete independence from the coding of PASS and so avoid the same latent errors, Rockwell was awarded the contract to develop BFS while PASS was being written by the Draper Laboratory of the Massachusetts Institute of Technology.

The first BFS developed for *Enterprise* and the ALT program was sufficient to hold the vehicle stable if the primary software were to lose control, allowing the crew to escape. As the ALT program neared completion, attention turned to the orbital test flights. This necessitated writing a new BFS, because in this case it was required to operate during both the ascent and descent portions of the mission and also be able to execute abort options. Thus BFS matured into a full guidance, navigation and control system that paralleled the primary software with full access to all sensor inputs and effector outputs.

BFS was organized in a similar manner to PASS, by being divided into system software and application software. The former managed timing, the PASS/BFS interface, uplink/downlink with Earth, engage/disengage control, and so on. The latter carried out GNC and SM functions. The GNC functions were the real soul of the BFS, since they were to take over after a failure in the PASS GNC, and either fly the vehicle into orbit or complete the descent. BFS GNC was not designed to manage orbital flight. In fact, the only functions available in this regard were for ascent, post-insertion and deorbit. System management functions were performed in the same way as PASS, as during launch and re-entry the four PASS GPCs were busy running GNC and did not have either the time or the resources to look after the other on board systems as well. BFS SM did that for them. On-orbit control of SM functions was restored to a GPC running PASS. Since BFS programming was much simpler than PASS, all GNC and SM functions could be executed together on a single computer. This simplicity also meant that all BFS would fit into that GPC's memory, eliminating the need to access an external memory.

Throughout launch and re-entry, BFS constantly listened to the data traffic over the flight-critical data buses in order to synchronize itself with PASS, ready to pick up at the exact point that PASS had reached when the crew decided to resort to the backup software. However, this synchronization had nothing at all to do with the synchronization enjoyed by the GPCs when running PASS GNC. In fact, the GPC with BFS was not in any redundant set, and by design PASS and BFS could not pollute each other with commands issued towards one another. This synchronization was just a way for BFS to monitor what PASS was doing. It was achieved by a three-step process known as "tracking".

In the first step, called "sync", BFS listened as PASS requested information or sent a command. In the second step, called "track", BFS had to receive important data transmitted over flight-critical data buses 5 to 8 by PASS specifically for it. This contained such things as the Orbiter's state vector. While receiving these data, BFS also verified that all four strings were providing the same information. If one string gave different values, BFS would assume that the GPC commanding that string had failed and would cease to listen to it. For BFS to track PASS properly, there had to be matching data on at least two strings. After these two steps, BFS then continuously carried out the third and final step of "listening". This was performed using the "wait for index" (WIX) mechanism by which BFS listened to all transactions occurring on the flight-critical buses but used only what it needed. Prior to sending a request to a unit on a given bus, PASS also transmitted a WIX "listening command" on the same bus. BFS compared this with a table index in its software. A positive match told BFS that it would need the data from the transaction that PASS was about to command, so it listened and retrieved the information. If the index was not in the table, BFS would ignore the transaction and await the index of the next PASS transaction.

If for some reason any of these three steps could not be performed, BFS would put itself into "standalone", a state which meant it would be unable to be engaged as backup software if the need arose. Since BFS needed to listen to PASS in order to be ready to take over, PASS had to be modified to make it more synchronous. The most important loss of data would be that from the inertial measurement units, since their outputs were essential for guidance and navigation during ascent and re-entry. Flight rules dictated that if BFS lost these data for more than 10 seconds during ascent or for more than 45 seconds during re-entry, it would become "no-go", meaning unfit to be engaged due to degradation of its state vector.

During ascent and re-entry the crew could view the information produced by BFS on any of their flight deck displays. If the need arose, BFS could be engaged simply by pressing a button on the hand controllers available to each pilot. Thanks to the tracking system described above, BFS could instantaneously take control without a dangerous transient. However, at no time during the Shuttle program was it ever necessary to engage BFS.

CREW OPERATIONS

Despite the complexity of the data processing system, the flight crew had available only a handful of controls for fully interacting with both software and hardware. In particular, a panel above the commander's seat on the left contained all the controls for activating, stopping, and loading system software into each of the five GPCs. In addition, three keyboards, one for each pilot and one on the aft station of the flight deck, enabled effective communication between crew and software. These keyboards were different from those of the Gemini and Apollo spacecraft because they were hexadecimal (base 16). Each was composed of a 4 × 8 matrix of 32 pushbutton keys and consisted of sixteen alphanumeric keys, two sign keys (+ and −), a decimal point, and thirteen special function keys. These pushbuttons enabled the crew to interact in a multitude of ways with both DPS hardware and software. To briefly mention a few, one of the first things a crew member might do would be to select a given display for either checking the status of a subsystem or changing a parameter value. The steps for selecting either a SPEC or a DISP page were as follows:

- *Depress the SPEC key.*[8]
- *Key in the SPEC or DISP number, omitting all leading zeros.* For example, a DISP number of 106 would be keyed in as "106" while a SPEC number 034 as "34". SPEC and DISP numbers always had three digits. Also, DISP pages had a number either starting with 9 as its first digit or ending with a digit in the range 6 to 9.
- *Depress the PRO key.*

Within a given selected display there could be as many as 99 sequentially ordered items or parameters that could be viewed or changed. The items that could be altered were identified by an item number that was a maximum of two digits placed in such a way that it was readily identifiable with the parameter to which it was associated. If the item numbering was obvious, the number could be implied and need not actually appear on the display. The keyboard operation to alter a parameter was referred to as an "item entry". In this regard, two different types of item entry could be performed: "toggle" and "data" entries. Toggling included selecting or deselecting a parameter, indicating or executing an action, and altering software configurations. The general procedure to execute the toggle item entry was as follows:

1. *Depress the ITEM key.*
2. *Key in the item number.*
3. *Depress the EXEC key.* At this point the parameter value was shown on the corresponding selected item line.

---

[8] This key was used for calling both SPEC and DISP pages.

Keyboard unit.

On the other hand, item data entries allowed the crew to load or change data in the software. Typical uses included initializing parameters, changing software limits, and specifying memory locations. The procedure to perform an item data entry was as follows:

1. *Depress the ITEM key.*
2. *Key in the item number.*
3. *Key in a delimiter ("+" or "−").* A delimiter served to separate item number codes from their corresponding data. The delimiter sign corresponded to the sign of the data to be used, but if no sign was associated with the data then it did not matter which delimiter was chosen. A "[]" after the data field on the display page indicated the item was sign-dependent. For item numbers that did not display brackets, the sign of the corresponding data was assumed to be positive. Hence to avoid improper data entry, it was good practice to always use the "+" key as a delimiter unless the item specifically required negative data.
4. *Key in the data.* As a general rule, leading and trailing zeros did not need to be entered.
5. *Depress the EXEC key.*

If it was necessary to undertake multiple data entries, the software allowed more than one data entry to be made with one command sequence. The procedure was the same as above except step 4, where consecutive data entries could be loaded by using a delimiter to separate each parameter. Item entries were incremented sequentially so that it was not necessary to enter the item number for each parameter following the one just entered. It was sufficient to hit another delimiter, and the next item number was ready to receive its associated data. To skip an item, the delimiter had to be hit twice. A sequence of item numbers could be skipped until the desired item number was reached. To illustrate: *ITEM 14 + 2 + + 7 EXEC* would mean that ITEM 14 was given the value 2, ITEM 15 was skipped, and ITEM 16 was given the value 7. It is worth remembering that multiple toggle entries were not permitted, as it would not make any sense to ask to view simultaneously more than one parameter. And since data entries were performed to modify the status or value of a parameter, they could be commanded and executed only in base pages and SPEC pages.

As the GPCs had such a limited memory, they had to be reconfigured on several occasions during a flight to process the software appropriate for that time. Hence at points during the mission it was necessary for the flight crew to perform what was called an OPS transition, loading and processing a new memory configuration in the lower memory of a GPC. This could be done with just a three-step keyboard entry as follows:

1. *Depress the OPS key.*
2. *Key in the three digits of the desired OPS.* While the first digit specified the OPS, the next two digits specified the major mode. Generally, a new OPS was called from mass memory by its first major mode.
3. *Depress the PRO key.*

If the transition was commanded successfully, the GPC(s) targeted for the OPS transition would stop processing their old software and load all the major modes of the new memory configuration, then begin to process the new operational sequence. In the time during which the new software was transferred, a GPC was in a so-called "pseudo OPS 0" state because it was temporarily in the OPS 0 state processing only system software. While loading new software, the GPCs were in transition, doing critical checks and processing. It was therefore imperative that these processes not be interrupted with inputs or switch throws. The rule was "everyone should be hands-off during OPS transition".

Once an OPS transition had been completed, it might have been necessary for the flight crew to perform an MM transition to change the major mode processed in the current OPS. This was done by the same sequence of keyboard inputs as for an OPS transition but, as explained earlier, during an OPS transition all of the major modes supported by that OPS were loaded into the GPC so it was not necessary to load new software during a major mode transition. A successful OPS or MM transition would be immediately indicated to the crew with the display of the appropriate OPS or MM page on one of the available displays.

C/W AND ON BOARD SYSTEM MANAGEMENT

Amongst other things, one of the goals of the Shuttle program was to lower operating costs by eventually reducing the size and scope of the ground support team required to run a space mission. Furthermore, with the Air Force promising to use the Shuttle for so-called "black missions" (secret operations such as retrieval or destruction of an enemy satellite), independence from ground control was mandatory. For this type of secret mission, the ideal would have been to have the crew perform their mission fast and without any particular assistance from the ground via communications that could easily be heard by the enemy. It soon became apparent that the only way to achieve such on board autonomy without overtaxing the crew would be to automate as many of the system monitoring tasks as possible. Since the computational requirements for this could only be grossly estimated, a tradeoff study was conducted to determine the relative merits of an integrated approach versus a separate, independent computer dedicated to system management. A corollary issue was the data acquisition process. On previous programs a great deal of data was telemetered to the ground, but only the information required by the crew to operate the spacecraft was made available on board. For the Shuttle, it would be necessary to make all the required data accessible on board as well.

These requirements resulted in a significant portion of the primary software being routines dedicated to the complex tasks of monitoring and providing the flight crew with interfaces to the Orbiter support systems, payloads, and the remote manipulator system. This software was called system management (SM) and it constituted one of the three major functions into which PASS was split. However, the designers did not want the management of such a complex machine as the Shuttle to be in the hands of software alone. The SM software was therefore accompanied by a hardware-based monitoring system called the caution and warning system (C/W) that made use of direct measurements and/or sampled the sensors and transducers without any software pre-elaboration. The C/W can be thought of as a coin with one face representing the hardware portion of the alerting system and the other face representing the software portion, with both relying upon the features of the other. If the avionics system could be likened to the brain, then the SM and C/W were the autonomic nervous system, functioning largely below the level of consciousness to control the visceral functions.

Categories of out-of-limit situations

Since not all out-of-limit situations have the same level of criticality, all the possible alarms were organized in four different classes, each requiring different actions to be undertaken by both the C/W and the flight crew.

Fire/smoke in the cabin and rapid cabin depressurization are the two most serious critical situation that could arise during a mission, so they were classified as Class 1 alarms and treated as Emergencies. There was no Class 1 alarm during the Shuttle program, but if it had occurred, the crew would have been immediately alerted by an aural alarm consisting of a siren activated by the smoke detection system or a klaxon activated by rapid loss or cabin pressure.

Class 2 alarms, simply dubbed Caution and Warnings, were associated with the monitoring of the parameters for many of the Orbiter's systems. They had their share of action during the program. Due the high number of parameters to be monitored and the need to promptly detect an out-of-limit situation, a hardware system (called Primary C/W) and software system (called Backup C/W, or further abbreviated to B/U C/W) were implemented. It should be noted, however, that the Primary C/W and Backup C/W operated in a manner that can be thought of as being symbiotic.

Alerts with a lower priority fell into Class 3, and they were managed only by the system management software to inform the flight crew of a situation leading up to a Class 2 alarm or an issue that would involve a long procedure (over five minutes) to fix. Alarms in this class were annunciated by a steady tone of predefined duration generated in the C/W electronics when activated by inputs from the computers. At the lowest level were the Class 0 alarms, also known as Limit Sensing. These were managed exclusively by software. A visual indication on DPS displays was provided to indicate when a given parameter was out of its operating range. These alarms were the only ones not annunciated to the flight crew by means of an aural alarm, since their only purpose was to inform the crew of possible problems with parameters that were not flight-critical.

To conclude this brief discussion of out-of-limit conditions, it is important to note the Class 1 and Class 2 Primary C/W alarms were exclusively driven by hardware, whereas all the other alarms were detected and managed by the system management software. This duality in the caution and warning system was simply due to the lack of faith in an exclusively software-controlled annunciation system alarm. In this way, the most important flight-critical parameters could be monitored by both hardware and software.

C/W hardware

Despite the different classes of alarm, a common feature for all but Class 0 was the crew annunciator matrix (CAM), an array of 40 light annunciators, each driven by one or more parameters. Each annunciator was illuminated by two bulbs that were easily replaceable, wired in parallel for redundancy. Their intensity could be adjusted by a dedicated brightness/variable switch. Once an annunciator had been lit, it would not be extinguished until all of the parameters driving it were back within limits. The only exception was the annunciator related to out-of-limit conditions detected by the Backup C/W system. The BACKUP C/W ALARM annunciator would stay on until the crew had acknowledged to the software that the alarm had been seen[9] and that recovery actions were under way. If the crew desired to avoid sudden distractions or excessive glare from the annunciator lights at key points in the mission, for example during a night landing, there was also the option of commanding the matrix not to illuminate when an alarm was raised.

One well-known feature was the master alarm (MA) pushbutton, which went off along with an aural sound to alert the flight crew to the occurrence of an out-of-limit condition.

---

[9] This was simply done by pressing the ACK button on the DPS keyboard.

C/W and on board system management

O₂ PRESS	H₂ PRESS	FUEL CELL REAC (R)	FUEL CELL STACK TEMP	FUEL CELL PUMP
CABIN ATM (R)	O₂ HEATER TEMP	MAIN BUS UNDERVOLT (R)	AC VOLTAGE	AC OVERLOAD
FREON LOOP	AV BAY/ CABIN AIR	IMU	FWD RCS (R)	RCS JET
H₂O LOOP	RGA/ACCEL	AIR DATA (R)	LEFT RCS (R)	RIGHT RCS (R)
	LEFT RHC (R)	RIGHT/AFT RHC (R)	LEFT OMS (R)	RIGHT OMS (R)
PAYLOAD WARNING (R)	GPC	FCS (R) SATURATION	OMS KIT	OMS TVC (R)
PAYLOAD CAUTION	PRIMARY C/W	FCS CHANNEL	MPS (R)	
BACKUP C/W ALARM (R)	APU TEMP	APU OVERSPEED	APU UNDERSPEED	HYD PRESS

Crew annunciator matrix.

The Primary C/W was hardware driven with the capability of monitoring up to 120 inputs, most of them received from transducers through signal conditioners. The remaining inputs were either received from the GPCs or were utilized by the C/W electronics for internal self-testing. In this last case, an internal failure would have illuminated the PRIMARY C/W annunciator on the CAM matrix. Inside a PROM memory in the C/W electronic unit, upper and lower limits for each parameter were stored and compared at a frequency of 80 hertz with the true values sensed. If after eight consecutive samplings the sensed value was outside its predetermined range, the parameter was considered to have been tripped and the master alarm light and tone were activated.

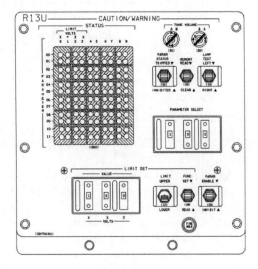

Panel R13U.

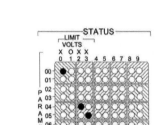 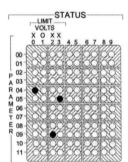 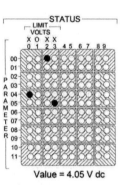

Limit values. (Courtesy of www.nasaspaceflight.com)

For a deeper interaction with the Primary C/W, the crew had at their disposal the controls of panel R13U, by which they could verify and change primary C/W limits, inhibit and enable parameters, and check their status. The main element of this panel was the STATUS light matrix by which, in a rather elaborate manner, all 120 of the monitored parameters could be displayed along with their upper and lower limits.

During the C/W hardware development, it was deemed desirable to use a single engineering measurement to represent all parameters. Because programs at that time used transducers with signals ranging from 0 to 5 volts, it was decided to use voltage as the common unit for all parameters when their values had to be displayed on the STATUS matrix. Hence the input transducer ranges were rescaled to match 0 volts as the minimum value and 5 volts as the maximum value. A calibration table was then provided to enable the crew to convert a parameter from this voltage reading into the appropriate engineering unit, and vice versa.

To read a parameter limit, it was first necessary to select the parameter of interest, specify whether the upper or lower limit was required, then read the position of the lights on the STATUS matrix along the first four columns, grouped under the LIMIT VOLTs marker. Values from 0 to 5 in increments of 0.5 could be shown. The value of each decimal point was the number written in the PARAMETER column. The second column was always blank to represent the decimal point.

System management software

Programmers assigned three different functions to the system management software. The first was a basic SM function whose task was to process and monitor data acquired from the various vehicle and payload subsystems, to provide fault detection and warning in the event of an anomaly. This triggered Class 2 Backup C/W alarms, and Class 0 and Class 3 out-of-limits conditions. Second, a number of special processes and mission-selectable processes provided specialized computations and processing sequences to determine and control the status and performance of various vehicle and payload subsystems. Finally, the display-and-uplink function provided the capability to initiate, alter, or terminate certain processing within some basic and special SM processes, as well as the on board capability to modify certain parameter tables used by the SM software.

The task of detecting anomalous situations triggering a Class 0, Class 2 or Class 3 alarm was assigned to the fault detection annunciation (FDA), a software module in the SM basic function. Up to three different limit sets or ranges could be assigned to each monitored parameter, derived from a set of up to four logical statements that defined the configuration for which a given limit set was appropriate. This used a process known as preconditioning. To illustrate, consider the water loop 2 pump out-pressure parameter that was preconditioned based on the configuration of the pump. If the pump was ON, then the preconditioning would have used the limit set to detect a failed pump. On the other hand, if the pump was OFF, the preconditioning would have used the limit set to detect a system leak. As another example, consider the fuel cell stack temperature parameter. Preconditioning was based on the fuel cell power level: the higher the power, the higher the temperature range. The preconditioning was software-based and the crew did not have any control over it at all, so if a system configuration was altered (e.g. a pump turned on) or system performances changed (e.g. decreasing the loads on the fuel cell, thereby changing its power output) then the software would select a different limit as required by the preconditioning.

Another key aspect of the FDA software was its capability to avoid a false alarm by means of a "noise" filter. This filter was an integer "n", with values ranging from 1 to 15 to represent the number of consecutive samples for which the parameter had to be sensed out-of-limits before the alarm would be annunciated. In this way, it was possible to avoid false alarms triggered by transients or by signals fluctuating around a limiting value. After having set off an alarm, a new alarm could not be annunciated until the parameter had been back within limits for at least "n" consecutive samples and again out of limits for "n" consecutive samples. For example, if a parameter was sampled once every second and had a filter value of two, it had to be out-of-limits for two seconds before annunciating an alarm. It would then have to be back within limits for at least two seconds and subsequently out of limits for two seconds before another alarm could be annunciated for that parameter.

Like any other software module in PASS, several display and SPEC pages were available to enable the flight crew to interact with the SM software. Two key displays were the SM SYS SUMM 1 and SM SYS SUMM 2, both available in SM OPS 2, which provided general system status information that could be accessed to rapidly diagnose a problem. For an alarm, the crew could call a system summary display to trace the problem to a given system and then continue troubleshooting it on system-specific SPEC, DISP and hardware panels. These two pages could be recalled by pressing the key SYS SUMM on any keyboard in the forward and aft flight stations, but because they were DISP pages their data was for display purposes only; it could not be modified. A more interactive opportunity was available with the SM TABLE MAINT (SPEC 60) available during SM OPS 2 and SM OPS 4. This allowed lower and upper limits, noise filter value, and enable/inhibit status for each PASS SM B/U C/W or alert parameter to be set. In this case, all the values were displayed in their engineering units, eliminating the conversion necessary when displaying a parameter on the STATUS matrix in panel R13U.

A useful function was recording into memory the value of a system parameter at a given instant in time, for future reference. In this regard SPEC 60 enabled so-called checkpoints to be created. Certain SM software control parameters were preloaded but could be changed in flight, with the changes being maintained in the SM GPCs. However, the most up to date values of these parameters would be lost if the software was overwritten or the GPC failed. To avoid the crew having to reset it all again and restore the changed parameters to the desired values, the checkpoint function allowed a few keystrokes to copy the parameter limit sets to a memory unit for possible future retrieval.

Fault detection and annunciation

As explained above, the C/W had both a hardware and a software interface, with the system management software handling the latter. For the hardware driven alarms (Class 1 and Class 2 Primary C/W) an anomaly was annunciated by an aural warning and the illumination of a lamp on the CAM specific to that out-of-limits hardware parameter, and by the illumination of the master alarm button, the pressing of which would silence the aural warning. If necessary, the parameter that trigged the alarm could also be viewed in the STATUS matrix on panel R13U.

Software driven alarms resulted in the generation of a C/W tone, illumination of the master alarm lights, and illumination of the BACKUP C/W ALARM light on the CAM matrix. In addition, a message would be displayed on the fault message line of the DPS display. Class 3 alarms did not illuminate the master alarm lights. Class 0 alarms did not illuminate the master alarm lights and did not generate a fault message line. Fault message lines were created for a fifth class of alarms, the so-called Class 5 Operator Errors. They represented the lowest priority alarm caused by a crew entry error and resulted in an ILLEGAL ENTRY fault message being displayed.

2

A skeleton for the Orbiter: structure and mechanisms

DESIGNING THE ORBITER STRUCTURE

Even before the Apollo program with its moonlandings reached its conclusion, the American aerospace community set about designing and building the next space program. Although with trips only in low Earth orbit it looked less exciting than the lunar missions, the new spaceship was to be something completely different from its predecessors. The keyword for the Space Shuttle was reusability, which translated into each vehicle having an operational life of at least 10 years, in which it would fly 100 missions. The engineers called upon to design the Shuttle were the same ones as had made Apollo feasible, and with the thrill and enthusiasm of seeing astronauts walking on the moon they tackled the problem head-on by deciding which material to use for the structure of the spaceplane.

The reusability requirement implied that each Orbiter had to be able to survive the fiery environment of atmospheric re-entry at least 100 times without a scratch either to the structure or, most importantly to the crew. This requirement meant designing a thermal protection system capable of preventing the structure from melting during re-entry. The ablative shield installed on the base of the Apollo capsule, although very simple in concept and very effective, was immediately ruled out because it would have been too difficult to create an ablative shield for the huge aerodynamic surface of the Orbiter that could be refurbished during a turn-around between missions of just two weeks. The Shuttle presented an opportunity for the former Apollo engineers to suggest a so-called "hot structure" in which the primary or load-bearing structure had some tolerance to heat, even if it were not able to directly face the fiery hell of re-entry. In other words, a "hot structure" is not only able to withstand mechanical loads but also be integrated into the thermal protection system. The highly successful X-15 and SR-71 "Blackbird" both proved this concept for high performance aircraft. Such structures require a material that exhibits high strength and mechanical properties in a high temperature environment. Titanium is one good candidate, with the amazing capability to retain its mechanical properties at 340°C. It was therefore the material of choice for building the Orbiter's structure.

36 A skeleton for the Orbiter: structure and mechanisms

Since the upper surface of the wings and fuselage of the vehicle would be exposed to much lower temperatures during re-entry than the belly, no additional thermal protection would be necessary in these areas. But for the belly, which would have to withstand 1,400°C, the titanium structure had to be covered by a thermal protection system. The first idea was shingles affixed to the primary structure in such a way as not only to withstand and transmit the airloads, but also to accommodate thermal expansion and contraction. This scheme promised not only to be robust but also to save a good deal of weight in the thermal protection system, as it was required only on the Orbiter's belly. As simple as it appeared, this configuration presented several significant problems. Firstly, the materials chosen for the shingles, molybdenum and columbium, are very susceptible to oxidation at high temperatures, severely impairing their mechanical properties. Anti-oxidation coatings were available, but even the tiniest scratch on the surface of a shingle could lead to a catastrophic failure in re-entry. Secondly, despite the excellent properties possessed by titanium, the aerospace industry had limited experience in the use of this metal; namely the SR-71. As this was a highly classified military project, it would not have been easy to gain access to the techniques for producing titanium structures. And, of course, given the limited budget for developing and building the Shuttle, cost was an important factor.

Since the early 1930s, aircraft have been built using aluminum alloys since these provide light and strong structures in an inexpensive way. With the option of creating a hot structure for the Orbiter diminishing, the question became: Could the vehicle be made of aluminum, for which a wealth of experience was already available in the aerospace industry? A positive answer to this question came on the day that a new thermal protection system made its appearance. In parallel with hot structure studies, the industry had developed an insulation made of interlaced fibers of silica that could be applied to the outside of the vehicle in the form of small tiles. This new material exhibited the same heat resistance as shingles with the additional advantage of being extremely light, only one-fourth the density of water. The tiles could be machined to a thickness that would prevent the temperature of the primary structure reaching its relatively low thermal limit. But an Orbiter with an aluminum primary structure and a thermal protection system made of ceramic tiles presented its own problems; in particular, the need to cover the *entire* skin of the vehicle with the tiles because the upper surface of the wings and fuselage would no longer be able to resist the heat of re-entry unprotected. Furthermore, for thermal protection comprising thousands of small fragile tiles, maintenance and refurbishment after each flight would be much more demanding and expensive. There were therefore pros and cons for both the hot structure configuration and the aluminum structure. As so often happens in a project of this magnitude, the deciding factor was money. While Congress allowed Apollo to get under way with a blank check, the Shuttle program was plagued by money woes that led designers to choose an Orbiter that had an aluminum structure covered by a thermal protection system of tiles. Although this design would be cheaper to develop, it was apparent that it would impose higher operating and maintenance costs once the new spaceship was in service.

The Orbiter made use of the well-known semi-monocoque structure, also known as a reinforced shell. Recognized very early on as the best for a flying machine, this kind of structure consists of an external skin (the shell) reinforced with longitudinal elements called *stringers* linked together by transversal elements called *frames* (for fuselage sections) or *ribs* (for wings). If one thinks of how easy it is to pierce an egg using a sharp object, it is easy to understand why such structures are configured this way. Pure shell structures (like an egg) are good at withstanding a load distributed over an area, such as the hydrostatic pressure inside a vessel, but are unable to resist the compression and shear of concentrated loads. An aerospace structure is subjected to a good number of loads that are applied in specific points, such as the attachment point of the landing gear to the fuselage or the attachment structure of a spacecraft on top of a rocket. Stringers, ribs and frames allow the structure to take and distribute all manner of loads, making it strong, stiff and lightweight. For the cross-section of the structure, every reinforcement element gets the specific shape and configuration deemed most suitable for that particular area. An aerospace structure, whether for an aircraft or for a spaceplane, must be as light as possible. The reinforced shell is inherently light, but this does not mean that engineers cannot devise ways to eliminate even more weight. This was particularly important for the Orbiter, because the heavier its structure the more powerful the engine to lift it, adding to the costs of engine development. Also the heavier the structure, the less the payload capacity, with the consequent reduction in the putative profits to be earned from operating the spaceship.

Thomas L. Moser, head of structural design for the Orbiter, remembers the head-on approach that he and his team took in developing the structure of the Orbiter, "We said we're going to be bold and aggressive." And bold and aggressive they were. The first decision was to use a lower than usual factor of safety. A structure is designed to withstand the maximum reasonable expected loads,[1] multiplied by a factor of safety to account for the inability to fully understand the behaviors of both materials and structures. For civil aircraft the accepted safety factor is 1.5, but Moser opted for a lower safety factor. As he remembers, "Now let's say that it takes a tenth of an inch of material to withstand the limit load. Say, 'Well, I want to be safe, so I'm going to put a factor of safety on top of that.' If it's a one-and-a-half factor of safety, instead of being 0.10 inch it'd be 0.15 inches. We said, 'We're not going to do that; we'll use 0.14 inches.' Well, that's weight, and when you take that over the entire vehicle it's a *lot* of weight." This approach drew criticism from the aerospace community, but as Moser says, "We didn't get to just do it because we wanted to, but for weight." And eventually it paid off.

Another way to save weight was by designing the Orbiter structure in terms of the so-called safe-life approach. Moser explains, "Some aircraft are designed so that you can have multiple load paths, but that added weight, so we said, 'We're not going to do that. We're going to have a safe-life design.' This means that if you have a piece of structure that has to carry the load and that structure fails, the structure fails; it doesn't have an alternate load path. We did it so that we were not adding any weight

---

[1] For an aerospace structure, these maximum loads are defined as ultimate or limit loads.

in the beginning, because once you add weight to a vehicle, it is very, very expensive and difficult to get it out."

Structural analysis is a complicated subject requiring a lot of analytical work and a good understanding of the mechanical processes acting on the structure. In a period in which computational power was far inferior to what is available nowadays, the structural engineers who designed the Orbiter had to really know their stuff. As Julie Kramer-White, who served as structural subsystem manager for the program in the early 1990s, explains, "A lot of it was done by hand-analysis, so they would take the wind tunnels and they would take their basic analysis and drive the external loads for the vehicle and then they would load the airframe." One mathematical tool created by NASA at the time of Apollo and nowadays widely used for applications unrelated to aerospace structures, was the finite element method (FEM). The structure is divided in a number of simple parts (finite elements) each of which has specific mathematical properties representing the mechanical characteristics of the material and the structural part that the finite element describes. FEM models were then processed by software developed by NASA, called NASTRAN, which calculated the load distribution over each component of the structure. In the case of the Orbiter, this would automatically propagate the airloads throughout the primary structure. "But," as Kramer-White continues, "when you reached a bolted joint or a certain aspect of actual structural design, all that was done by hand, and you can see that in the stress reports when you look at them."

It is usual in the aviation industry to verify the structural integrity of the aircraft structure by building one or more so-called test articles in order to determine whether the structure will really withstand the loads for which it was designed. Because of limited funding, Moser and his group came up with a really bold proposal for the Orbiter. As he recalls, they said, "We think that what we can do is we can load the test article, the entire Orbiter, to 120 per cent of the maximum load and predict what we think it's going to do. Then we'll extrapolate to 140 per cent." In other words, the static test article was instrumented with some 3,000 strain gauges placed all over the structure, and how it would respond to 120 per cent of the limit loads was predicted. If the predictions were correct, it meant the calculations done to design the structure with a safety factor of 1.4 were correct and that the structure could withstand all the loads for which it was designed, including the safety factor. The second departure from a normal program of structural tests consisted of eliminating the fatigue test article. As Moser explains, this was based on an educated assumption. "We didn't think fatigue was an issue for the Shuttle, since it wasn't going to have that many flights – each Orbiter was designed to fly 100 times. So we said, 'That's probably not going to be an issue. We will check it and make sure it's okay.'" The decision not to consider fatigue saved $100 million from the Orbiter development, eliminated further weight from the structure, and by reducing the loads imposed on the static test article enabled this to be made spaceworthy to serve as the second Orbiter of the fleet, *Challenger*.

SHUTTLE ORBITER STRUCTURE: FUSELAGE

With the Orbiter developed as a reusable spaceship capable of landing on a runway, its configuration was necessarily similar to an airplane with a fuselage, wings and empennage. The fuselage itself consists of the forward fuselage, mid fuselage and aft fuselage.

Orbiter Fuselage: forward section

The forward fuselage design was dictated not only by the requirement to incorporate in its structure the crew compartment module, but also to improve the hypersonic pitch trim and directional stability and to reduce re-entry heating on the body sides. It had also to react to basic body-bending loads[2] and to nose landing gear loads. From a construction point of view, it was built in upper and lower fuselage sections in order to accommodate the crew module in the manner of a sandwich. For both sections, conventional 2024 aluminum alloy was chosen for the skin-stringer panels, frames and bulkheads. The forward section also held a pod for the forward reaction control system, secured to the fuselage by means of 16 fasteners for ease of

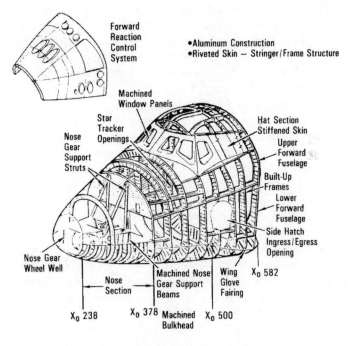

Forward fuselage structure.

---

[2] A body-bending load tends to change the radius of a curvature of the body.

installation and removal during maintenance, and the landing gear and nose cap. To provide for an increase in strength, large machined beams and struts were included in the nose next to the forward reaction control system pod and the landing gear well. For weight and center of gravity control, there was lead ballast in the landing gear well and on the aft bulkhead.

Regarding the crew module, Moser recalls that the initial idea was "to make it just like an airplane, so there isn't a separate pressure vessel, if you will, for the crew". Examined more closely, this configuration raised an important flaw. In space, in case of damage to the forward fuselage such as a crack or a hole in the skin, the safety of the astronauts would be seriously jeopardized by depressurization of the module. The engineers led by Moser then opted to create the module as a simple pressure vessel to sit inside the fuselage and be attached to it at discrete hard points. As Kramer-White explains, "The crew module is hung on a series of swing links, so you think of it as a basket that's hung inside the forward fuselage. As the fuselage bends and warps on-orbit due to thermal loads or bends due to landing loads, it's hanging in there and it doesn't pass any of that load onto the crew module." From a structural point of view, this meant that the only load that the module had to withstand was its pressurization. It was then constructed of 2219 aluminum alloy plates, stiffened with internal welded stringers and frames. Some 300 openings through the vessel, sealed for tightness, allowed cabling and wiring to connect the compartment with the rest of the Orbiter. As Moser recalls in a colorful manner, "That simplified the heck out of a very, very critical part of the Orbiter." But its deceptive simplicity did not mean that it was not safe in case of a failure such as a crack in the vessel skin. As Moser explains, "We designed it such that a crack would grow but it could not reach critical length and grow catastrophically. We would detect a leak in there." To achieve this capability, "We designed the pressurization system with the environmental control system, so it could accommodate a leak of the size we thought would be the maximum we could stand and get home safely." His conclusion, "Again we simplified the design, but we made it conservative enough that we knew that we were safe."

From a layout standpoint, the crew compartment was arranged as a flight deck, a middeck, and a lower equipment bay. The flight deck was the uppermost part of the crew compartment, housing the commander's and pilot's work stations forward and side by side. These stations had controls and displays for maintaining autonomous control of the vehicle throughout all mission phases. In the aft section of the flight deck there were the two mission specialist seats. The aft flight deck also contained several stations for executing attitude or translational maneuvers for rendezvous, station-keeping, docking, payload bay door operations, payload deployment and retrieval, and closed-circuit television operations.

An aperture in the left side of the flight deck floor led to the middeck, where there were three avionics equipment bays, two on the forward side and one on the aft part of the deck. Depending on the mission requirements, bunk sleep stations, exercise equipment and a galley could be installed on the middeck as well several modular lockers to store the crew's personal gear, personal hygiene equipment, experiments, and mission-necessary equipment. For launch and re-entry, four crew members were

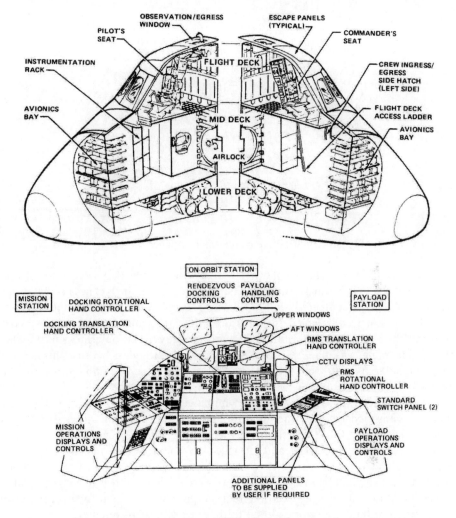

Crew compartment (top) and flight deck aft station (bottom).

seated on the flight deck and the others on the middeck. The middeck floor allowed access to the lower equipment bay that contained the major components of the waste management and environmental control life and support systems, such as pumps, fans, lithium hydroxide canisters, absorbers, heat exchangers and ducting.

In space the flight crew required external visibility for rendezvous, docking and payload-handling operations, and for the descent they needed forward, left and right viewing. The flight deck had six windows in the forward section, plus two overhead windows and two payload bay windows at the aft station for rendezvous, docking and payload viewing.

The six forward-facing windows were at that time the thickest pieces of glass ever produced of high optical quality. These windows had to maintain the cabin pressure

42 A skeleton for the Orbiter: structure and mechanisms

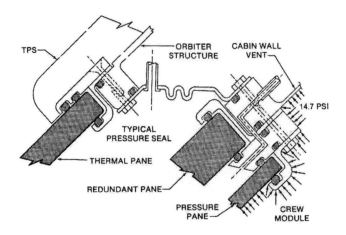

Structure of the window panes.

and to withstand the heat of re-entry. To satisfy both needs, they were made of three glass panes for the forward windows. The innermost, with a thickness of 0.6 inches, was constructed of tempered aluminosilicate glass to withstand the cabin pressure. For this reason it was called the pressure pane. Its outer surface was coated to transmit the visible spectrum but reflect infrared rays. The outermost pane, called the thermal pane, was made of low-expansion fused silica glass because this had high optical quality and sufficient thermal shock resistance to withstand temperatures up to 430°C. Whilst the exterior was uncoated, the inner surface had a high efficiency anti-reflection coating. A third pane, 35 millimeters thick, also made of fused silica glass, was inserted between the pressure and thermal panes in case of damage to either other pane. For this reason it was called the redundant pane. While the thermal pane was mounted and attached to the forward upper fuselage, the redundant and pressure panes were mounted on the crew module. The overhead windows and payload bay viewing windows were made with the same layout, but the panes had different thickness and the thermal pane was omitted for the aft-viewing windows since they were not exposed to the intense heat of re-entry.

The crew would ingress/egress the Orbiter at the middeck level through a circular 40-inch hatch on the left side that could be operated either from inside or outside. It was attached to the cabin by means of hinges, a torque tube and support fittings, and would open outwardly 90 degrees down with the Orbiter horizontal or 90 degrees sideways with the Orbiter vertical. In this way during ingress on the launch pad, the crew would not risk damaging the hatch by stepping on it. The hatch had a circular window 10 inches in diameter. In an emergency, the hatch was the primary means of exit.

A second emergency exit was provided by the left-hand overhead window, which could be jettisoned by a pyrotechnic charge. The jettison system could be initiated by pulling a ring handle on the forward flight deck center console. The outer pane would be jettisoned upward and aft, followed 0.3 seconds later by downward

Shuttle Orbiter structure: fuselage 43

Ingress/egress hatch in the horizontal orientation.

rotation of the inner pane which would then be held in place by a capturing device. Once open, the crew could egress using a mission specialist's seat to climb up through the window. Seven emergency ground descent devices (Sky Genies) were stowed on the overhead aft flight deck outboard of each overhead window, one for each crew member. With these devices, astronauts could lower themselves to the ground via the starboard side of the Orbiter. As such an emergency egress would be declared soon after landing, descending with ropes and rubbing against the skin of the vehicle, which would still be very hot, would have posed a risk.

Extravehicular activity (EVA), popularly known as spacewalking, is one of the most fascinating activities involved in space exploration. Prior to the advent of the Shuttle, NASA had gained a good amount of experience of EVA during the Gemini and Apollo programs. With the Shuttle, a great many activities

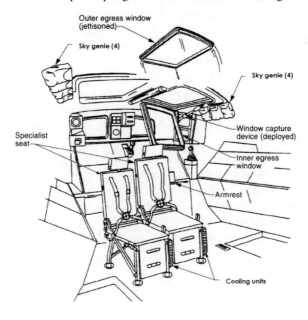

Overhead egress window jettisoned.

44 A skeleton for the Orbiter: structure and mechanisms

Inner airlock hatch open.   Outer airlock hatch open.

involving astronauts working in the open were envisaged. For this reason an airlock was added to the rear wall of the middeck to allow spacesuited astronauts[3] to pass through into the payload bay. It was a cylinder 63 inches in diameter and 83 inches high, with two 39.4-inch diameter D-shaped openings with pressure-sealing hatches. It provided space for all EVA gear, checkout panels and recharge stations. Both hatches were mounted directly on the airlock structure. The inner hatch was mounted on the external wall to isolate the airlock from the crew cabin. To open this hatch, it was pulled into the crew cabin approximately 6 inches, then pivoted up and to the side. It was closed by reversing the procedure. The outer hatch isolated the airlock from the unpressurized payload bay when closed and gave access to the bay when open. During opening, it had to be first pulled into the airlock and then forward at the bottom and rotated down until it rested with the inner side facing the airlock floor. Each hatch was provided with six equally spaced latches, a gear box and actuator, and two pressure equalization valves and hatch opening handles, one set on each side. Each hatch also had two pressure seals on the hatch side of the structural interface, one on the hatch cover and one on the structural interface. In addition, there were differential pressure gauges and relief valves.

For missions that had a Spacelab pressurized module in the payload bay, a tunnel adapter was added to connect the airlock to the Spacelab tunnel. If an EVA became necessary, the spacewalkers would access the payload bay using a hatch built into the top of the tunnel adapter. For missions involving dockings with space stations, the airlock was relocated into the payload bay, mounted on a truss, and equipped with a docking mechanism. The only incident involving the airlock occurred in November 1996 when a loose screw in the gearbox prevented *Columbia*'s airlock from opening and required the two EVAs programmed for the STS-80 mission to be canceled.

---

[3] Normally astronauts spacewalk in pairs. During STS-49 in May 1992, *Endeavour*'s maiden voyage, a contingency situation resulted in three crew members going outside to retrieve an Intelsat communications satellite.

Orbiter fuselage: mid fuselage section

The mid fuselage was the biggest structural section of the Orbiter, with a length of 60 feet, a width of 17 feet and a height of 13 feet. Built using conventional aluminum, it was open at both ends. It interfaced with the forward and aft fuselage sections and the wings, and also supported the payload bay doors, door hinges, robotic manipulator arm, Ku-band antenna, and various other vehicle components. It was divided into thirteen bays, each made of vertical side machined elements and horizontal elements made of boron/aluminum tubes that incorporated bonded titanium end fittings. Its external contour was defined with numerical control machined skin panels stiffened with longitudinal and vertical stiffeners. Two machined sills and longerons running its full length were to respectively take the bending and longitudinal loads imparted by the payload and the payload bay doors. Another vital component was a lateral trunnion support structure located on the side wall forward of the wing carry-through structure to respond to main landing gear loads.

After detailed analysis of actual flight data of the descent stress thermal gradient loads, torsional straps were added to the lower stringers in bays 1 through 11 in order to tie all the stringers together in a manner similar to a box section. By eliminating torsional deformation, this would provide a positive margin of safety. In addition, room-temperature vulcanizing silicone rubber material was bonded to the lower section in bays 2 through 14 to act as a heat sink and evenly distribute temperatures across the bottom of the fuselage, thereby reducing the thermal gradient and again ensuring a positive margin of safety.

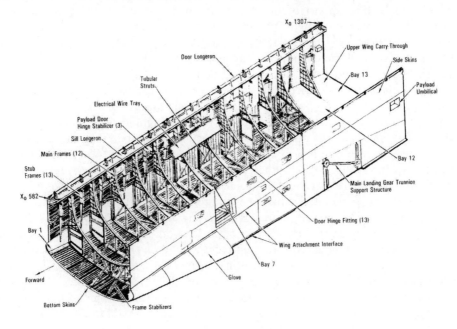

Mid fuselage structure.

The mid fuselage section accommodated the *raison d'être* of the Orbiter, that is to say the cavernous payload (or cargo) bay. Thirty years of operational service proved that the payload bay was what enabled the Shuttle to serve as a stable platform from which to make scientific observations, release commercial satellites and telescopes into orbit, and send probes into interplanetary space. Astronauts spent many hours spacewalking in the payload bay to outfit or refurbishing satellites, in particular the Hubble Space Telescope. Several payloads were carried to investigate how to build large space infrastructures like the International Space Station, and again many hours were spent developing the skills required to assemble the various elements of a space infrastructure. And in its last thirteen years of use, the Shuttle assembled the largest space infrastructure ever, the International Space Station.

At this point, it is worthwhile reviewing the reasons for the Orbiter having such a large cargo bay. The design of the Shuttle was undertaken in an era when NASA was sending men to the Moon and proposals for a large space station in orbit were on the drawing board. Already under development was the modification of the third stage of the mighty Saturn V rocket with facilities and experiments. This Skylab "Orbital Workshop" was launched in 1973 and successively occupied by three crews of three astronauts. But even before the end of the Apollo program, the production line for the heavy-lift Saturn V had been shut down. As a result, NASA planners started to think of a future space station as a large structure that would be assembled in orbit piece by piece. Its assembly would require a vehicle capable of transporting the various component. For this reason, NASA wanted the Shuttle to have a payload bay 17 feet in diameter. In the face of Congressional opposition to the program, NASA sought the support of the Air Force, which agreed but demanded some changes to the overall configuration of the Orbiter. In particular, it being the Cold War, the Air Force wished to place into orbit ever bigger and heavier spy satellites. To carry these satellites, the payload bay would have to be greatly enlarged. In fact, the Air Force demanded that it be at least 60 feet in length.

To give the Shuttle the incredible versatility that it demonstrated throughout the program, it was necessary primarily to design a proper means of accommodating the payloads in the cargo bay. Doing this was a rather complex assignment, since, as Moser says, "We had to know the characteristics of the payload that was going to go in that mid fuselage. We had to know its size, what it weighed, where the center of gravity was, how many payloads there were going to be, where they were going to be attached, how stiff they were." The problem was that due to the variety of missions the Shuttle was expected to conduct, the requirements for payloads were either vague or did not exist. Nevertheless, the payload bay had to be designed. As Moser continues, "We looked at multiple types of payloads, different orientations, different centers of gravity, different positions, different weights, everything else. We designed the mid fuselage to accommodate ten million types of payloads. The Orbiter has never had a problem accommodating any payload."

Due to the variety of possible payloads, as well as their number and position in the cargo bay for a given mission, particular attention was paid to the design of the payload retaining fittings. In order to avoid having a very stiff payload bolted to the mid fuselage structure, the attachment fittings were designed in such a way that they

Shuttle Orbiter structure: fuselage 47

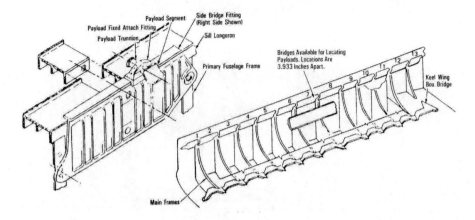

Payload retention system.

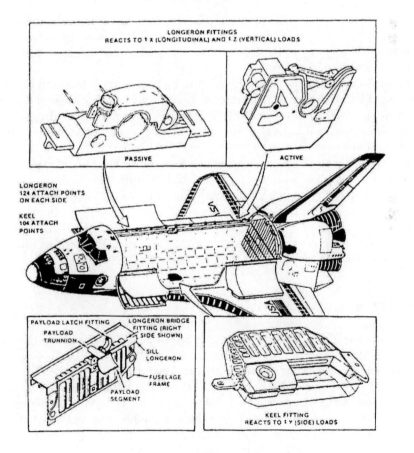

Active/passive payload retention systems.

48 A skeleton for the Orbiter: structure and mechanisms

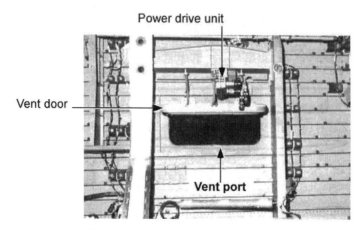

A view of the vent port and door from the payload bay.

could slide. Hence, as Moser explains, "The Orbiter didn't care how stiff the payload was, and the payload didn't care how flexible the Orbiter was. The payload can be designed independently of the Orbiter and vice versa. We isolated, we decoupled the design by making it a statically determined system."

For non-deployable payloads intended to be kept in one place during the entire mission, so-called passive fittings were used, over which neither Mission Control nor the crew had any control. For deployable payloads, active motor driven retention mechanisms were used, activated by the crew to release a payload or to secure it into the bay. Both types of retention mechanism could be installed on the mid fuselage longerons by means of bridges, and to the bottom of the bay by keel bridges. In total there were 13 longeron bridges per side and 12 keel bridges available per installation, but only those bridges required for a particular cargo were flown. Because of main frame spacing, varying load capability and subframe attachments, the bridges were not interchangeable. Longeron bridge fittings were attached to the payload bay frame at the longeron level and at the side of the bay. Keel bridge fittings were attached to the payload bay frames at the bottom of the bay.

ORBITER ACTIVE VENT SYSTEM

For all sections of the fuselage, seven vent ports on each side made up the so-called active vent system (AVS) to equalize the unpressurized compartments to the ambient environment as the Orbiter traveled from the pressurized atmosphere of Earth to the vacuum of space. Some of the vent doors also had intermediate positions to purge the unpressurized compartments with either dry air or nitrogen. Purging operations were performed on Earth to provide for thermal conditioning and moisture control, and to prevent hazardous gases or contaminants from entering the compartments. Each vent port aperture was sized according to the volume that it had to vent.

The vent doors were left in their purge position until the countdown got to T-28 seconds, then opened sequentially at intervals of approximately 2.5 seconds. If at T-7 seconds any door was out of configuration, the countdown would be interrupted. For the full duration of the ascent, the doors remained open to allow their compartments to vent as the vehicle entered the vacuum of space. They remained open for the full duration of the orbital flight in order to vent gases released by vehicle insulation in the unpressurized compartments. While preparing for atmospheric re-entry, the crew commanded all the doors shut and soon thereafter opened only the four doors on the port side to vent any hazardous gases that had accumulated during the deorbit burn. During re-entry, the doors had to remain closed to prevent hot plasma from entering the vehicle. On having slowed to a speed of 2,400 feet per second, the door opening sequence was automatically commanded, allowing the unpressurized compartments to match the ambient environment as the Orbiter returned to Earth. For post-landing operations, the doors were again placed in their purge positions to permit conditioning of the unpressurized compartments. While during ascent the crew could not intervene to manually open or close the doors, for all other phases of the flight, including re-entry, the crew could command the opening/closing sequences.

Orbiter fuselage: aft section

The aft fuselage was the most densely packed part of the Orbiter, housing all of the actuators, turbopumps and propellant lines necessary to feed and gimbal the three main engines. It also had the three auxiliary power units with associated hydraulic lines, engine and vehicle avionics, and ammonia boiler and flash evaporator used for active thermal protection during ascent and re-entry. It supported and interfaced with the left-hand and right-hand aft orbital maneuvering system/reaction control system pods, both wings, the body flap, the vertical tail, the launch umbilicals, and the three main engines. This section of the fuselage bore the immense stress of the 1.5 million pounds of thrust generated by the main engine cluster and transmitted it to the rest of the fuselage without catastrophic failure. To accomplish this the 28 truss members of the internal thrust structure of what was essentially a box 18 feet long, 22 feet wide and 20 feet high were made of titanium and fabricated by a process called diffusion-bonding. As Moser explains, "What we did, is we would take two pieces of titanium that were going to go together, and instead of welding them we'd put them in high temperature in a vacuum and push them together until they bonded. Just molecularly they became one and the same." The result was a single hollow, homogeneous mass that was lighter but stronger than a forged part. For selected areas of the structure the diffusion-bonded truss element were reinforced with boron/epoxy to further reduce weight and increase strength.

An outer shell made of frames and integral-machined aluminum skin panels was wrapped around the thrust structure and fitted with penetrations to provide access to the installed systems, and mountings for the secondary structure that was made of aluminum. An intriguing feature was the aft heat shield seal (AHSS), which acted as both structure and mechanism at the three main engine interfaces, and played a major role in the structural integrity of the aft compartment. In fact, it served to

50 A skeleton for the Orbiter: structure and mechanisms

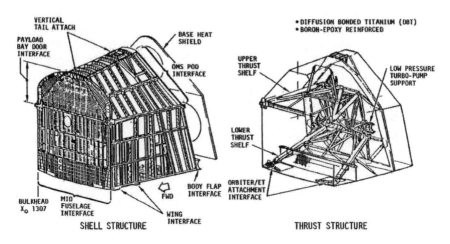

Aft fuselage structure.

minimize compartment pressure and loads generated during ascent venting/engine gimbaling, and re-entry repressurization. Seal leakage could cause an increase in pressure, with possible resultant failure of other aft fuselage structural components. Each aft shield consisted of a stationary conical dome heat shield fastened to the base structure, a hemispherical engine heat shield made of Inco 625 mounted on and moving with the engine itself, and a seal mounted to the dome heat shield. This last one bore against the engine heat shield, and accommodated the forward motion between the engine and aft fuselage compartment. The seal mechanism had a mean diameter of 7 feet and consisted of two sliding and flexible seals. The former was to accommodate the gimbaling of the engine and the latter the forward motion between the engine and the compartment structure. From a mechanical standpoint, the sliding seal was a series of graphite blocks held in a retainer with 48 spring cans uniformly spaced about the periphery to exert pressure on the graphite blocks to make them slide and seal against the spherical surface of the engine heat shield. Support and reaction to the spring cans was provided by the heat shield, while the ring was held in position by three articulated links anchored to the dome heat shield, allowing the sliding seal to follow the forward translation of the engine heat shield as the thrust built up.

The dome heat shield was made of bonded aluminum honeycomb with reusable surface insulation tiles on the outside and thermal insulation blankets on the inside. It consisted of a pair of half-cones structurally joined by bolts at the meridian vertical splices and by a row of outer peripheral bolts attaching the dome heat shield to the base structure. Additional bolts attached the insulation tiles at both the meridian and peripheral joints. The engine heat shield was made initially of René 41 then changed to Inco 625, protected by a thermal coating on the outside and thermal insulation blankets on the inside. The shield consisted of two spherical segments structurally joined by dual rows of bolts at the vertical meridian and by a row of bolts at the heat shield to engine nozzle inner peripheral joint.

Payload bay doors

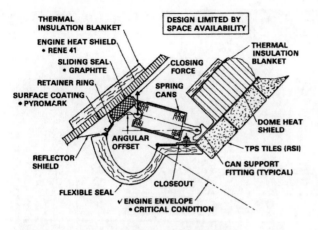

Aft heat shield seal installation.

The aft fuselage section also housed two umbilical cavities on the underside, each containing a structural attachment with the external tank as well as the electrical and propellant line disconnects. Associated with each cavity, there was a so-called ET door that was kept open during ground operations to allow mating of the umbilicals between the external tank and the Orbiter and also during powered flight. Once the external tank had been jettisoned, both ET doors were commanded closed in order to seal the cavities and shield the underlying structure from re-entry heating. During ground operations and ascent, both doors were held open by means of two centerline latches that engaged a fitting on the outboard edge of a door.

Nominally, the doors were closed manually by the crew via a series of switches. Prior to closing the doors, the centerline latches had to be disengaged from the doors and stowed. To do so, the latches rotated and retracted into the body of the Orbiter so that they were flush with the thermal protection system mold line. At this point, the crew could command electromechanical actuators to drive the doors closed through a system of bellcranks and pushrods. As the doors moved within two inches of being closed, two rollers mounted on the outboard edge of each door contacted two ready-to-latch paddles located in the umbilical cavity. This condition was signaled to the crew, who then commanded other actuators to drive uplock latches to engage three rollers located on the inside face of the door, pulling it firmly against the Orbiter. Compression of the aerothermal barrier resulted in a seal that protected the umbilical cavity from the heat of atmospheric re-entry. If an ascent abort were made, ET door closure would be achieved automatically through sequential software commands sent by the GPCs. If the manual procedure failed at the end of a nominal ascent, Mission Control would have closed the doors by a series of commands, monitoring the state of each mechanism prior to sending the next command. Real-time commands were preferred over the automatic closure, because the automatic software did not receive position feedback. Soon after landing, the doors were commanded open by the crew in order to allow the ground crews to access the umbilical cavities and the aft support points.

PAYLOAD BAY DOORS

With a length of 60 feet, a diameter of 20 feet and a combined area of approximately 1,600 square feet, the payload bay doors of the Orbiter remain the largest fairing ever

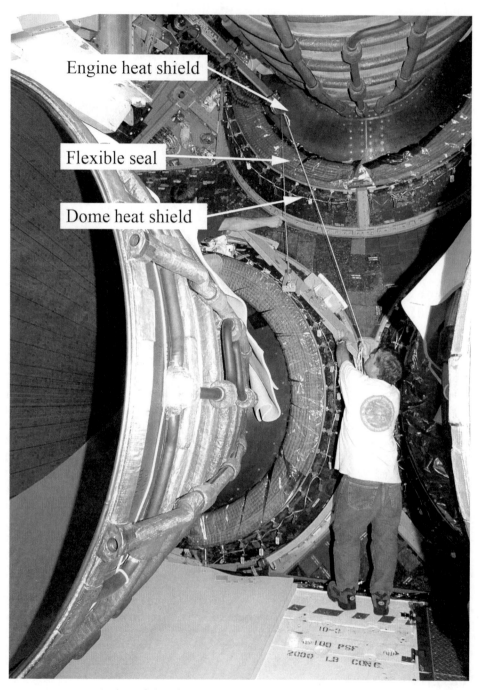

A view of the aft heat shields at the base of the SSMEs.

A view of one umbilical cavity on the aft underside of the Orbiter.

built for a launcher. As a fairing, these doors played the two-fold role of being both an aerodynamic cover to provide a non-turbulent airflow over the fuselage in order to reduce drag, and to prevent contamination of the payload by the external terrestrial environment. For the aerospace industry this was the first time composite materials were used for such a large component. But composite materials can be considerably lighter than metallic materials, so it is easy to understand why they were chosen for the Shuttle. As Moser recalls, the issue was weight, "We had to cut more weight out the vehicle. I think we were looking to save 600 pounds of weight in the payload bay doors."

The initial material chosen for the payload bay doors was aluminum sandwich, a sort of composite material made by sandwiching an aluminum honeycomb between sheets of aluminum. This had been used in building the Apollo command module. But to eliminate 600 pounds something else would have to be used. The solution was to use modern composite materials.

Structurally speaking, a payload bay door was made of five skin panels each of which was constructed in a honeycomb sandwich fashion with the faceplates and the honeycomb made using graphite-epoxy and Nomex respectively. The panels were connected by means of shear pins and stiffened by means of an internal framing of ribs, intercostals and longerons, all made from solid graphite-epoxy laminates.

The payload bay doors were to be opened shortly after the vehicle achieved orbit, and left so until several hours before deorbit. For the full duration of the orbital flight, the thermal stress of passing through the Earth's shadow would cause expansion and contraction, warping the door in a more or less marked way. The degree of structural deformation of the doors would depend on the attitude held by the Orbiter. While the warping of the doors would not impair the mission in orbit, it

54 A skeleton for the Orbiter: structure and mechanisms

Detail of one ET door latch.

could prevent the doors from closing for the return to Earth. If they could not be properly closed, then the hot plasma of re-entry would penetrate the Orbiter with potentially catastrophic results. To allow for this effect, engineers decided to provide the doors with a high degree of flexibility. As Moser remembers, "We said, 'In the structural design we're going to do something to alleviate that issue. What we're going to do is make the payload bay doors very flexible so that once they start to close on-orbit you can zip them closed.'" Thus the doors mimic a shirt with a zip closure. This was achieved by having three sets of latching mechanisms, one on the aft bulkhead of the forward fuselage, another on the forward bulkhead of the aft fuselage, and the others along the centerline of the doors.

Each door was opened and closed by power drive units consisting of two electric motors for redundancy, rotating a shaft which ran the full length of the mid fuselage beneath the sill longeron. Due to the length of the cargo bay, this shaft was split into several smaller sections interlinked by means of couplers that evenly distributed the torque the full length of the shaft. Rotation of the shaft activated six rotary actuators to create the linear motion required to operate the opening/closing mechanism. This consisted of a push-pull rod, a bellcrank and a link. Vertical motion of the push-pull rod caused the bellcrank to pivot around its hinge point and pull the link connected to the door and rotate it on its hinges. The six push-pull rods available per door could be easily seen by the crew viewing through the rear windows of the flight deck. They were color-coded to assist the crew in opening and closing the doors. As a push-pull rod acted to open or close a door, certain colored bands were displayed to indicate the degree of rotation, and thus the amount the door was open or closed. This information would indicate whether a door was warped or jammed. For example, if the door was completely closed then only one gold band would be partially visible at the top of its push-pull rods, but in the event of jamming silver bands would be visible.

Two hours and 40 minutes prior to deorbit, the drive system of the left-hand door would start to close that door. As the door moved through the 4-degree open

Payload bay doors

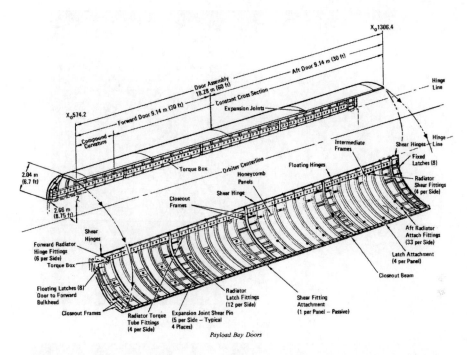

Payload bay door structure.

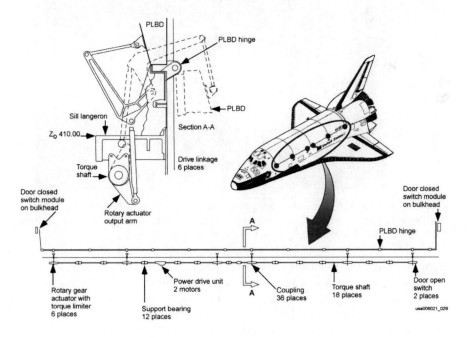

Payload bay door drive system.

56 A skeleton for the Orbiter: structure and mechanisms

Payload bay door push-pull rod.

A view of the payload bay door closing mechanism on the forward bulkhead. It is possible to see part of the linkage mechanism of the short side of the door and the rollers of the bulkhead that are engaged by the door latching mechanism.

position, ready-to-latch switches on the forward and aft bulkheads sent a signal to initiate the bulkhead latch actuators located on both short sides of the door. As the door reached the 1.75-degree open position, contact with the forward and aft bulkhead door-closed switches shut off power to the door drive system. At this point, the forward and aft latch rotary actuators transmitted power to each of the four latches in the gang on both short sides of the door and were connected through a

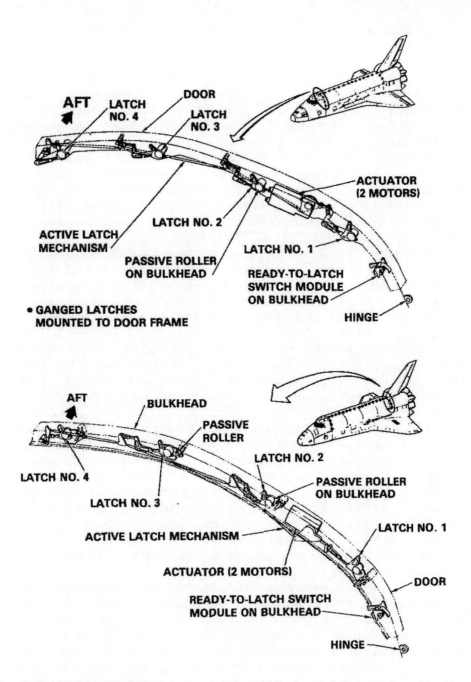

Typical bulkhead circular latch system on forward (top) and aft (bottom) short sides of a payload bay door.

mechanical linkage. The kinematics of the ganged linkage were such that the first latch engaged the first roller on the bulkhead, pulling it to a position within reach of the second roller. The latch had a hook that would rotate from the open position to engage the roller, pull it, drawing it between the latch arm and a scalloped mating surface. As the door was being pulled to the roller and the roller made contact with the scalloped portion of the latch, this action had the effect of aligning the door.

This sequence was repeated for the other three latches until the door was finally pulled into total contact with the bulkhead structure. At this point, power from the left-hand door latch actuators was removed and switched to the right-hand door. It is important to understand that this staggered four-latch-gang latching sequence was made possible by a particular disposition of the linkage that allowed the latches to be latched and unlatched in ascending and descending order. This phasing allowed the closing latches to help the remaining latches to move into the range of engagement of the rollers on the bulkheads.

A command was then sent to latch the 60-foot-long centerline of the two doors. This was achieved by means of four gangs, each with four latches. Each latch had an active portion on the right-hand door that consisted of a hook to engage the passive portion, consisting of a roller, on the left-hand door. Each latch also had provisions to allow the door to close properly. Since the latching mechanism was strong enough to cause one door to overlap and ride on top of the other, a backing plate oriented in the X/Z-plane was placed on either side of the latch hook's passive roller. A roller on the active side of the latch was positioned opposite each plate. The contact between the rollers and the plates prevented overlapping from occurring. A third roller, located between the other two, made contact with a fork positioned outside of the passive roller of the latch's passive side to constrain the door in the Z axis. To constrain the doors in the X axis, the centerline was provided with passive shear fittings consisting of one roller placed on the right-hand door that slid during closure through a claw on the left side. This prevented translation in the X axis during the latching sequence of the gang. Each gang drove its set of four latches simultaneously, but closure of the centerline was done in a staggered manner with the forward-most and aft-most gangs actuating first because at that point in the operation the door ends were latched to the forward and aft bulkheads and those sections were fairly close to each other. Upon latching these two gangs, the remaining 30 feet of centerline was close enough for the remaining two gangs to complete the operation.

Opening of the doors was simply the inverse of the closing sequence. Both door opening and closure could be performed in auto (preferred) or manual mode. In auto mode, software monitored the latches and power units for premature out-of-sequence operations, such as a door opening prior to being unlatched. If anything was out of configuration the auto mode halted the sequence, withdrew power from all elements, and issued a fault message. These and other safety checks prevented the doors and latches from suffering damage. When operating the doors manually, the crew would carefully monitor limit-switch indications and motor drive times to verify that the doors and latches were opening or closing properly. Afterwards, a quick check of the push-pull rods would indicate if any warping had occurred.

Payload bay doors 59

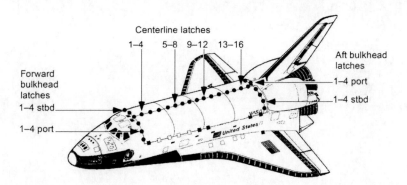

Payload bay door latches locations.

The flexibility of the doors was achieved not only by the materials used but also by the way in which they were connected to the fuselage. As Moser explains, "We said, 'We're going to let those payload bay doors behave like they are not part of the fuselage structure, so that we can make sure that they're flexible enough to close on orbit'." Fixed and floating hinges were designed. The five fixed hinges prevented any movement on the three axes and permitted rotation only on the hinge axis. However, if all the hinges were of this type, external stresses would be imposed on the doors by the fuselage structure. The floating hinges allowed free movement along the hinge's longitudinal axis. In this manner, the Orbiter could expand, contract, bend and twist without imparting any loads on the payload bay doors. In designing flexible doors, as Moser put it, "What you do, is you give up that part of the structure. Think about the structure of the Orbiter fuselage as just being a big tube. If you take a tube and you bend it, it's pretty stiff, but if you cut half of it away and bend it, it's not very still at all.'" To compensate, the mid fuselage structure had to be designed to be stronger and therefore heavier. But at least this would avoid the problem of the doors warping and jamming, preventing the vehicle from safely returning to Earth.

Given the extreme effort to save every single pound of weight, the payload bay doors were designed to be able to support only the loads encountered in orbital flight and their drive system was designed with the torque to open and close the doors only in the weightlessness of space. As a result, when the Orbiter was being processed on Earth, under full gravity, the doors could not be opened using the drive mechanism and were not able to hold their own weight in an unlatched horizontal configuration. During ground processing, support fixtures had to be attached to the doors to provide the necessary force or torque to keep them from deforming. For opening the doors, a special "zero g" apparatus was used to simulate the weightlessness of space. This merely applied a force equal to the weight of the doors (and attached ground fixtures) in the opposite direction to gravity so that the drive mechanism could open/close the doors.

The "processing flow" called for installing the payload into the Orbiter on the launch pad. Even in a vertical orientation, the doors would warp when open. For each door, the weight and curvature would bend the forward-most end in and the

60 A skeleton for the Orbiter: structure and mechanisms

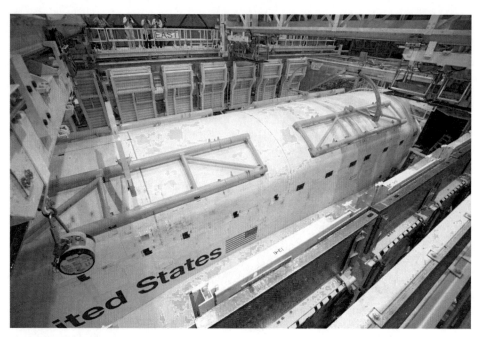

The "zero G" apparatus used for opening/closing the payload bay doors on Earth.

Torque tubes to prevent door warping during vertical processing.

aft-most end out, preventing the doors from closing and latching. Torque tubes were installed on each segment of a door to counteract this warping. The torque applied varied from segment to segment according to segment weight and warping. The torque applied to the forward-most arm of each torque tube pulled the centerline of the forward part of that segment out whilst pushing inward the centerline of the aft part. The equilibrium produced in this way enabled the latches to remain within latching distances. Since in the vertical orientation the drive system did not require to overcome gravity, opening and closing could be performed without external assistance.

Heat generated on board the Orbiter was transferred to an active thermal control system that consisted of two independent Freon-21 loops running through radiators affixed to the inner sides of the payload bay doors. Each radiator had four panels, numbered from forward to aft. The two forward-most panels could be deployed to an angle of 35.5 degrees to shed heat from both their upper and lower surfaces, thereby providing a 20 per cent increase in heat rejection. The other two panels were fixed to shed heat from their upper surfaces only. The decision to deploy the radiators was based on mission heat loads and thermal requirements. In most cases the additional heat transfer capability to be derived from deploying the forward panels was not required.

Each panel was an aluminum sandwich coated with a silver Teflon adhesive tape that reduced solar energy absorption and increased heat transfer from the Freon-21 to space. Coolant fluid from the Orbiter passed through the tubes of the two loops inside the panels, and ran in series from aft to forward in each panel prior to flowing back into the vehicle.

ORBITER WING

At the start of the Shuttle program, it was envisaged that the whole system would be reusable and be capable of landing on a conventional runway. This required the new spaceship to be capable of generating lift when re-entering the atmosphere. This was easier said than done, since defining the correct shape, dimension, position and airfoil of a wing is one of the most difficult issues in designing an aircraft. And all the more so for a vehicle that was to leave Earth as a rocket, accomplish its orbital mission as a spaceship, and return to Earth as a glider.

The Shuttle was meant to be a complete departure from all the previous spaceship designs. Very advanced concepts would be transferred from the drawing board into a real flying machine. NASA engineers, and companies competing for the contract to develop the Shuttle, started with the idea of building a so-called lifting body; a dream and nightmare at the same time for aerodynamicists. The main advantage of such a configuration was that the fuselage and wing are blended together. This offered the advantage of a smaller structure, compared to a conventional machine of the same performance, and it eliminated the irritating aerodynamical wing-body interfaces that would have created drag. Tests with experimental aircraft such as the HL-10, X-23 and X-24 demonstrated the viability of this approach but also highlighted issues that could not be neglected. A lifting body does not generate as much lift as a winged aircraft and the increased drag raises the landing speed, which is a rather undesirable handling quality to any pilot. In addition, the complex aerodynamics of this kind of vehicle are such that even the smallest mistake in defining the right configuration can seriously affect the whole design. And to safely pilot a lifting body it is necessary to develop sophisticated software to enable an on board computer to stabilize a machine that it is intrinsically unstable. Although very appealing as a very new and innovative design, the idea of making the Shuttle a lifting body was soon abandoned in favor of a more conventional solution with a fuselage and wings.

If using a conventional solution appeared to be a step back in the development of something really different, it brought with it its own set of issues. The most important issue was the shape of the wing. This was related to the re-entry profile. At the start of the program it was intended that the Orbiter would return to Earth by executing a ballistic re-entry. This was based on work undertaken in the early 1950s by Julian Allen and Alfred Eggers, aerodynamicists at the Langley Aeronautical Laboratory of the National Advisory Committee for Aeronautics, the predecessor of NASA. Allen and Eggers had proved that the best shape for a re-entry vehicle is not a sharp one but rather a blunt one, because when a blunt shape penetrates the atmosphere from space the compressed air creates a curved shockwave. By "standing off", this shockwave protects the skin of the vehicle and by flowing around the vehicle it carries away the heat. Combining this effect with a high angle of attack to reduce the time spent in the upper part of the atmosphere, reduces the heat that reaches the vehicle's structure. In turn this reduces the specifications for the thermal protection system, which can be lighter and less sophisticated. It was for this reason that all spacecraft built before and since the Shuttle have been capsules that re-enter the atmosphere with their protected base facing the direction of travel.

To give the Shuttle a blunt shape during re-entry it would have been sufficient to have a pair of short straight wings, similar to the X-15 rocket-powered hypersonic research aircraft. The idea was to make a ballistic re-entry, maintaining a high angle of attack until the vehicle had descended to an altitude of about 40,000 feet, at which point its speed would have slowed to 300 feet per second. Then the pilot would push the nose down and dive for 15,000 feet to achieve the speed required for subsonic flight. The landing would occur at a modest 130 knots; half that of a lifting body. With the wings required to provide lift only in the final part of the atmospheric flight, they would be fairly simple to design because they could be configured for subsonic flight.

But the Air Force had in mind something completely different. Firstly, they disliked the idea of having to perform a deep dive to bring the vehicle to an attitude appropriate for subsonic flight, arguing this would increase the risk of losing control. But perhaps the most important reason was the Cold War. In the late 1960s and early 1970s the two superpowers aimed to achieve as much control as possible over "outer space". In a James Bond-like fashion, one possible mission for the Shuttle called for it to be launched and promptly rendezvous with and snatch an enemy "spy satellite", returning home 90 minutes later. Many such satellites are launched into polar orbits, meaning that the plane of the orbit is perpendicular to the equator and they pass over both poles in traveling a full orbit. As the Earth rotates below, such a satellite has the opportunity to view the entire globe in each 24-hour period. But launching into polar orbit from the Kennedy Space Center was not allowed since it would have required passing over densely populated areas and posed a threat in the event of an accident during the ascent. Flying south was ruled out by the prospect of an ailing Orbiter coming down over communist Cuba. To accomplish the Air Force missions, it would be necessary to launch from Vandenberg Air Force Base in California. This site had the additional advantage that it was not accessible to members of the public, meaning that launch operations would be able to be conducted in secret. Polar missions would be able to launch to the south over the Pacific without posing any risk. But launching from Vandenberg introduced a new parameter into the Shuttle program, namely the cross-range capability of the Orbiter during re-entry.

The Air Force also wanted the vehicle to return to the launch site after a one-orbit polar mission, but due to the rotation of the Earth this would be 1,100 miles off the orbital track. This required the Orbiter to be capable of gliding during all phases of its re-entry in order to steer towards the landing site. In addition, NASA and the Air Force were concerned that a Shuttle might have to abort its mission and come back as soon as possible. For launches from the Kennedy Space Center, heading eastward, this would not be a problem since it would be possible to find a good selection of emergency landing sites across the country. But for missions in polar orbit, the only place to land, even in an emergency, was Vandenberg itself. Thus a cross-range of up to 1,100 miles was deemed essential.

This meant the wings could not be short and straight, but a delta shape capable of producing a considerable amount of lift at hypersonic speeds to glide far to the left and to the right of the initial direction of flight. But flying with a delta wing to obtain cross-range would expose the structure of the Orbiter to thermal heating for a longer period of time. This would require the thermal protection system to cover the entire exposed surface. And since a delta wing would have a considerably greater area than a small straight wing, this would further increase the weight of the thermal protection system. Nevertheless, the delta wing was chosen to satisfy the Air Force's need for cross-range because, without Air Force support, the program would have stalled in Congress.

If the shape of the wing was dictated mainly by the Air Force's requirement for cross-range and for satisfactory control throughout the entire flight regime, the size of the wing was mainly determined by landing requirements. Irrespective of whether a flying machine is an aircraft or a spaceplane, a slower landing is safer. The larger the area of a wing the greater the lift that it can generate at low speed and hence the

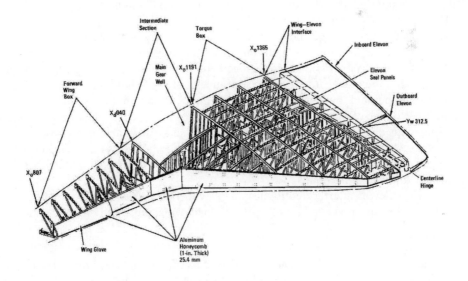

Wing configuration.

64 A skeleton for the Orbiter: structure and mechanisms

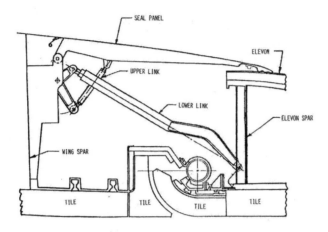

Elevon seal configuration.

lower the landing speed. Another requirement concerning landing is the length of the main gear struts. Aircraft with swept wings, such as a delta wing, have to land with a high angle of attack in order to increase the lift at low speed. Since this angle is also a function of the sweep of the leading edge of the wing, the greater the required angle of attack the longer the landing gear struts must be to prevent the rear of the aircraft from scraping on the runway. The longer the struts, the heavier the main landing gear and the greater the volume required to accommodate it inside the vehicle during flight. By combining all of these requirements, the optimized shape proved to be a "double delta" wing with an 81-degree sweep angle for the forward section and a 45-degree sweep angle for the main section. The wing was approximately 60 feet in length at its intersection with the fuselage and had a maximum thickness of 5 feet.

Each wing had four major subassemblies: the forward wing box, the intermediate section, the torque box, and the wing-elevon interface. The forward wing box was a conventional configuration of aluminum ribs, aluminum tubes and tubular struts,[4] with upper and lower skin panels in stiffened aluminum and the leading edge spar in corrugate aluminum. The intermediate section consisted of aluminum multi-ribs and tubes with upper and lower skin panels consisting of aluminum honeycomb. Almost half of the volume of this section was occupied by the main gear well. All torsion and bending loads applied to the wing were combined and distributed to the rest of the vehicle structure by means of the torque box section, which was composed of four corrugated aluminum spars plus a series of tubular struts, with upper and lower skin panels in stiffened aluminum.

The wing/elevon interface contained the elevon hinge mechanisms and the seal panels on the upper skin. Aircraft with a delta wing are often called "tailless"

---

[4] Generally speaking, for all sections of the Orbiter's wing these tubular struts acted as ribs to maintain the airfoil of the wing, whereas in a conventional aircraft the ribs are holed webs. The tubular strut configuration was chosen to save weight.

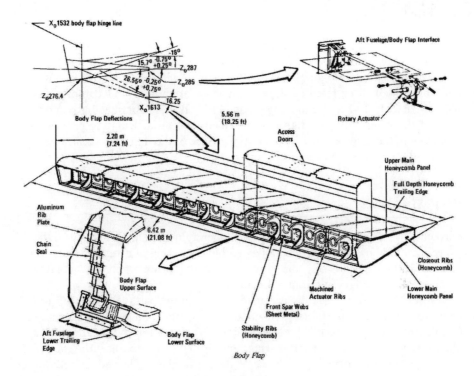

Body flap structure.

because they do not have a horizontal empennage for stabilizers and an elevator. In this case the aerosurfaces of the trailing edge of the wing are called elevons because they act as an aileron (if moved together in opposite directions) and as an elevator (if moved together in the same direction). For the Orbiter, there were two elevons for each wing made of conventional aluminum multi-rib and beam construction with aluminum honeycomb skins for compatibility with both the acoustic environment and thermal interaction. Each elevon could travel 40 degrees up and 25 degrees down by means of hydraulic actuators located at points along their forward extremities. The fact that the elevon hinge line was established near the lower wing surface meant that elevon motion resulted in large changes in the position of the upper elevon forward edge, opening a gap between the elevon and wing that had to be sealed to prevent hot gas from penetrating the interior during re-entry and damaging the aluminum structure and the hydraulic systems of the actuators. In addition to thermal sealing, the upper wing/elevon junction required a fairing to provide aerodynamic smoothness in an area subject to high vibration, pressure differentials, and buffet/flutter loads. This gap was sealed in the wing/elevon transition section, and consisted of a series of hinged panels made of Inconel honeycomb sandwich construction for the outboard elevon and of titanium honeycomb sandwich construction for the inboard one. A titanium honeycomb rub strip was placed on the upper surface of the elevon leading edge to act as a sealing surface area for the seal panels.

66 A skeleton for the Orbiter: structure and mechanisms

BODY FLAP

Although not a flap in the sense commonly used for aircraft, the aerosurface at the rear of the aft fuselage called the body flap had the function of protecting the main engine bells during re-entry and providing the Orbiter with pitch control trim during atmospheric flight after re-entry. It was made of aluminum and comprised ribs, spars, skin panels and a trailing edge assembly. Like the thermal barrier on the elevons of the wing, the body flap was provided with an articulating pressure and thermal seal in the forward area on the lower surface of the flap to block heat and air flow from the structure during ascent and re-entry.

VERTICAL TAIL

To provide directional stability during atmospheric flight following re-entry, the aft fuselage section housed the vertical tail assembly consisting of a vertical stabilizer, a combined rudder/speedbrake, the housing for the power unit to actuate the aerosurface, and the landing drag parachute.

The vertical stabilizer structure was made of aluminum, with the main torque box constructed of integral-machined skins and strings, ribs, and two machined spars. The full structure was attached by two tension tie bolts at the root of the front spar to the forward bulkhead of the aft fuselage and by eight shear bolts at the root of the rear spar to the upper structural surface of the aft fuselage. The rudder/speedbrake

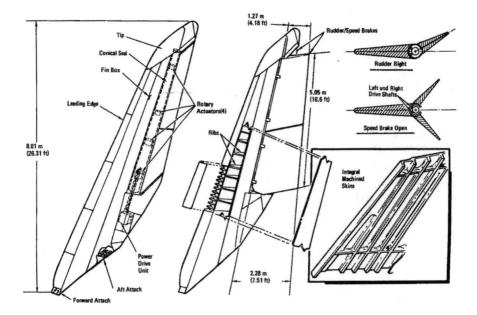

Vertical tail structure.

control surface consisted of two diverging halves, each made of aluminum ribs and spars with aluminum honeycomb skin panels attached to the fin through rotating hinge parts. The aerosurface could be used both as a conventional rudder for yaw control with both halves moving in the same direction, and as speedbrake by opening the two halves in opposite directions.

ROBOTIC MANIPULATOR ARM

Most of the missions flown by the Shuttle relied on a truly iconic system installed in the payload bay. Previous spacecraft did not perform activities such as deploying and retrieving satellites, servicing telescopes, and assembling a space station. From the inception of the program it was understood that for the Orbiter to be able to conduct such tasks it would require a robotic manipulator.

At that time, Canada was developing an interest in a topic that is quite common nowadays but was then a novelty, namely teleoperations. This involves the use of remotely controlled manipulator systems to perform risky jobs in hostile environments, such as underwater operations, nuclear power plant maintenance and mining. NASA approached the government of Canada proposing to use their skills in developing a robotic manipulator system. A memorandum of understanding was signed in July 1975 with the National Research Council of Canada for the design, development, test and evaluation of the Shuttle remote manipulator system (SRMS). The cost of the development phase was estimated at US$78.5 million, paid in full by Canada. On the other hand, NASA committed to purchase each production unit for US$66 million and promised that Canada would have preferred access to the Shuttle for its payloads and astronauts.

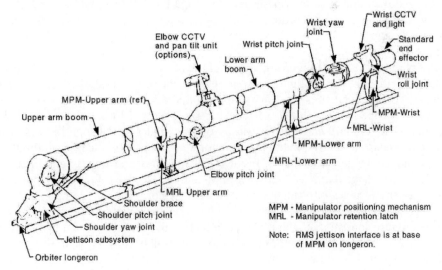

RMS stowed position and general arrangement.

The SRMS was developed to stringent requirements. In particular, its envelope could not exceed 15 inches in diameter and its maximum length was 50 feet. The overall weight, including the display and control subsystem, closed-circuit television (CCTV) and manipulator controller interface unit could not exceed 994 pounds. The system was required to maneuver a "design case" payload of 32,000 pounds having a maximum diameter of 15 feet and a length of 60 feet, and be capable of retrieving a payload of 65,000 pounds in a contingency operation. The articulation of the robotic arm very closely resembled the human arm, with three electromechanical joints, six degrees of freedom, and two boom segments. To maximize its capability to move in any direction and cover the largest possible envelope, the three joints were placed (by analogy with human anatomy) on the shoulder (the point of connection with the Orbiter), the elbow (the joint between the two booms) and the wrist (the connection with the end effector). The shoulder was capable of pitch and yaw actions, the elbow swung in pitch only, and the wrist facilitated pitch, yaw and roll. The initial plan for hydraulic motors was soon discarded due to the hazard that a fluid leak would pose to the payload, vehicle and crew. Small, lighter and hazard-free reversible frameless brushless direct current motors with high ratio gear boxes were instead selected as actuators for each degree of freedom.

The frameless configuration allowed a single housing to include the motor, brake and tachometer. The choice of a brushless motor allowed for some weight saving due to the absence of commutator brushes, and also reduced wear and tear of the motor. With all joint motors of the same type and low power, the torque values required for each joint were provided by an extremely high precision, high reduction, two-stage gearbox with an epicycle/planet system. The first stage provided for a gear reduction ratio that was variable from joint to joint to meet specific joint torque requirements. The second stage was identical for all joints, and consisted of a planetary design that maximized the number of load paths and provided high stiffness and low backlash. A key requirement of the gearbox was back-drivability in order to reduce the scope for overstressing the structural integrity of the arm by excessive loads resulting from high forces and moments.

But perhaps the most interesting feature of the robotic arm was its "hand", the so-called end effector. The configuration selected was the one that best satisfied the two stringent requirements of a two-stage grappling operation in which there was a large capture envelope and the initial soft docking was followed by a rigidization of the interface. The standard end effector was a hollow cylinder installed on the wrist roll joint at the end of the arm, made primarily of aluminum with a length of about 21 inches and a diameter of 13 inches. Inside the cylinder was the grappling mechanism consisting of a large diameter snare drive ring and a mechanism for pulling this ring and three snare cables into the body of the end effector.

To grapple a payload, it was necessary to have installed on its structure a so-called grapple fixture consisting of a grapple pin, a target, and three alignment cams which would mate with cam lobes on the end effector. The task of the grapple fixture was to transmit torsional bending and shear loads between the end effector and the payload within specified limits for each load to be applied either singularly or in combination. The standard grapple fixture was called the "flight releasable grapple fixture"

Robotic manipulator arm

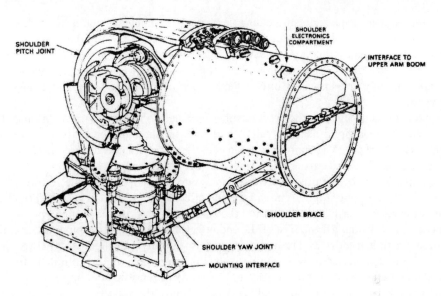

Overall configuration of the shoulder joint.

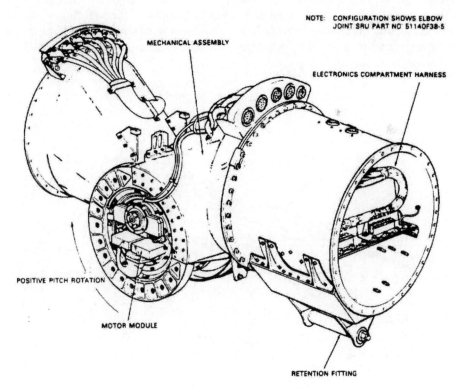

Overall configuration of the elbow joint.

because it had a mechanism to enable a spacewalking astronaut to release the grapple pin in the event of the end effector failing to release the payload. For payloads that had either to be provided with power or to receive/send data whilst grappled, the "electrical flight grapple fixture" was employed. This sported a male connector plug for insertion into a female connector on a "special purpose end effector".

To capture a payload, the crew member assigned to maneuver the arm (and thus known as the robotic arm operator) had first to align the end effector over the grapple fixture pin. The snare inside the end effector consisted of three cables which had one end attached to a fixed ring and the other end to a rotating ring. The sequence began by rotating the three-wire snare around the grapple fixture pin in order to center the grapple fixture in the end effector. The payload was then rigidized by drawing the snare assembly inside the end effector using a jackscrew, pulling the payload tightly against the face of the end effector so that the arm/payload assembly could be fully rigidized. During this process, capture misalignment between the payload and end effector was eliminated as the three cams on the grapple fixture sat in the end effector. Some 4,000 to 7,120 newtons of tension were applied overall to the grapple fixture as the snare cables were driven to the rigid position. To release a payload, the snare mechanism was moved outward until the force pulling the payload against the end effector was nulled, derigidizing the payload. Then the snare of the end effector was rotated open to release the grapple fixture. During grappling and ungrappling operations, current limit commands were sent to each joint motor to allow the arm to move and compensate for misalignment errors.

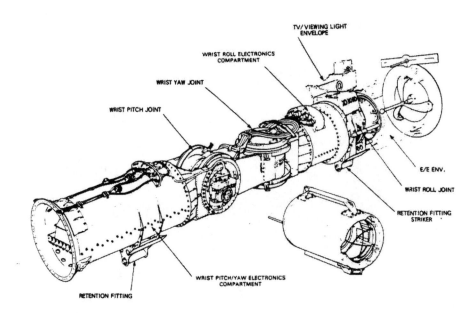

Overall configuration of the wrist joint.

Robotic manipulator arm 71

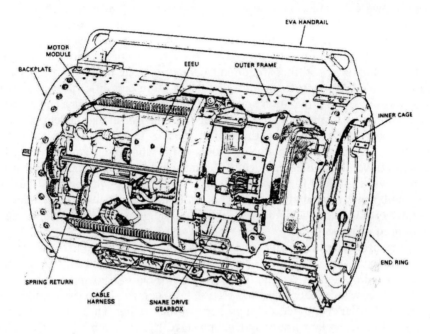

End effector arrangement.

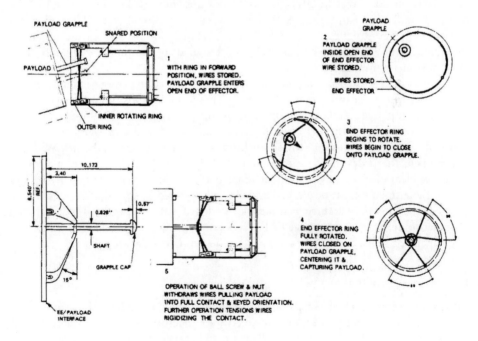

End effector capture and rigidize sequence.

The elbow joint connected the lower and upper booms; the human equivalent of forearm and arm. During development, dynamic rigid body analysis revealed that joint torques, end effector speed, and stopping distance requirements contributed significantly to the manipulator's weight and operation. In addition, it was found that the manipulator's stiffness requirements had a strong influence on its weight. So for simplicity of construction a circular section graphite/epoxy composite design was chosen. To maintain stability of the thin-walled booms, intermediate stiffening rings were placed internally at regular intervals along the length of each boom to prevent the flattening that could cause a thin-walled tube to collapse under bending. Another advantage was to place the vibration frequency of the tube wall well above the high energy acoustic range experienced during Shuttle liftoff. To limit susceptibility to impact damage, a protective bumper of crushable honeycomb was wrapped around each boom. In a high velocity impact, such as a micrometeoroid strike, the damage would be visible on the bumper, indicating the area in need of inspection.

Thermal protection was by means of both active and passive control systems. The passive one consisted of multi-layered insulation blankets attached with Velcro to the booms and to each other. And because areas around the moving parts could not be blanketed, white coating paint was used to control hardware temperatures exploiting the solar energy reflection property of the coating. The active part used 26 heaters in two redundant heater systems powered by two main buses. The heaters were placed mostly at the arm's joints and at the end effector to warm the electronics and electric motors. Weight-efficient wire cables were mounted on the exterior of the booms by adhesive Kapton tape. Sufficient slack was allowed for relative thermal expansion and contraction between the cables and arm booms, and for movement of the joints. The electrical harnesses were designed to minimize cross-talk and electromagnetic interference.

For launch and re-entry, the robotic arm was stowed stretched out in a line on the port sill longeron, sitting on three so-called manipulator positioning mechanisms that were simply pedestals containing a latch for securing the arm. For launch, re-entry, and any other period in which the payload bay doors were closed, the pedestals were rolled into the stowed position, some 11.88 degrees inboard. With the doors open, the pedestals were rotated 19.5 degrees outboard in order to provide sufficient clearance for payload unberthing, berthing, and maneuvering. If for any reason the arm could not be cradled and stowed, it could be jettisoned to enable the doors to be closed. All three pedestals provided a jettison point, along with the shoulder joint attachment on the Orbiter where there was a redundant pyrotechnically operated guillotine to sever a wire bundle prior to separation.

Robotic arm operations

SRMS operations were conducted at the aft flight deck workstation by at least a pair of astronauts. The astronaut that controlled the trajectory of the arm was stationed on the port side while the other worked on the starboard side to control inputs, payload retention latch assemblies, and cameras.

Robotic manipulator arm

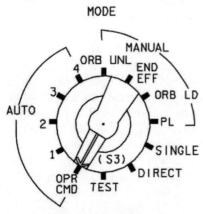

Robotic arm MODE switch.

The robotic arm could be operated in several different ways. Some of them were computer-assisted, moving the joints simultaneously as required to bring the point of resolution (POR) to the desired location.[5] Other modes moved only one joint at a time. During training for a mission, the astronaut assigned to operate the arm used a specific simulator to become familiar with the various modes and thus learn when a given mode should be selected as part of an overall plan to perform an operation. A rotary switch was used to select the desired arm mode in two major classes of auto (or computer-assisted) and manual-augmented modes.

Generally speaking when the arm was used in the auto mode, it was commanded to automatically move the POR to a desired location and then hold this position. The software automatically calculated the velocity commands for the joints in a manner that would not exceed the maximum permissible joint rates, and safely halt the arm within 2 feet of the desired point. Auto mode was present in both auto commanded and in operator commanded submodes.

In the operator commanded submode the end effector was moved from its present position and orientation to a new one defined by the operator via the keyboard and robotic arm display. After keying in the data, the operator had to command a check to verify the availability of a set of joint angles that would put the arm at the desired point. It is important to recognize that this check did not verify the trajectory of the arm, so each command had to be verified as safe prior to use. If the destination point was valid, a "good" flag would appear on the display and the operator would put the mode switch to OPR CMD, remove power to the brakes, and press ENTER to start the arm moving. The auto commanded submode was very similar to the operator commanded sequence. The only difference was that instead of having the robotic arm operator enter the data into the computer, the arm was maneuvered to one of two hundred points that were recognized by the software. These points could be grouped in up to 20 different sets for commanding complex operations. Once a set was chosen and ENTER was pressed, the software continuously calculated the straight line from the current POR to the next point. An example of using this submode was the orbital inspection of the Orbiter's thermal protection system using

---

[5] The point of resolution was a software-defined point about which all translations and rotations occurred. For an unloaded arm this corresponded with the tip of the end effector, but for a loaded arm it was usually defined to be at or close to the geometric center of the payload.

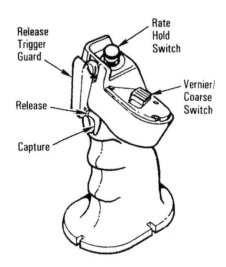

RMS rotation hand controller.

a special sensor boom. The auto mode allowed this complex inspection to be performed without the sensor boom coming into contact with the Orbiter. It was performed by all Shuttle missions after *Columbia* was lost as a result of damage to its thermal protection system.

By selecting one of the four manual augmented modes available, the robotic arm operator could take full control and use a translational hand controller (THC) and a rotational hand controller (RHC) to control the arm's POR. For each hand controller, the manual inputs provided were converted in a proportional way via software into velocity commands. A rate-hold button on top of the RHC allowed the arm operator to maintain the translational and rotational rates of the POR at commanded values.

To assist the operator in commanding the desired input, a pan/tilt CCTV camera was installed on the elbow joint and a fixed CCTV camera was installed on the wrist joint. Other cameras in the payload bay enabled the operator to view the arm relative to the payload or surrounding structures. This information was presented to the crew at the aft flight deck workstation by two monitors with split screen capability.

The general principle common to each manual augmented mode was that rate commands from the THC would cause motion at the tip of the end effector parallel to the Orbiter-referenced coordinate frame and compatible with the up/down, left/right and in/out directions of the THC itself. Similarly, commands from the RHC caused rotations at the tip of the end effector in the Orbiter-referenced coordinate frame.

For a grappling operation in conjunction with the wrist-mounted camera, the operator would select a manual augmented end effector submode which maintained compatibility at all times between the rate commands of the THC and RCH and the instantaneous orientation of the end effector. That is, the operator could imagine he was sitting at the tip of the end effector. The up/down, left/right and in/out motions of the THC commanded a corresponding motion of the end effector, which manifested itself on the monitor by the background in the scene moving in the opposite direction. The operator had therefore to use a fly-to control strategy and apply commands to the THC and RHC to steer towards the target area presented on the monitor.

If, having grappled the payload, the task was to berth it on a structure such as the International Space Station, the operator would use the manual augmented Orbiter loaded submode. In this case, THC and RHC inputs would command movement of the full arm as seen from the aft viewing windows in terms of the Orbiter's reference system. Thus, if the arm had to be translated to the left, with the operator looking aft in the cargo bay, he was required to move the THC to the left. For this submode the POR was a point inside the payload. To move an arm in this manner with no payload attached, the Orbiter unloaded submode would be selected. The only difference with the previous one was that in this case the POR would be the tip of the end effector. It is important to understand that the Orbiter loaded/unloaded and the end effector modes differ solely in the reference system employed; namely, Orbiter reference for the Orbiter modes and end effector reference for the end effector mode. Let us examine further why this is important. If a payload needs to be grappled, the end effector has to be brought to the grapple fixture. All commands imparted to the arm will have the purpose of translating and rotating the end effector so that it can arrive in the correct attitude for engaging the grapple fixture. If we now suppose that the end effector is directly "in front" of the fixture, a simple translation command will bring it into contact with the grapple. But along which axis, should it translate? To answer this question, it is necessary to remember that in order to translate and rotate something a reference system must be defined together with an origin point. Due to the disparate shapes and sizes of the payloads, grapple fixtures could be placed anywhere with an orientation that was usually not parallel to any of the Orbiter body axes. Returning to the case where the end effector is ready to be translated towards the grapple fixture, and supposing that the grapple fixture is not oriented to any of the Orbiter body axes, if we retain the Orbiter body reference system then a single translational command will not be enough; it is necessary to perform a series of translations, all parallel to the Orbiter axes, before the end effector can catch the fixture. If, instead, commands are given in the end effector's own reference system, the X axis will coincide with the longitudinal axis of the effector. Hence, regardless of the grapple fixture orientation, a simple translational command on the X axis of the end effector will bring it to the desired target. This is equivalent to standing on the tip of the end effector and viewing the approaching target, and it explains the rationale for placing a camera on the wrist joint.

The body reference system of the Orbiter was invaluable when the payload had to be moved around the payload bay. For an arm operator looking at the payload bay and a payload from the aft flight deck station, it would be more natural to mentally process movements of the entire arm relative to the Orbiter's reference system. If, for instance, the payload had to be moved 10 feet from the bay, such a movement would be commanded as a simple translation along the Z axis. In this case it would not have been sensible to use the reference system of the end effector because the task was to move the payload rather than the end effector. Once a payload has been grappled, the end effector ceases to be part of the arm and can be considered as an extension of the payload. In other words, in the end effector mode, commands are given to move the end effector, whereas in the Orbiter loaded/unloaded modes commands are giving to move the arm and, with it, the payload.

Finally, a fourth submode was provided, called payload mode. It was equivalent in concept to the Orbiter loaded submode, but the hand controller movements were in reference to a predefined software coordinate system, such as the International Space Station reference system when the arm was to berth a payload to one of the existing modules.

A skeleton for the Orbiter: structure and mechanisms

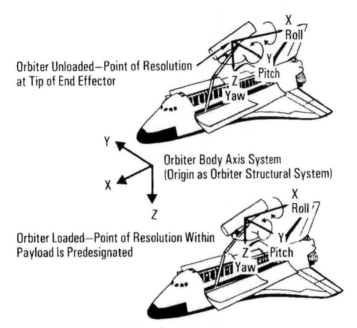

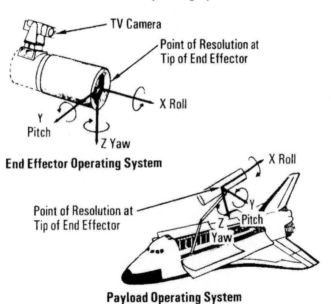

Control coordinate system.

Robotic manipulator arm

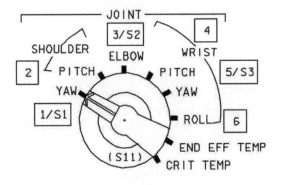

Joint MODE switch.

To unstow and stow the arm on-orbit, another manual augmented mode was used. In this so-called single joint mode the operator would move the arm on a joint-by-joint basis. On selecting this mode, the operator had from time to time to select which joint to activate and key in the desired amount of translation or rotation. While the selected joint was brought in motion, all other joints would hold their positions. Two contingency modes were also provided. The first was called the direct drive control mode and could be used after a software or hardware failure. In this mode, the drive command would bypass the robotic arm's computer to go directly to the joint motor. This meant that the arm software could not check whether the commands imparted would exceed the joint limits, so great care had to be taken in maneuvering the arm. In case of software degradation, another contingency mode, called the backup drive control mode, allowed joint-by-joint use of the arm. Despite having degraded joint performance in both contingency modes, these provided a means of maneuvering a loaded arm to a safe payload release position, or to maneuver an unloaded arm to its storage position.

Early on in the Shuttle program, with successive missions having ever heavier payloads to deploy, it was found that commanding a straight line movement of the arm produced uncommanded translations and rotations. This annoying phenomenon was attributed to a number of factors, including torque disturbances due to dynamic coupling between the joints, friction and so on. It was also noted that when any joint rate crossed zero and the friction torque changed its sign, the uncommanded motion increased. Furthermore, this motion was more significant when the commanded rate was low. This problem became really annoying for the crew of STS-31 (24-29 April 1990) when deploying the Hubble Space Telescope. Afterwards, the arm's software was upgraded with the new position orientation hold submode (POHS). While the arm was in motion, every 80 milliseconds the software compared the actual position of the POR of the arm with its projected position. The difference was processed to automatically generate the small commands required to move the actual position into the projected position. After this software upgrade was introduced, uncommanded motions did not trouble any subsequent crews, regardless of the size of the payload being manipulated by the arm.

Off-nominal arm operations

Although the robotic arm was conceived for deploying and retrieving satellites, it performed a variety of tasks during the Shuttle program. For example, it was used to apply force to lock the antenna of a payload bay experiment in place during STS-41G (5-13 October 1984) and to remove a large icicle that formed on the Orbiter's waste nozzle on STS-41D (30 August-5 September 1984), and configured as a "fly swatter" to engage a jammed switch lever on a satellite on STS-51D (12-15 April 1985). The *Columbia* tragedy also introduced two new possible scenarios in which the robotic arm would have played a pivotal role. The first scenario, referred to as Orbiter repair maneuver (ORM), would have seen the robotic arm grapple the ISS to position the Orbiter so that an ISS crew member could use the space station's own robotic arm to effect a repair. The possibility of an ORM also allowed STS-125 (11-24 May 2009), the final Hubble servicing mission, to take place. Following the loss of *Columbia* in February 2003, all subsequent missions were launched under the premise that if there were to be a problem that ruled out attempting to return home, the ISS would serve as a safe haven. This was fine for missions that visited the ISS. However, owing to the inclination of the Hubble Space Telescope's orbit, this mission would not be able to use the ISS as a safe haven. For this one-off mission, another Shuttle was readied as a rescue vehicle. One orbiter would grapple the other to allow the astronauts of the crippled Orbiter to transfer across. Another scenario, tested for real during STS-121 (4-17 July 2006) showed that an astronaut could ride on the end of the robotic arm's sensor boom extension and make repairs to the thermal protection system.

But perhaps the most off-nominal operation undertaken with the robotic arm was serving as a stable platform for a spacewalker. As five-time Shuttle astronaut and robotic arm operator Steve Hawley notes, "The arm gives you an incredible amount of capability to make an EVA very efficient, either by moving the crew member to a worksite quickly, compared to how long it would take him to go on his own, or by being able to convey hardware to him." The spacewalker could "stand" on a portable foot restraint attached directly to the arm, or he could use a manipulator foot restraint that was grappled by the end effector. In this way, the stable platform represented by the arm allowed the spacewalker to perform tasks much easier than if he were freely floating. It was also possible for a crew member to simply grab the EVA handhold on the end effector and be pulled into position.

Moving an astronaut around in EVA was not as simple a task as it may seem. For example, how rapidly should he be moved? Based on his experience as a robotic arm operator, Hawley explains, "The arm moves either in what they call vernier control or coarse control, and that determines the rates you can command. You like to use coarse rates if you can, because you save time, but that might not be comfortable for the guy on the end of the arm, particularly if he's holding something. I found that in some cases you could use coarse rates and he wouldn't even know."

In riding the robotic arm, an astronaut had to provide clear instructions to the arm operator. Recalling the missions on which he served as an arm operator, including the second Hubble servicing, Hawley says, "We developed a coordinate system to communicate in. He, on the end of the arm, could be in a variety of orientations and if he says, 'Take me to the right', you need to know if he means his right or Orbiter right. So we had a coordinate system that was based on the Orbiter and another one that was based on his body as he stood on the end of the arm. We spent a lot of time practicing being very disciplined in our communication. He might say, 'Take me forward in the bay', which is very clear. It means whatever orientation he's at, he wants to go towards the nose of the Orbiter. Then he would say, 'Okay, I'm going to switch to body coordinate now,' as you might do if you were inside a bay in Hubble. If he's not exactly sure where the nose of the Orbiter is, he'd say, 'Take me head up,' or 'feet down' or 'left' and my job would be to know with respect to his body which way he wanted to go."

During an EVA, spacewalkers would talk to each other, so some discipline was required in directing the arm operator. As Hawley explains, "There is a rule that only the guy on the end of the arm can command the arm. That was a nice little technique to keep any misunderstanding from happening. The only command that the guy that wasn't on the arm could give to the arm operator would be 'Stop'. If he said that then you stopped, because maybe he could see something that nobody else could."

ORBITER STRUCTURAL MAITENANCE

The original requirement that drove the design of the Orbiter's structure was that it should last for a decade or 100 missions, whichever came first, without any need to perform maintenance to the structure. As Kramer-White explains, "There were things done initially to try to make sure that you could inspect it, and there were some inspection doors, but it really wasn't designed to be maintained for thirty years. It wasn't really intended to be flown for thirty or forty years, have people climbing around it doing inspections or have them climbing around it doing modifications." By the mid-1980s this limitation was already surfacing, because it was clear that all the Orbiters would serve well beyond the 10-year life initially thought in the design. So it was necessary to start to draw up a maintenance plan for the primary structure along with a manual for performing structural repairs.

The famous American airline Pan Am was contacted to provide hindsight of the field of aircraft structural maintenance and, as Kramer-White recalls, "They figured out an inspection scheme. It was never intended to be the forever inspection scheme. There was just no reference. They took what was available as an aircraft inspection program and morphed it to the Shuttle and laid in a set of inspections which wound up starting in the down period after *Challenger*." Going deeper into details, Kramer-White adds, "The idea was that it would be a living document, that as we learned we would change it. We specified a set of intervals. Then we'd go do those inspections, and if we found anything we'd adjust the inspection intervals. If we inspected several times and never found anything, then we'd lengthen the intervals for the subsequent vehicles."

Clearly, with such a small fleet of Orbiters flying at a rate well below the initially envisaged one flight per week, drawing up and modifying such a maintenance plan took a long time. Furthermore, each new inspection technique, each new tool, had to be thoroughly certified before it was authorized for use on the Orbiters.

Not having been built to be maintained, getting access to the innermost parts of the Orbiter's structure could be quite difficult, and it also posed a risk to the integrity of the structure itself. One good example was represented by the visual inspections inside the wing structure. To gain access, technicians could use large holes built into the wing spars for manufacturing purposes. Once inside, great care had to be taken in moving around the intricate net of tubes that formed the primary structure. As Kramer-White explains, "They were so fragile. They were in great tension and compression, which is what they were designed for. But you'd be in there banging around doing maintenance, and God forbid, you'd bump a tube because you could actually literally almost squeeze it like a Coke can. If you grabbed it, you could dent it and then it had to be replaced."

Upon finding some kind of damage to the primary structure, the next step would be to repair it. And this brought a whole new series of problems. "Over time, we had to learn to deal with and classify certain types of defects as what we used to call 'fair wear and tear'. It's ok if you scratch it. Just go touch it up. But if the scratch is deeper than this, you call a stress analyst. Or if the corrosion is like this, here's the standard repair." By analogy with the aviation industry, a structural repair manual was built up which contained allowable limits for damage, and the repairs to carry out if the limits were exceeded. For damage not contemplated by the manual, a dedicated repair had to be designed and performed, in the process adding a new section to the manual.

3

Power to orbit: the main engines

INTRODUCTION

When a Shuttle ascended to orbit, it did so with the help of a cluster of three engines. The Space Shuttle main engine (SSME) was the most powerful, reusable, high-performance, liquid propellant rocket engine with variable thrust ever developed. Fed by liquid hydrogen and liquid oxygen, and operating for just a little bit short of nine minutes for each flight, they delivered a combined output of 37 million horsepower, the equivalent of two dozen Hoover Dams. While operating, an engine had to endure temperatures ranging from –423°F (the temperature of the liquid hydrogen used to cool the engine) to 6,000°F (hotter than the boiling point of iron) in the combustion chamber. It could be throttled between 67 and 109 per cent of its rated power level in 1 per cent increments. The value of 100 per cent corresponded to a thrust level of 375,000 pounds of force at sea level and 470,000 pounds of force in a vacuum; 104 per cent (the setting used for a nominal ascent, after the initial flights which were made at 100 per cent) corresponded to 393,800 pounds of force at sea level and 488,800 pounds of force in a vacuum; 109 per cent corresponded to 417,000 pounds of force at sea level and 513, 250 pounds of force in a vacuum (despite being certified to run at this thrust level, the engine was baselined to use such a power level only during an emergency, which fortunately never occurred). Mounted on the rear of the Orbiter, the engines were called engine 1 (center), engine 2 (left) and engine 3 (right). Each was 14 feet long, 7.5 feet in diameter at the nozzle exit, weighed around 7,000 pounds, and was designed to operate for a total of 15,000 seconds over a life span of thirty starts.

MAIN PROPULSION SYSTEM

At its rated power, an SSME consumed 155 pounds of hydrogen and 934 pounds of oxygen per second. Multiply this by three and expand the time scale to that for the climb to orbit, and it is easy to appreciate why one of the most visible elements of the

82 Power to orbit: the main engines

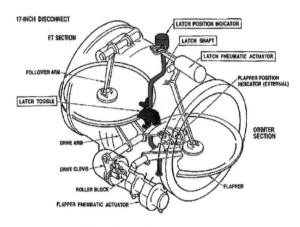

Propellant disconnect valve.

"stack" was the external tank attached to the Orbiter's belly. This will be discussed in a later chapter; all that need be said here is that its aluminum shell was 27.5 feet in diameter and 154 feet in length, and had a forward tank for some 143,000 gallons of liquid oxygen and an aft tank for 385,000 gallons of liquid hydrogen.

The propellant entered the underside of the aft fuselage through a pair of feedline manifolds 17 inches in diameter. There, two disconnect valves provided a physical link between the external manifolds and the internal feedlines which delivered the propellants into the combustion chambers. Each disconnect valve consisted of a pair of flappers, one on the Orbiter's side of the interface and the other on the external side of the interface. During ascent, both flappers remained open for a clean and flawless flow of propellants to the engines. Prior to separation of the external tank at engine shutdown, both flappers in each mated pair of disconnects were commanded to close in order to prevent propellant discharge either from the external tank or from the Orbiter upon physical separation and also to preclude contamination of the Orbiter's propulsion system during landing and ground operations. Since inadvertent closure of either flapper during engine operation would stop propellant flow with potentially catastrophic consequences, after the *Challenger* accident a latch mechanism was added to the Orbiter's side of the disconnects to serve as a mechanical backup to the force of the flowing fluid. The latch was placed on a shaft in the flow stream in order to overlap both flappers and thus prevent closure for any reason. At external tank separation, the latch would disengage from the external tank's flapper and allow the Orbi-

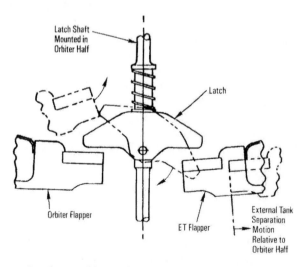

Latch assembly on the propellant disconnect valve.

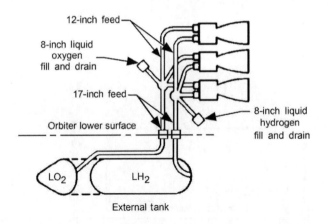

Main propulsion system schema.

ter's disconnect flapper to toggle the latch and thereby close the flappers on both the tank and Orbiter.

Inside the Orbiter, three 12-inch-diameter feedlines split the propellants equally for delivery to each of the individual engines. Both the 17-inch manifolds and the 12-inch feedlines were part of a very complex maze of feedlines inside the aft fuselage. Along with the lines feeding propellant, there were other lines for safe operation of the engines from the time when the external tank was filled to when the engines were shut down on attaining orbit. The role of these other lines will be explained in detail later, so for the moment, to provide a general picture of the main propellant system, it is sufficient to list them as the fill/drain manifolds, relief manifolds, backup liquid hydrogen dump line, and external tank pressurization lines.

Also essential to the propellant system was the helium system, which consisted of seven 4.7-cubic-foot helium supply tanks, three 17.3-cubic-foot helium supply tanks, associated regulators, check valves, distribution lines and control valves. The purpose of this system will be described below, but at this point it is worth noting that the system was divided into four separate subsystems, one for each of the three main engines and a fourth for the pneumatic systems that operated all of the propellant valves.

SSME PROPELLANT FLOW

As with the saying that all roads lead to Rome, inside an SSME the two propellants followed different paths through a maze of feedlines, turbopumps and preburners on their way to the combustion chamber.

The first milestone on the journey, for both fuel and oxidizer, was a prevalve on each 12-inch feedline. Beyond this, each propellant was directed to a low-pressure pump in which the delivery pressure would be boosted well above the tank storage pressure. From a configuration point of view, the low-pressure fuel turbopump (LPFTP) was an axial-flow pump driven by a two-stage axial flow turbine powered by gaseous hydrogen, and it increased the pressure from 30 psi to 276 psi. The low-pressure oxidizer turbopump (LPOTP) was also an axial-flow pump driven by a six-stage turbine powered by liquid oxygen, and it increased the liquid oxygen pressure from 100 psi to 422 psi. Next were the high-pressure turbopumps, which had the task

84 Power to orbit: the main engines

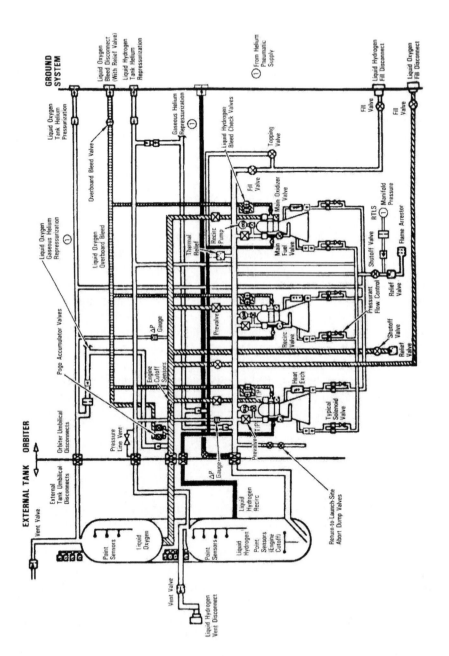

Main propulsion system fluid system.

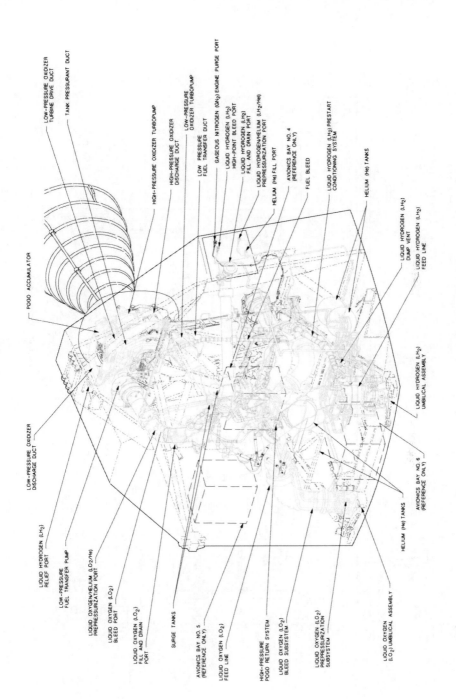

Propellant feedline arrangement in the Orbiter aft section.

86 Power to orbit: the main engines

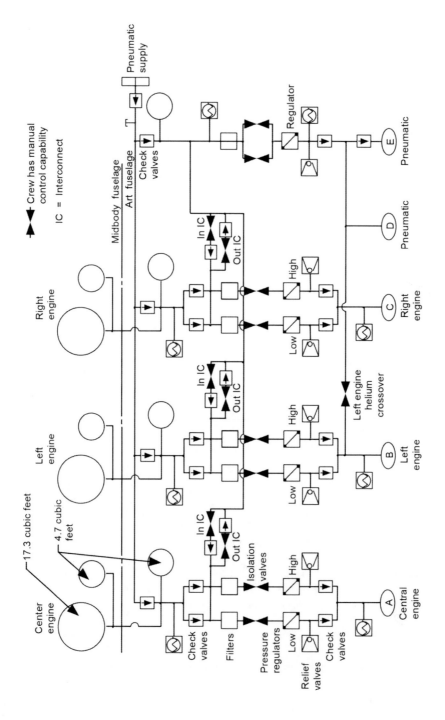

Helium system supply and storage.

SSME propellant flow 87

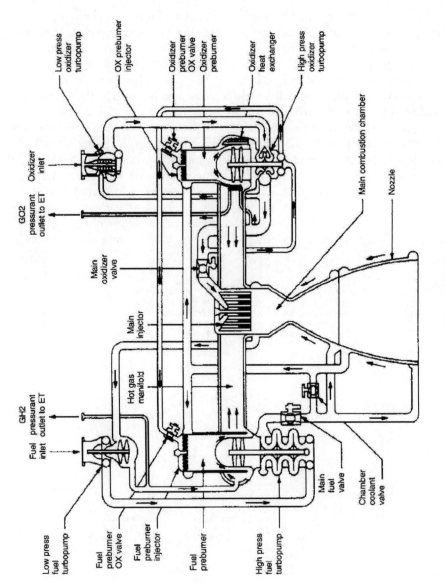

SSME schematics.

88 Power to orbit: the main engines

of further raising the pressure to levels suitable for permitting the flow to continue into the chamber for combustion sufficient to produce the necessary thrust. From a configuration point of view, the high-pressure fuel turbopump (HPFTP) was a three-stage, hot-gas turbine which operated at about 35,360 rpm and boosted the liquid hydrogen pressure from 276 psi to 6,515 psi. The high-pressure oxidizer turbopump (HPOTP), however, consisted of two single-stage centrifugal pumps mounted on a common shaft and driven by a two-stage, hot-gas turbine. At an operating speed of 28,120 rpm, it would boost the liquid oxygen pressure from 422 psi to 4,300 psi.

Although a process using two different turbopumps adds complexity and weight, and reduces reliability, it must be remembered that in order for an engine such as an SSME to work efficiently and deliver the highest possible thrust, the propellants must be delivered to its combustion chamber at pressures much higher than they are stored in the external tank. One of the most worrisome issues with turbopumps is that if the difference between the pressure at the inlet and at the outlet is too high, low-pressure cavities can form in the liquid, causing the formation of bubbles. These bubbles are short lived, since they implode immediately after formation. Amongst other things, this phenomena, called cavitation, can cause a rapid degradation of the moving part of the pump that at best results in a loss of efficiency and at worst causes catastrophic failure. Since pumps for rocket applications run at very high speeds, cavitation can destroy a pump within seconds. Staging the increase in

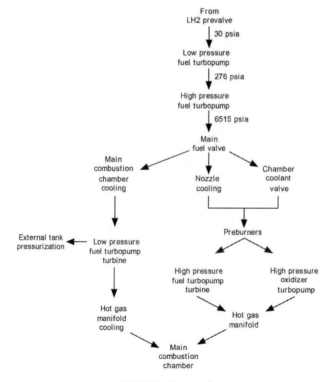

SSME hydrogen flow.

pressure enables the high-pressure pump to work in conditions that eliminate the possibility of cavitation.

After exiting the high-pressure turbopumps, the fuel and oxidizer had to follow different paths to provide the high efficiency and high thrust for which the engines were designed.

In the case of the liquid hydrogen, the flow discharged from the high-pressure turbopump was routed first through a main fuel valve which split it into three flow paths. In one, liquid hydrogen was sent to flow around the combustion chamber as coolant to prevent the chamber from melting due to the high internal temperature. As the hydrogen picked up heat, it became gaseous and its expansion was used to drive the turbine of the low-pressure fuel turbopump. From there, part of the gaseous hydrogen was directed to flow around the so-called hot gas manifold in order to pick up further heat prior to being fed into the main combustion chamber. The remaining gaseous hydrogen was sent back to the external tank to maintain the pressurization. Following the second path, liquid hydrogen was made to circulate around the nozzle walls as coolant to draw the heat from the high-temperature exhaust and prevent the nozzle from exceeding its maximum working temperature. Entering the nozzle from the bottom, the hydrogen climbed up to the top, where it joined the third flow, which was the hydrogen that flowed through the chamber coolant valve and bypassed the nozzle coolant circuit. This united flow was promptly split and sent to the fuel and oxidizer preburners, small combustion chambers where hydrogen and oxygen were burned in order to create the gas required to drive the turbines of the high-pressure pumps.

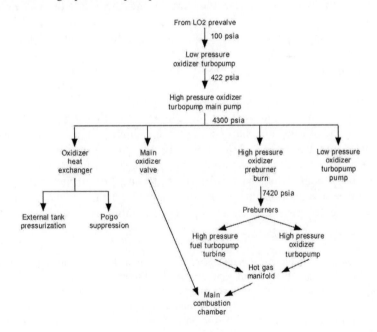

SSME oxygen flow.

In a similar way, the liquid oxygen flow was split into multiple paths, but in this case with four paths. In the first path a small portion of the liquid oxygen was sent to the oxidizer heat exchanger and the resulting gas was used to pressurize the external tank and the POGO system. In the second path, most of the oxygen passed through the main oxidizer valve and directly into the main combustion chamber. A third path routed oxidizer to the two preburners via the high-pressure oxidizer turbopump that further raised the pressure to 7,200 psi. Finally, on the fourth path, a small flow of oxidizer was sent to the turbine driving the low-pressure oxidizer pump and from there back to the main stream for the high-pressure pump.

It is this very complexity that accounts for the incredible performance delivered without a glitch during every launch of the Shuttle program.

In designing the SSME, engineers opted for a so-called staged-combustion cycle, a close-loop cycle in which all propellants are used to produce thrust. To understand this, it is necessary to remember that any rocket engine requires the combustion of propellants at high pressures. In the case of a high-performance engine the pressure can be several thousand psi. To avoid building incredibly robust and heavy tanks, propellants are stored at pressures far lower than those necessary for combustion and turbopumps are used to increase the pressures to the required level.[1] The theory of a turbopump is a well-known discipline, and designing such machines is nothing really exotic. But when it comes to rocket applications, a big problem arises concerning the way in which the turbopumps are driven. As the name implies, they have to be run by turbines. This requires a working fluid that can be created in several ways. One way is to burn a small proportion of the propellants in a gas generator, use the exhaust to drive the turbine and then vent the gas overboard. Known as the gas generator cycle, this approach was used for the F-1 and J-2 engines that powered the mighty Saturn V rocket, and it is used by modern launch vehicles such as the Delta IV, Ariane V and Atlas V. Despite providing very good performance, a gas generator cycle has the far from negligible drawback that not all of the propellants carried on board are used to produce thrust, since that proportion used to drive the turbopumps is lost by being vented.

The high performance specified for the SSME meant that the propellants had to be supplied to the combustion chamber at pressures higher than any previous engine. This required more powerful turbopumps, driven by more powerful turbines. It was realized early on that if a gas generator cycle approach was used, then a prohibitive amount of propellant would be consumed in driving the turbines. The inevitable result would be a reduction in the amount of payload that could be carried into orbit. It was decided instead to develop a staged combustion cycle which, although

---

[1] For the sake of completeness it is necessary to remember that for some applications where low performance is requested, turbopumps can be dispensed with and propellants are fed by the internal pressure at which they are stored in the tanks. In this case, the tanks are slightly heavier because they have to withstand higher than usual pressures, but this increase in weight is compensated by the simplicity and cheapness of a system that has no complex and dangerous turbine-driven pumps.

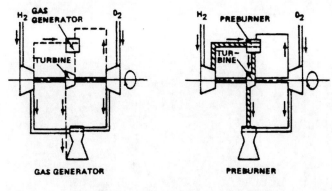

Gas generator (left) and preburner (right).

more complex, would provide the desired high performance without increasing the amount of propellant to be carried. A staged combustion cycle retains the gas generator, but instead of being vented overboard the exhaust is "made to work" by being fed into the engine and burned in the main combustion chamber.

The SSME was the first liquid rocket engine developed by NASA to use this cycle, and the configuration adopted used two preburners in which a fuel-rich combustion created the gas to drive the high-pressure turbopumps. The fuel-rich combustion was chosen to deliver an exhaust at a relatively low temperature that would not damage the turbine blades. It is interesting to notice that while part of the liquid oxygen was sent directly to the main combustion chamber, all of the hydrogen was first passed through the preburners so that none was sent directly to the combustion chamber.

The reason for having two preburners stemmed from the fact that the SSME had to be capable of being throttled to vary its thrust. This required the oxygen mass flow rate to the combustion chamber to be capable of being varied, whilst at the same time maintaining the mixing ratio constant in order to avoid an early depletion of either propellant. When a change in thrust was required, the engine controller commanded the oxygen valve on both preburners to allow the necessary oxygen to flow in order to generate the amount of gas to drive the high-pressure turbopumps at the right speed. In this way, the amount of hydrogen would match the amount of oxygen needed to produce the commanded thrust level.

SSME MAIN COMPONENTS

Powerhead

To achieve the desired propellant flow paths, the SSME was mainly built around the so-called powerhead, an assembly of eight major units that can be thought of as the backbone of the engine. Specifically, the hot gas manifold served as a structural base for the main injector of the combustion chamber, the two preburner injectors, the oxygen heat exchanger, the high-pressure turbopumps, the main combustion chamber and the two preburners. The injectors were welded to the top of the manifold, and the combustion chamber and turbopumps were bolted to its bottom. This configuration created a compact, efficient package wherein the

92 Power to orbit: the main engines

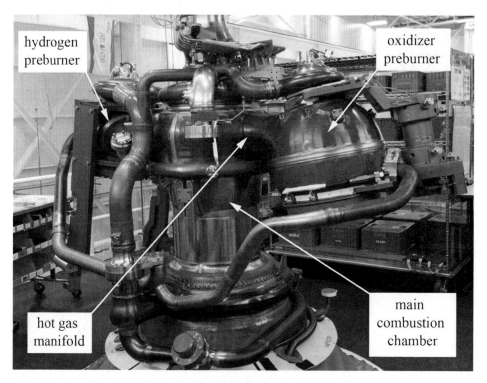

Powerhead.

turbopumps were closely coupled to the preburner and the duct lengths and losses could be minimized. To simplify their replacement, both turbopumps were canted from the vertical by ten degrees. Along with being a structural support for all major engine components, the hot gas manifold also acted as a hot gas passageway from the high-pressure turbopump turbines to the combustion chamber injectors.

Main combustion chamber

The combustion chamber is the roaring pulsing heart of every rocket engine so far developed. Not only because it is normally buried well inside the bundle of piping, valves and wiring of a rocket engine, but also because the inferno generated within it creates the vital energy that will be converted into thrust. Generally speaking, a combustion chamber is a metal cavity that does not appear very interesting. But getting its shape right can give engineers a lot of headaches and involve many hours of experimental trial and error. This is because the process of combustion in the chamber is actually a "controlled explosion".

It is necessary for the propellants to remain inside the chamber for sufficient time to enable the chemical reaction to run efficiently. This time has to be longer than that required for the reaction to complete, to ensure that all the propellant is converted to hot exhaust gas at high pressure for passage through the exit nozzle.[2] Although it would be interesting to explore the equations used in sizing a combustion chamber, we need only note that these mathematical relationships show that the stay time is dependent on the volume of the chamber but, perhaps surprisingly, not its geometry. This means that as far as the equations and combustion chamber performances are concerned, for a specified volume the chamber can be literally any shape. In practice, however, the choice is quite limited. For example, a long chamber with a small cross section would subject the exhaust flow to a lot of friction, with the resultant pressure drops yielding unreliable thrust. In addition, a long chamber requires a larger space envelope which imposes space limitations on the injector design to accommodate the required number of injector holes. On the other hand, in a short chamber with a large cross-section the atomization or vaporization zone would occupy a relatively large portion of the chamber volume, while the mixing and combustion zone would be too short for efficient combustion.

Three shapes have been widely explored and used for rocket engines: spherical, nearly spherical, and cylindrical. Compared to a cylindrical chamber, the spherical or nearly-spherical chamber with the same volume offers the advantage of less cooling surface and weight.[3] The first advantage is because a sphere has the smallest surface-to-volume ratio, while the second is because the structural walls of the spherical chamber are about half the thickness of those of a cylindrical chamber for the same strength of material and chamber pressure.

Nevertheless, spherical chambers are nowadays rejected since they are difficult to manufacture and, under most circumstances, they have poor overall performance. For this reason, every modern rocket engine has a cylindrical combustion chamber. Once again, it is necessary to stress that it is of paramount importance to calculate the right volume of chamber to maximize the thermal energy of the exhaust. Since kinetic energy is proportional to the square of velocity, it is essential to achieve the highest exit velocity from the nozzle of the combustion chamber. The chamber of the SSME is a cylinder with a throat section at the bottom of it creating a convergent-divergent conduit that enables the exhaust to smoothly accelerate while discharging into the nozzle. A casual inspection of the inside and outside of the chamber would suggest that its walls are solid metal, but this is not the case.

During operation at full power, the combustion chamber of the SSME contained a reaction that occurred at about 6,000°F. This temperature was far too high for even the most thermally resilient material nowadays available to withstand. Nevertheless, during all 135 launches of the Shuttle program, the combustion chambers survived in excellent health and, with refurbishment, for more than one

---

[2] It must be added here that in reality, due to several factors, it is not actually possible to have all the propellant reacted and transformed to gaseous exhausts, so there is a loss in performance of the chamber. The aim of the combustion chamber designer is to minimize such losses, since they translate into a thrust lower than that theoretically possible.

[3] A sphere has the smallest surface-to-volume ratio. As will be shown, combustion chambers require to be constantly cooled during functioning.

94 Power to orbit: the main engines

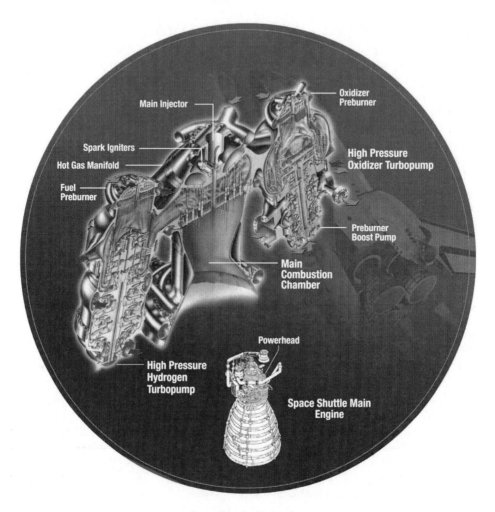

Powerhead schematics.

flight. The secret was the internal structure of the chamber walls which, rather than being solid, were hollow. The structure of the chamber was a spin-forged Narloy-Z liner, machined to form the inner diameter of the chamber. On the outer surface of this liner, 390 axial coolant channels were made by a milling process, at the completion of which molten wax was poured into each of the channels while the liner was being continuously rotated horizontally. After cooling to room temperature, the liner was wet sanded by hand to expose the Narloy-Z on the part between the channels. Next a layer of copper was electrodeposited to close the channels and serve as a hydrogen embrittlement barrier and then a layer of nickel was added by electrodeposition. The result was machined to the desired dimensional requirements. As the last manufacturing step, the wax was melted and washed out of the channels by a complex cleaning process that assured no damage to the channel walls. In this

SSME main components

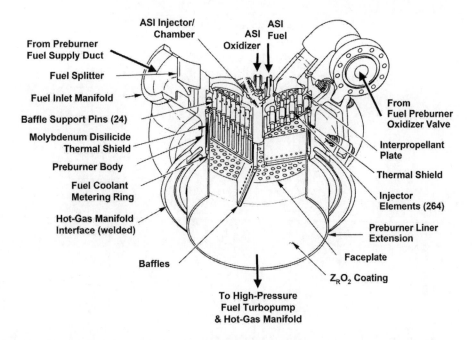

Fuel preburner.

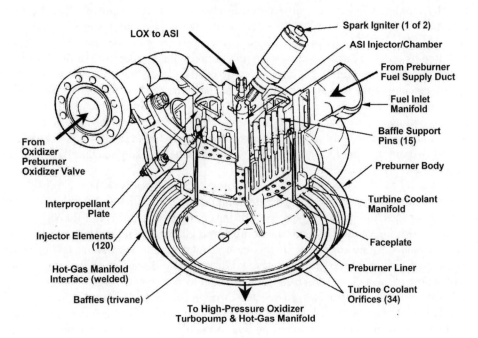

Oxidizer preburner.

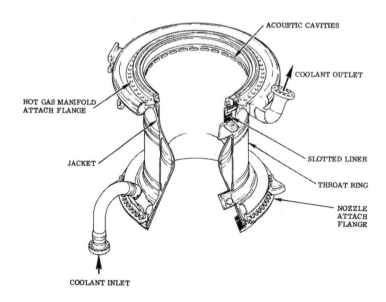

Main combustion chamber.

way, the internal structure of the chamber wall was made of hundreds of small channels within which cryogenic liquid hydrogen could be made to flow as coolant, thereby making the wall of the chamber serve as a heat exchanger. The heated hydrogen turned to gas and was made to do work prior to being fed into the chamber. In fact, the hydrogen would carry away a good portion of the heat acquired by the wall during combustion, maintaining the temperature of the material inside its operative limits. To maximize the efficiency of the heat exchanger, the hydrogen entered the channels from the throat section going up to the injector plate. Each milled channel had a length of 24 inches with varying rectangular cross sectional dimensions ranging from 0.035 by 0.093 inches at the throat to 0.060 by 0.247 inches at the top of the liner. To strengthen the chamber, a two-half jacket was placed around the liner, together with a throat ring for supplemental structural support and reinforcement.

Just as a heart receives blood by means of a series of blood vessels, a combustion chamber must be constantly fed with essential fluids. It has already been shown how liquid hydrogen and liquid oxygen reached the combustion chamber of the SSME by a rather winding path. To enter the combustion chamber both propellants had to pass through an injector plate, in this case a device made with some 600 coaxial elements. Cold liquid oxygen from the oxidizer manifold was injected through the center post of each element. At the same time, cold gaseous hydrogen and hot fuel-rich gas from the preburners were injected through the space surrounding each post, coaxially with the oxygen. Flow shields bolted to the outer row of elements helped to protect them from damage and erosion due to the high speed gases coming from the preburners on the hot gas manifold.

SSME main components

In general terms, to start operating a combustion chamber an appropriate device will ignite the propellants and then be turned off once the reaction is self-sustaining. Since the invention of the modern rocket engine, a number of ignition devices have been developed and utilized. Regardless of the method used, there is one overriding requirement: the minimum ignition delay. If the propellants entering the chamber are not promptly ignited, an explosive mixture can form and lead to a detonation known as a "hard start" with damaging results. Some propellants are so reactive that they ignite on coming into contact, but a hydrogen and oxygen mixture requires a spark.

For the SSME, the ignition system chosen was the so-called augmented spark igniter (ASI), permanently located in the center of the injector. This consisted of two redundant spark igniters that fired inside a small chamber where fuel and oxidizer were injected in a precise manner in order to establish specific conditions. Oxidizer was injected into the chamber through two different holes to form streams of slow-moving liquid that tended towards the center of the chamber and created an oxidizer-rich environment at the point of ignition. Fuel was injected through the sides of the chamber, as eight tangential streams of light, fast-moving gas whirlpools around the oxidizer to allow a thorough propellant mixing and cooling of the walls. The small chamber was essentially a nozzle that lacked a throat and dispersed the combustion products widely across the face of the main injector. After 4.4 seconds, the igniters were turned off, while the ignition flame continued in order to prevent intermittent and possibly damaging blowback from the main combustion area. The igniter was capable of an unlimited number of starts because the sparks were located so deeply in the injector plate that the combustion did not seriously affect their lives.

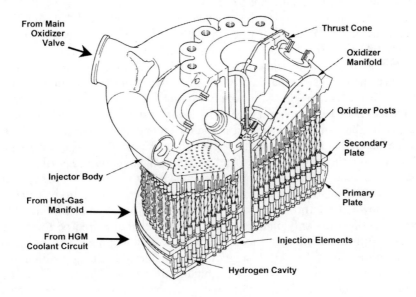

Main combustion chamber injector.

98 Power to orbit: the main engines

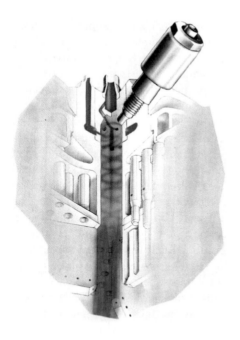

Augmented spark igniter.

Nozzle

If a rocket engine were a circus, the combustion chamber would be the lion and the nozzle the lion's tamer. As a tamer must control a wild beast, the nozzle must tame the incredible energy of the gas produced in the combustion chamber to produce the desired thrust. As a result of the third law of motion that for every action there is an equal and opposite reaction, thrust is generated by the ejection of gases through the nozzle of a combustion chamber at high velocity, thereby imparting a momentum to the nozzle and hence to the engine and the structure on which the engine is mounted. In other words, the thrust experienced by a rocket engine is the result of the reaction created by the ejection of high velocity matter. Since the gas exiting the combustion chamber has a supersonic speed, in order to force it to accelerate faster and faster the nozzle is designed as a divergent tube because, as supersonic flow theory explains, in such a conduit a fluid can expand by trading its static pressure for increased kinetic energy.

Bearing in mind that the flow in a rocket nozzle accelerates while losing pressure, the fundamental equation for rocket thrust says that (if all other parameters are fixed) the highest thrust occurs when the pressure of the gas flow at the exit of the nozzle is exactly the same as the external (ambient) pressure. If the exit pressure is lower, then the flow is said to be overexpanded because it has expanded more than necessary. If it is higher, the flow is underexpanded because it could have expand a little bit more. In either case, the result is less thrust than if the exit and ambient pressures are the same. Since the aim of the rocket engine designer is to get the most from the engine, and because thrust is created in the nozzle, particular attention is paid to creating the correct nozzle. And based on the equations describing the flow inside a supersonic nozzle, the key is the so-called area ratio; that is to say, the ratio between the area of the nozzle exit section and the smallest section of the nozzle, which is known as the throat section. From the thrust equation, it is evident that for a given level of thrust only one area ratio will deliver the desired performance. But there is a drawback, in that the ambient atmospheric pressure is not constant throughout the ascent. In fact, the highest thrust achievable changes with altitude. If one wishes a nozzle always to deliver the highest thrust, then the area ratio would be required to constantly change. Unfortunately, a nozzle cable of constantly changing shape and dimension has yet to be invented. For this reason, it is essential for the designer to know the trajectory of the vehicle, or more specifically a profile of ascent altitude versus time. Armed with this information, the designer can pursue an iterative process for selecting the nozzle area ratio that will give the best overall result for that trajectory.

Although the main engines would ignite at liftoff, the solid rocket boosters would lift the Orbiter above the lowest (and thickest) part of the atmosphere. As a result, the SSMEs would spend most of their time firing with an ambient pressure of almost zero. For this reason, the nozzle was designed with an area ratio of 69:1, a very large value that allowed the flow to almost fully expand at high altitude for a performance very close to optimum. In contrast, at low altitude, such as at liftoff and ascending through the troposphere, the engines operated in an overexpanded condition because the high area ratio forced the flow to expand to pressures below ambient whilst still inside the nozzle. When this point was reached, a shockwave formed in the nozzle, further reducing the thrust. As the Orbiter climbed, the shockwave inside the nozzle moved progressively towards the exit section, ultimately vanishing when the exit pressure matched the ambient pressure. Thereafter, the flow exiting the nozzles had a higher pressure than the external environment, which was exponentially approaching zero. The underexpanded flow kept expanding beyond the nozzle but this expansion, being in the external environment, made no contribution to the thrust delivered by the engines.

If one watches video footage of a Shuttle launch and ascent, it is easy to see the formation and motion of waves inside the nozzles. A shockwave implies an abrupt change in fluid density and this is clearly visible in the flow due to condensation of water vapor. If the video zooms in on the SSMEs, it is also possible to observe the shockwave progressively move towards the nozzle exit as the vehicle climbs. After the shockwave reaches the nozzle exit, it is possible to see how the plume below gets larger and larger due to the expansion continuing outside the nozzles. Regarding this phenomena, Daniel C. Brandenstein has an amusing anecdote. For his first flight into space, as pilot of STS-8, he had the right-hand seat aboard *Challenger*. Immediately behind him was mission specialist Dale Gardner. As Brandenstein remembers, "We weren't very far into the launch, and Dale says, 'Dan, how do the engines look? You know, the instrumentation on the engines, are they running all right?' I said, 'Yes, they look fine.' Thirty seconds later he says, 'Dan, how do the engines look?' 'Fine.' A minute later, 'Dan, how do the engines look?' 'Fine.' I don't know how many times this happened." What Brandenstein had evidently forgotten was that during the development of the SSME some engines had exploded, and moments before they did so their flame would turn solid and flutter. "Well, as you get higher in altitude and from the perspective *he* had, the flames from the engines were fluttering, so his connection was, well, when the flames flutter, the engine blows up." As Brandenstein elaborated, "You just have a different perspective as you get higher in altitude. The air pressure goes way down and you get into a vacuum, so basically what holds your flame real tight is the atmospheric pressure factor in that. Well, you get outside atmospheric pressure, they expand and they flutter a little bit more. Once again, from his perspective, from the inside looking back out at it, the flames were fluttering and he was concerned about that."

While from a theoretical point of view what matters for nozzle performance is the area ratio, having to deal with a real fluid requires that a nozzle be designed in such a way as to bring the fluid from the throat section to the exit as smoothly as possible. A too abrupt divergence in the nozzle would cause the flow of gas to separate from the walls, with resultant high energy loses and reduced thrust. In addition, a nozzle with a large divergence angle can create strong shockwaves that cause the fluid to detach from the walls and in some instances even produce damage. On the other side of the coin, a nozzle cannot be too long since the fluid would be subjected to higher losses due to friction with the walls, and that could also cause fluid separation. In any case, it is best to limit the envelope of a nozzle in order to reduce weight. Choosing the right shape for a nozzle is therefore more complex than simply calculating the best area ratio.

For many applications, one of the best nozzles is the so-called conical nozzle with a 15-degree divergent half angle because it is a good compromise of weight, length, and performance. But for the SSME the nozzle is bell shaped. It has a contour with a very high expansion section (30 to 60 degrees) soon after the throat section, and then there is a gradual reversal of the contour slope so that at the exit section the angle is almost zero. In such a nozzle, the flow coming out the throat section rapidly expands in the high expansion section, but since this expansion occurs very rapidly the fluid does not have the time to detach from the wall before it gradually and softly starts to expand in the reversal and exits in a direction almost parallel to the centerline of the nozzle. A bell nozzles performs much better than a conical nozzle, whilst at the same time being shorter with weight and cost benefits. It is therefore not surprising that the high performing SSME was equipped with this kind of nozzle.

When a flow expands, it also gets colder. But due to the high kinetic energy, the flow expands so rapidly in the nozzle that the expansion can be considered adiabatic, meaning that the gas does not lose heat but remains at the same temperature as when it emerged from the combustion chamber.[4] Like the chamber, the nozzle is equipped for regenerative cooling. As explained earlier, one of the routes taken by the liquid hydrogen inside the engine delivered some fuel to three transfer ducts that ran the full length of the nozzle down to the aft section, where they split in two smaller ducts that entered the fuel manifold placed on the perimeter of the nozzle exit section. The fuel in the manifold therefore made a single up-pass through 1,080 stainless steel tubes that formed the nozzle, making it a huge counterflow heat exchanger. The warmed fluid was collected on the upper section of the nozzle in order to continue its journey to the combustion chamber. A structural jacket and nine hatbands were provided for additional strength, and to keep the tubes together and maintain the nozzle's shape. Apart from STS-93, no nozzle ever suffered a problem. STS-93 lifted off on 23 July 1999 with Eileen Collins in the left-hand seat, the first mission to be

---

[4] In reality there is some heat loss, mostly due to chemical reactions of the combustion products while still inside the nozzle. It is worth noting that rocket engines employed as the reaction control systems by satellites that use compressed gas as the working fluid can experience drastic cooling and run the risk of ice formation.

SSME nozzles.

commanded by a female astronaut. Telemetry revealed that the main engines were shut down 0.16 seconds early. Whilst this might seem to be a trivial issue because the underspeed of 16 feet per second was easily made up when the smaller orbital maneuvering engines were fired for orbital insertion, an investigation discovered that during the ascent the turbine temperatures for the right-hand SSME were approximately 100°F higher than the pre-flight prediction.

In addition, post-launch review of the data showed that about 5 seconds into the start, numerous parameters suffered small anomalous shifts. Based on the models available, this anomaly was consistent with ruptured nozzle tubes. A review of the video showed that just after SSME ignition, and prior to liftoff, there was a streak in the exhaust plume of that engine. And just after liftoff, the streak was observed to be emanating from the nozzle wall. This confirmed the "nozzle leak" hypothesis. Visual inspection soon after landing revealed that three nozzle tubes had been ruptured. A hardware processing review found that two injector posts in the combustion chamber had been deactivated during servicing by inserting pins into the orifices located in the interpropellant plate of the main injector. As soon as the engine was available for inspection, one of these pins proved to be missing and there was a ding on the main chamber in line with the post at a location 2.5 inches upstream of the throat section. Each pin was approximately 0.9 inch long by 0.1 inch in diameter, weighed a mere 1.5 grams and was gold plated. The ruptured tubes were sectioned from the nozzle and sent for analysis. The key finding was that the gold discoloration at the ruptured tubes was indeed gold, thereby confirming that the ejected pin had

102 Power to orbit: the main engines

caused the tube ruptures. The nozzle was bolted to the bottom of the combustion chamber aft of the chamber throat section. The area ratio mentioned above for the nozzle of the SSME was in reference to this section.

POGO system

To a rocket engineer, POGO does not refer to the famous "pogo stick" with which children used to jump around in kindergarten, the term refers to a nasty vibrational problem encountered in powered flight. The first time it was encountered was during development of the Titan II launcher intended to orbit the Gemini spacecraft of the second NASA space program. Some 90 seconds into the first test flight, the missile developed a longitudinal vibration in the range 10 to 13 hertz. It lasted for roughly

The damaged right-hand SSME on STS-93.

30 seconds and reached a maximum amplitude of plus/minus 2.5 g at about 11 hertz. If it had been carrying astronauts, such a level of vibration and acceleration would have been quite painful and would have severely impaired their ability to carry out piloting tasks or to react in an emergency. It was not by chance that NASA had specified that such vibrations must not exceed 2.5 g! Although this gained the nickname of POGO by analogy with the bouncy action of the children's jumping toy,[5] to the engineers it was a very serious issue.

After analysis of the available data, it was suggested that this longitudinal motion was due to pressure oscillations in the propellant feedlines caused by fluctuations in the combustion chamber cyclically forcing pulses of fuel into the turbines. If such an oscillation were to match one of the natural frequencies of the structure, a resonance would increase the magnitude of the vibration and induce stress, perhaps sufficient for catastrophic failure. Data gathered during several flights, some of which were lost due to POGO oscillations, validated this theory and prompted the installation on the engines of accumulators or cavities designed to prevent the vibration of the fluids in the feedlines from synchronizing with the resonant frequencies of the structure. But POGO could not be totally eliminated because the oscillations would come and go as the propellant was depleted. The worst case occurred on Gemini V. For 13 seconds towards the end of the first stage of the ascent, a 10 hertz oscillation superimposed an amplitude of 0.38 g on the 3.3 g acceleration and gave Gordon Cooper and Pete Conrad an unpleasant ride.

POGO also affected almost every flight of the mighty Saturn V. The most serious case was during the second stage burn of Apollo 13, when two episodes occurred on the center engine, followed by a third that imposed an estimated peak load of 34 g on the attachment structure of the center engine before the combustion chamber low-level pressure sensor commanded a shut down. Post-flight analysis showed that only one more cycle of amplitude growth could have been sustained without catastrophic structural failure!

When SSME development started, the Apollo 13 experience was still fresh in the minds of the engineers. Whilst for all previous rockets the POGO problem had been fixed by the addition of some sort of remedy after having completed the vehicle and engine design, for the Shuttle it was decided to "design out" the problem at its root by fitting a so-called POGO suppression system between the low-pressure and high-pressure liquid oxygen turbopumps. It was initially intended to install an accumulator at the inlet of the low-pressure pump, but this would have left a long oxidizer duct to interact with the downstream high-pressure pump, and studies had shown that despite complicating engine development the optimal location for the suppressor was within the SSME at the high-pressure turbopump inlet, operating as a broad band filter.

The suppression system was a spherical container about the size of a basketball. It was charged with hot (400°F) gaseous oxygen for most of the flight, and separated from the cold liquid oxygen by a baffle plate which prevented the liquid propellant

---

[5] Although this term is written as POGO, it is not an acronym.

104 Power to orbit: the main engines

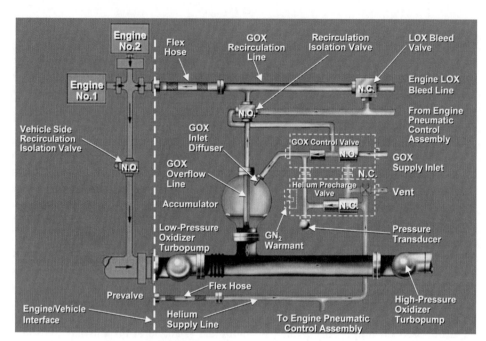

POGO suppression system schematics.

gasifying and forming bubbles. Operating as a Helmholtz resonator, it attenuated the flow oscillations into the oxygen high-pressure turbopump in the frequency range of 5 to 50 hertz for a "smooth" flow rate in the high-pressure liquid oxygen turbopump. During powered flight, the amounts of liquid and gaseous oxygen were continuously varied in response to the dynamics of the system, to prevent the onset of POGO.

Engine controller

All rocket engines have a heart in the form of the combustion chamber, but not all of them have a brain as well. The engines developed prior to the SSME were readily commanded by the vehicle's on board computers and did not require an autonomous computational capability. For the SSME the story was different. To understand why, let us recall some of the features of this remarkable engine. It was the most complex and "hottest" engine ever built. The complexity was tied to the mission requirements, which said the engine must be capable of being throttled; a common feature of an internal combustion engine or a jet engine, but not of such a large rocket engine. The SSME runs "hotter" than any other rocket engine because at any given moment it is closer to destroying itself. Earlier engines were overbuilt because they were designed to burn at full power throughout. Their combustion chambers and cooling systems were better than optimum, with the result that the engines were heavier than a less-protected design. They operated below their theoretical best performance. And they did not have a means of maintaining a high performance in the event of an anomaly. In contrast the SSME was capable of sensing how close to exploding it was, and would adjust the propellant flow levels to maintain maximum performances at all times. In many ways, the SSME was a real leap forward in rocket engine technology.

In 1972, when Rocketdyne won the contract to develop this new beast of the rocket zoo, they decided that in order to avoid complex interfaces between the engine and the vehicle the operation of the engine would have to be managed by a dedicated computer. In effect, this engine controller would be the SSME's brain. It was agreed by Rocketdyne and NASA that for three reasons a digital computer would be better than an analog one. There would be more flexibility at the software level. It would allow for future upgrades of the software.[6] And failure detection functions would be simpler. For a fail operational/fail-safe approach, the engine controller was made of two computers. Failure of the first computer would not impede operational capability because the second would take over automatically. If the second computer failed, the engine would be gracefully shut down. The main flight control system would be able to apply an appropriate abort scenario.

The engine controller software would use an open-loop strategy during the start and shutdown phases, but control the chamber pressure and propellant mixture ratio in a closed-loop manner in flight, issuing instructions to the engine control elements fifty times per second. This high frequency was necessary to satisfy the requirement to control a rapidly changing engine environment. Similar to what happened with the PASS, the engine software grew too large for the memory that could be incorporated into the controller. For this reason, a set of pre-flight checkout programs were stored in the PASS mass memory units and rolled in during the countdown. A controller was fitted directly on the engine, and operated in conjunction with vehicle command and recorder channels, engine sensors, pneumatic valves, hydraulic actuators, spark igniters, and electrical harnesses to provide a self-contained system for checking out, controlling and monitoring an engine. Being installed on the engine itself, it is easy to infer the high temperatures and vibrations in which these controllers had to reliably operate. From a hardware point of view, each controller was a sealed, pressurized, and thermally conditioned package of highly reliable electronics.

In the 1990s, an important program for upgrading these controllers was started by NASA and Boeing. To understand this, it is necessary to remember that the engine's turbopumps were the key elements in providing the combustion chamber with the pressure required to generate the desired thrust. The high-pressure turbopumps had to operate at extreme speeds, 34,000 rpm for the fuel pump and 23,000 rpm for the oxidizer pump. This required highly specialized bearings and precisely balanced components. In the blink of an eye, a flaw in the turbopump structure would induce failure. To detect imminent failures, some redlines were introduced. "The high-pressure pumps had two redlines that were considered as a basis for shutting the engine down before it came apart. One of them is the temperature redline," explains George D. Hopson, the SSME Project Manager at the Marshall Space Flight Center.

---

[6] NASA had already realized that it was easier to upgrade software than hardware.

Redlines based on temperatures allowed the detection of the onset of most of the failure modes of the engine, but they could not catch failure arising from excessive vibrations in the high-pressure turbopumps.

Studies of a means of detecting anomalous vibrations begun in the 1980s were abandoned because, as Hopson says, "We found out that if you got some moisture in the wiring connectors, you would get noise that would be interpreted by the redline system to be pump vibrations…the noise was spurious signals that could trigger a shutdown."

The advanced health management system (AHMS) upgrade program started in the 1990s solved this problem by examining it from a different perspective. "The big difference between the AHMS and the first vibration shutdown system was that the first system used composite vibration. In other words, the old system used the whole spectrum of frequencies, so any noise was considered along with real vibration. The AHMS only considered synchronous vibration," Hopson explains. Specifically, "The pump turns at 'x' number of rpm. If you have a vibration that matches the speed at which the pump is running, then that is synchronous. And if it is synchronous it is real."

Along with the introduction of the vibration redline monitoring, the AHMS led to a number of changes to the engine controller, notably the addition of advanced digital signal processors, radiation-hardened memory, and new software. As capable as the original controller was, it was not capable of diagnosing or correcting many engine anomalies and failures that could affect a wide range of engine parameters without necessarily exceeding any of the single-parameter redline thresholds. Thanks to the AHMS upgrade, the engine controller acquired a companion in the form of the health management computer (HMC) which collaborated with the engine controller to provide a more "intelligent" system capable of diagnosing and either correcting or mitigating the effects of various anomalies and failures. Because there was no space available on the already cramped engine envelope, the HMC was placed nearby in the aft fuselage.

The AHMS made its first flight in monitor-mode only on one engine during STS-116 in December 2006 and operated in active mode[7] on one engine on STS-117 in June 2007. Data collected during these two missions showed that the AHMS could safely operate during a real ascent, and starting with STS-118 it was active on all three engines.

Hydraulic system

All of the propellant valves[8] of the SSME were hydraulically actuated by means of three hydraulic systems, with all the valves on a given engine receiving hydraulic

---

[7] Active mode meant it was authorized to shut down the engine if anomalies were detected.
[8] The propellant valves were the main fuel valve, main oxidizer valve, fuel preburner oxidizer valve, oxidizer preburner oxidizer valve, and chamber coolant valve.

pressure from the same hydraulic system. In the event of a hydraulic system failure, all the hydraulically actuated valves could be "blown" closed by the pressure of the helium supply of that engine.

One of these failures was the so-called hydraulic lockup, a condition in which all of the propellant valves on an engine were hydraulically locked in a fixed position. This was a built-in protective response by the propellant valve actuator/control circuit that took effect whenever low hydraulic pressure or loss of control of one or more of the propellant valve actuators rendered closed-loop control of engine thrust or propellant mixture ratio impossible. In response, the engine controller would hydraulically isolate all five engine valves at their most recent position, preventing the valves from drifting. This strategy allowed the engine to continue to operate at approximately the throttle level that applied when the lockup occurred. Hydraulic lockup was therefore a means of allowing an engine to continue thrusting in a safe manner in a state that normally would have required its shutdown. The only drawback was that the engine was no longer able to be throttled. Hydraulic lockup did not affect the capability of the engine controller to monitor critical operating parameters or issue an automatic shutdown if an operating limit were subsequently exceeded.

Another important user of the engine hydraulic system were the servoactuators used for thrust vectoring control. Each engine could be gimbaled to direct its thrust in order to hold the vehicle on the ascent trajectory calculated by the on board guidance and navigation software. Each engine had two servoactuators, one for pitch and the other for yaw. They were fastened to the thrust structure in the aft fuselage and to the powerhead, enabling the entire engine to be gimbaled. Each servoactuator received hydraulic pressure from two of the Orbiter's three hydraulic systems, with one acting as the primary supply and the other as the secondary or backup supply.

SSME DEVELOPMENT

The development of the SSME was particularly challenging and innumerable tests were performed at both component and engine level, using several test facilities built specifically for the purpose. One problem that required a lot of troubleshooting and imposed a considerable delay in the program, was the detection of a subsynchronous whirl in the high-pressure fuel turbopump.

Otto R. Goetz worked on the SSME, and explains this as a situation in which "the rotor vibrates and whirls at a speed lower than the actual rotational speed". This seriously set back the program. "We had redline vibration sensors, accelerometers, on the turbopump and they cut us off. It was something that was not expected. We got consultants from universities. We asked even a consultant from England what to do about the subsynchronous whirl. We struggled for almost a year in order to solve it." As J. R. Thompson, who participated in managing the developmental program, remembers, the cause proved to be "inadequate cooling of the bearings. They became soft and allowed the rotor to orbit subsynchronously within the bearing cage itself." And the remedy happened to be quite straightforward, as Thompson

108   **Power to orbit: the main engines**

SSME low-pressure turbopumps.

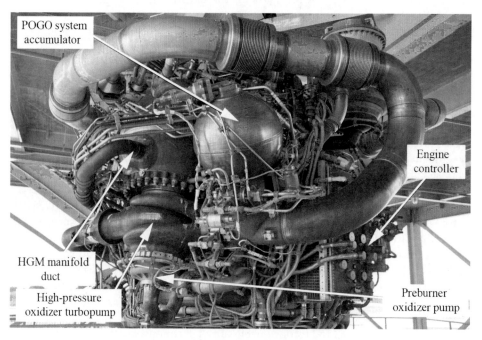

SSME systems.

SSME systems.

continues, "We ended up solving that problem by adding a little paddle – it wasn't any bigger than your thumbnail. A vortex was being created without that paddle where the coolant flow dumped into the bearings, and the introduction of the paddle disrupted the vortex and allowed the cooling to get there."

But Rockwell and NASA engineers did not have time to rest and enjoy the thrill of solving this issue. Another obstacle was already lurking on the horizon. As Goetz says, "We detected cracks in the turbine blades in the high-pressure fuel pump. That was the next big challenge that needed to be solved, because any turbine failure is catastrophic." In the aeronautics industry, turbine blades were produced from high strength materials with crystals growing in the direction of the centrifugal force. In order to make the blades even stronger, as Goetz explains, "We went and had a very intense materials development program and converted those directionally solidified turbine blades into single-crystal blades. One crystal was a whole turbine blade." A further strengthening provision had to be developed due to the incredible thermal load seen by a blade at the moment of engine ignition, when the blade was essentially at room temperature while the gas driving it was at 2,000°F. "Within two seconds the blade saw a delta-temperature of two thousand degrees. Its core was still at ambient temperature while the surface, the skin, went to 2,000 degrees. That caused cracking, because the delta internal to the material caused high stresses." The solution adopted was to apply a coating to the surface of the blade to act as an insulator and reduce for a while the delta temperature.

110   Power to orbit: the main engines

SSME EVOLUTION

When the SSME was developed in the early 1970s it was a leap forward in terms of rocket propulsion, and it still represents the best engine for human space exploration ever developed. As Goetz remembers, "At the beginning of the Shuttle program most people were afraid of the engine because of the extreme power density that it has at high pressures. Everybody expected that if there is a failure of the Shuttle, it may be caused by the engine." Hence NASA and Rockwell engineers worked to increase its power to enhance its reliability and make it safer to operate. The first five flights saw *Columbia* running its main engines at a maximum of 100 per cent of the rated power level. Then, as Goetz recalls, "We went into Phase I, which allowed the flight to be at 104 [per cent] and we flew up to the *Challenger* accident with this." This improved version of the engine allowed the program to carry out more complex missions with heavier payloads, reaping a string of successes.

Even more demanding missions were on the horizon, in particular the so-called "black missions" for the Department of Defense with national security payloads. As these missions were to be launched from Vandenberg Air Force Base in California, mainly into orbits that would be more demanding in terms of thrust, work began to run the engines at 109 per cent of the rated power level. This was to be achieved by means of improved durability, with changes to several critical components such as the high-pressure turbopumps, main combustion chamber, hydraulic actuators, etc. As Goetz explains, "It was a Phase II engine. That was primarily controller software and controller reliability, taking into consideration some of the things that we learned from *Challenger* and other lesson learned. We also did some minor improvement to other components, mainly how they were built, not how they functioned or how they performed." However, Phase II failed in achieving the goal of bringing the SSME to 109 per cent of its rated power level. Nevertheless, Phase II engines were flown after *Challenger*. In fact, the desire to continue to improve the engine persisted, and was eventually implemented in three parts.

As Goetz says, one of the main goals was "to replace the turbopumps, because the turbopumps had to be removed from the vehicle after every flight. In order to do that, the engine had to be removed from the Orbiter. That was very costly. We wanted to get higher reliability. We wanted a longer time between overhauls." Other objectives were to enhance safety and reliability by eliminating critical failure modes, lowering recurring maintenance costs, and increasing capability by increasing margins relative to the operating environment. These initiatives resulted in major upgrades to several engine components, incrementally introduced into the fleet as block changes. It is interesting that this design approach followed a different methodology to previous modifications, since they were not obliged to be performance neutral. This meant that a small decrease in performance would be accepted if it would deliver a significant improvement in safety. The loss of specific impulse in engine performance and/or an increase in weight was offset by weight reduction in other elements to ensure that the overall performance of the vehicle was not compromised. Thanks to the numerous modifications made during this developmental program, turnaround was significantly reduced by eliminating the need to remove hardware between flights for inspection and/or maintenance.

The launch of *Discovery* on 13 July 1995 for STS-70 marked the first time that an SSME was flown in the Block I configuration.[9] Although several vital parts of the engine were completely different from those flown previously, the layout remained the same.

In line with Goetz's explanation, work on the high-pressure oxidizer turbopump saw the welded casing structure replaced by a fine-grain casting. This eliminated 293 welds, 250 of which had lacked the capability of inspection from both sides. This structural change greatly enhanced safety since, as Goetz says, "These welds caused a lot of problems. They had to be X-rayed, dye-penetrant inspected, and what have you. That was very, very expensive and time-consuming...We never knew: did we, or did we not get everything by X-raying these welds? Were there left over cracks or not?" To the relief of the engineers and flight crews, switching to a fine-grain casting banished these worrisome questions.

When the SSME was conceived, engineers had neither the sophisticated computer programs nor the sophisticated analysis techniques available today. This led to a sort of misunderstanding in assessing pressure, flow and temperature distributions of the exhausts discharged from the preburners into the hot manifold. One consequence was a significant impact on the turbine blade life, especially on the fuel side. The Block I developmental phase presented an opportunity to fix this. While the original hot gas manifold sported three small hot gas transfer ducts on the fuel side and two on the oxidizer side, the new design had only two ducts on the fuel side. Eric S. Ransone spent many years overhauling the engines. As he explains, this modification "mainly reduced turbulence, and with the reduced turbulence you had reduced pressures and temperatures". This simplified design had 52 fewer piece parts and 74 fewer welds, resulting in a 40 per cent reduction in the fabrication and assembly time and a 50 per cent reduction in rework costs. In addition, the new powerhead had high margins for main injector assembly and less unit-to-unit variation. On the downside, the weight increased by 170 pounds, but this was compensated by generating a higher specific impulse by a combination of eliminating the baffles in the main injector and reducing the boundary layer coolant flow in the main combustion chamber.

As noted earlier, another feature of the powerhead was the heat exchanger placed inside the oxidizer preburner. The original configuration consisted of a small primary coil and a coil of two larger parallel tubes joined by a welded bifurcation joint. Such complex welded piping had to work when completely immersed in a high-pressure hydrogen-rich environment with inlet and outlet penetrations in the pressure shell of the powerhead. Remembering how important it was to eliminate welds in the high-pressure oxidizer turbopump casing, it is not difficult to guess how

---

[9] STS-70 was only an evaluation flight for the new engine configuration. For this reason only the center engine was a Block I. The other two were still the Phase II version. The first mission to fly with all three main engines of the Block I type was *Endeavour* in May 1996 as STS-77.

112 Power to orbit: the main engines

designers set out to develop a better configuration for the heat exchanger. In fact, it was redesigned as a single-tube that eliminated the interpropellant welds in the hot gas region. At the same time, the thickness of the tube wall was increased for improved wear resistance and to reduce susceptibility to foreign object damage.

As the Block I engines were being flown successfully, other improvements were identified and tackled with passion. The most important examples were related to the main combustion chamber and the high-pressure fuel turbopump. Other components that would clearly benefit from redesign were the low-pressure oxidizer turbopump, low-pressure fuel turbopump, purge check valves, and controller software. All but the high-pressure fuel turbopump were fairly easily improved in time for *Endeavour* to launch for STS-89 with three Block IIA engines. "Of all the IIA changes, the most significant was that we decreased the nozzle area ratio about 10 per cent," explains Hopson. "Basically what that amounted to was increasing the nozzle throat area. And what that did was to lower the outlet temperatures and pressures of the preburners that supplied the gas to run the pumps. The hydrogen pump was running pretty close to its limit on turbine temperature and pressure, and we wanted a bit more margin. When we decreased that area ratio, it reduced the temperatures of the gas that drove the pumps by about 100 to 200 degrees Fahrenheit, which gave us extra margin." In addition, the new combustion chamber was redesigned as several large casting pieces instead of many small parts to be welded. In all, 46 welds were eliminated, including 28 that lacked the capability of inspection from both sides. By switching to castings, the fabrication/assembly time was greatly reduced and the quality of the components increased.

However, it was proving difficult to redesign the high-pressure fuel turbopump to

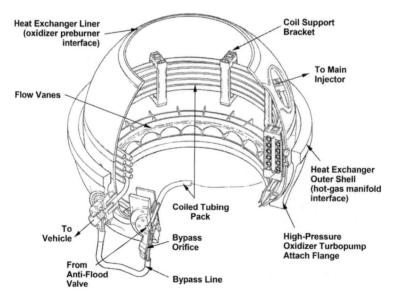

Oxygen heat exchanger.

make it more reliable and safe. Hopson provides us with some essential background to the issue. "We have something we call a redline. Almost anything that goes wrong in an engine will increase the turbine temperature, so this redline was there in case a temperature increase indicated a serious problem. When it hits the redline, the engine controller shuts that engine down." So the thing to do was to remain on the safe side of the redline for the high-pressure fuel turbopump. But on several occasions in tests it was found that the turbopump could fail well before reaching its redline, and if that were to occur in flight it would be catastrophic. "It would explode before we ever got there [to the redline], which would mean you'd lose both the crew and the vehicle. Because those engines are clustered together, one pump explodes and it wipes out all nearby engines and components." The root problem was "the pump was designed for minimum weight, and testing showed that if it were damaged it never would survive to reach the redline". It was evident that some change into the turbopump structure would have to be made in order to improve this sort of Sword of Damocles situation. Attention turned towards making the rotating parts of the pump more robust and stiff. "The real sensitive part inside a pump is the rotating part," Hopson continues. "The rotating part of the old pump included a shaft and a turbine disk, bolted together. It was the best pump that you could design that has the lowest possible weight and can also do the job...but if you lost half of a turbine blade you bought the farm, because the next thing you knew the rotor would be into the case because of the unbalance. You'd have an explosion."

In the new design the shaft and disk were forged as one piece, making the rotor assembly very tolerant of damage/unbalance compared to the bolted configuration. In addition, a third bearing was added to the forging to provide some extra stiffness. As Hopson reports, with these changes, the turbopump "could withstand unbalance better than the other one could. We put 400 extra pounds into that pump. Even if

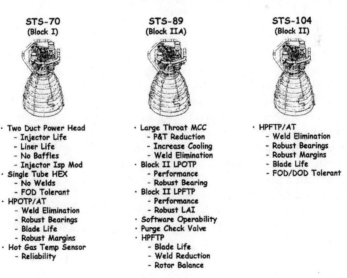

Variants of the SSME.

you lost half the turbine blades, the pump would contain the damage until the redline shut down the engine safely." Obviously, for this pump too, the welded construction that required meticulous process control, inspections and frequent repairs, was replaced by a unique cast in order to provide an additional margin against burst failure modes. The turbine blades were also improved using a different material and thin thermally compliant airfoils to deal with the rapid thermal transients at ignition and shutdown. With the final development and testing of this new pump, the Block II configuration was born.

When launched as STS-104 in July 2001, *Atlantis* became the first Orbiter to fly with one Block II engine. It was also the first to fly with three such engines during STS-110 the following year. From that point on, Block II engines flew up to the end of the program.

It is important to understand that in the evolution from Block I to Block II, each configuration retained all the improved features developed for the previous versions. Hence, the Block II engines had all the new components designed for the Block I and Block IIA version plus the final addition of the new high-pressure fuel turbopump. It is also necessary to recognize that originally there were no plans to have a Block IIA configuration. All the changes implemented in the Block IIA were to be added to the Block II version, plus the new high-pressure fuel turbopump. It was the difficulties in developing the fuel pump that led to the decision to have the interim version with all of Block II changes except the fuel pump. When the fuel pump was finally ready for use, the Block IIA engines were upgraded to Block II and thereafter any new engines were manufactured in the Block II configuration.

SSME OPERATIONS

Prelaunch operations

Despite being used for just short of ten minutes on a mission lasting several days, the SSMEs were without a doubt amongst the most complex systems to be operated on board the Orbiter. Ground processing operations began several hours prior to liftoff.

Of course, to be able to use the SSMEs the external tank had to be fully loaded with both liquid hydrogen and liquid oxygen. But before propellant loading could begin, the engines had to be purged. This preparation phase consisted of purging the oxidizer side of the engines with dry nitrogen and the hydrogen side with dry helium in order to remove any air and moisture that could have accumulated inside the lines, since it would turn into ice upon initiating the engine thermal conditioning. Once this was completed, loading operations would start with about six hours remaining in the countdown and continue to just a few minutes prior to liftoff. Both propellants were supplied by ground facilities that connected with the Orbiter at the left (for hydrogen) and right (for oxygen) ground support equipment umbilicals. The propellants entered the Orbiter via the outboard/inboard fill and drain valves, followed the appropriate feedline manifolds, and exited the Orbiter at the disconnect valves into the massive external tank.

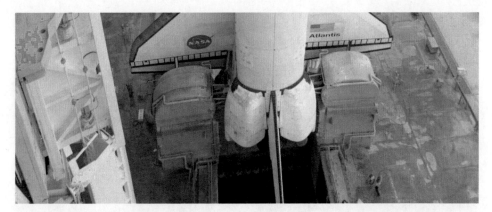
The two T-0 ground umbilicals by which propellants were loaded into the ET.

As the external tank was filled, all three SSMEs were thermally conditioned by the circulation of the cryogenic propellant around their main components. This was to chill the engines in order to ensure a safe ignition. Conditioning was performed by drawing liquid hydrogen from the feedline manifolds, passing it around the closed fuel prevalve, and then through the low-pressure and high-pressure turbopumps and, via a recirculation pump, to a topping manifold (also called a recirculation line) that fed the propellant into its tank. The liquid oxygen followed a similar path in order to chill its lines and turbopumps. In this case though, once the propellant had passed through the engines the three flows were manifolded together and sent back to the production facility at a flow rate of 18 pounds per second using the outboard bleed valve and the T–0 ground umbilical. However, not all oxygen circulating into an engine returned to the ground facility, because a very small amount flowed out of the turbopump seal drain lines and was vented overboard through three drain lines that ran along the nozzle and opened near the bottom. In footage showing the Shuttle on the launch pad it is possible to see "white smoke" emerging; it is the drained oxygen. Unlike the hydrogen, the oxygen chilling circuit did not have a recirculation line into the external tank. Since liquid oxygen is harmless in the atmosphere (around 20 per cent of air is oxygen) it could be returned to its production facility without a problem. Being more dangerous, hydrogen was routed into the tank in order to minimize risks. Thermal conditioning continued until T-9 seconds, in preparation for the engine start sequence.

The fuel system purge began 4 minutes from liftoff. Forty seconds later there was a test to confirm the proper functioning of each engine gimbal actuator by cycling it through a predefined profile of extensions and retractions. If all the actuators passed the test, the three engines were gimbaled into a predefined position at T-2 minutes 15 seconds and held there until ignition. With nine seconds left, the GPCs commanded the liquid hydrogen prevalve to open.[10] Meanwhile, onlookers were able to observe two interesting processes occurring on the pad.

---

[10] The liquid oxygen prevalves had been opened during external tank loading, in order to permit engine chill-down.

116   Power to orbit: the main engines

Gaseous oxygen vented from the engines during thermal conditioning.

Water started to flood the launch pad prior to engine ignition.

In order to protect the Orbiter and its payload from damage by the acoustical energy that reflected off the mobile launcher platform during liftoff, a sound suppression water system was installed. At T-16 seconds this began to flood the pad with thousands of gallons of water per second. Initially, water was poured down from sixteen nozzles atop the launch pad flame deflector and from outlets in the SSME exhaust hole of the mobile launcher platform. When the solid rocket booster were lit at T=0, a deluge of water started to flow from six large nozzles known as "rainbirds" atop the platform. Water was also sprayed into the solid rocket booster exhaust holes of the mobile launcher platform in order to provide overpressure protection to the Orbiter. This water system continued to operate for

Burnoff preigniters in action prior to engine ignition.

10 seconds after the Shuttle lifted off, since in terms of acoustical reflection the worst moment was at T + 5 seconds.

At the beginning of the final 10 seconds in the countdown, another quite visible action occurred, this time directly below the main engine nozzles. During the thermal conditioning, some hydrogen would inevitably have found a way through the engines and into the combustion chambers. This posed a hazard, since at engine ignition the hydrogen-rich atmosphere in the engine bell would ignite, causing a small explosion capable of damaging the engine. To preclude this, six hydrogen burnoff preignitors were installed on the platform, positioned to face the main engines. Just prior to main engine ignition they sent thousands of hot, luminescent balls into the area below the nozzles to ignite any free hydrogen, producing a fascinating pyrotechnic show. Main engine ignition began at T-6.6 seconds. Whilst it might appear to a casual observer that all three engines lit simultaneously, it actually occurred in a staggered pattern at an interval of 120 milliseconds, with the right engine starting first, then the left one and finally the center one. If at T-3 seconds the engines had not achieved at least 90 per cent of the rated power level, a launch pad abort was automatically initiated and the engines were shut down.[11] If everything was nominal, the solid rocket boosters would ignite at T = 0.

Engine start sequence

Igniting a rocket engine is not at all like turning the key on the car dashboard to start the engine. Any rocket ignition sequence consists of a complex and precisely timed

---

[11] Launch pad aborts occurred several times during the Shuttle program, and these will be discussed in Chapter 11.

118 Power to orbit: the main engines

opening and closing of valves in order to achieve as rapidly as possible a safe stable combustion that can be ramped up to the maximum power level for liftoff.

The start sequence of the SSME cluster began at T-4 minutes with a final purge using helium introduced downstream of the main fuel valve in order to displace any gas that might remain in the lines and would freeze when hydrogen was allowed into the lines. At this point the engine controllers ran health checks of their engines to verify that each parameter was within the prescribed range. If no failures were found, the controller adopted an "engine ready" status.

As soon as the start command was received, the main fuel valve was fully opened in two-thirds of a second, allowing hydrogen to flow into the combustion chamber and nozzle channels. The latent heat of the hardware was sufficient to warm up this liquid, which flashed to gas. The gaseous hydrogen then flowed to the fuel preburner and started the high-pressure turbopump. This eliminated the need for an auxiliary power unit simply to start the engine. But is also caused a headache in developing the startup sequence. In fact, the rapid expansion of the hydrogen caused flow blockage and momentary flow reversal, a phenomena called "fuel system oscillations" which gave rise to a pulsating flow rate with an unstable pressure oscillation at a frequency of approximately 2 hertz. The oscillations continued to increase in magnitude with dips (reduction in pressure) occurring at approximately 0.25, 0.75 and 1.25 seconds into the starting sequence.

At the same time, the augmented spark igniters (ASI) in the two preburners and the main combustion chamber were energized ready to trigger the inferno as soon as the fuel and oxidizer were present in the correct mixing ratio. Fuel was fed into each ASI as soon as the main fuel valves were opened, but the oxidizer had to wait for the oxidizer valves to be opened. The proper mixture ratio for ignition was achieved by the second dip in pressure caused by the fuel system oscillations. Due to the fuel system oscillations, the oxidizer valves could not be opened straight away, but had to

Staggered ignition sequence.

follow a precise opening/closing sequence timed with the dips in the fuel system in a process called priming, the objective of which was to fill the preburners and main combustion chamber. An oxidizer system was said to be "primed" when it was filled with liquid such that the flow rate entering the injector was equal to the flow rate discharged by the injector into the combustion chamber, resulting in a rapid rise in chamber pressure. The fuel preburner was the first to be primed at 1.4 seconds into the starting sequence, followed one tenth of a second later by the main combustion chamber and one tenth of a second after that by the remaining oxidizer preburner. The oxidizer valve of the fuel preburner was the first to be opened, 0.10 seconds after the main fuel valve opened. The valve was first commanded to open to 56 per cent, then at 0.72 seconds into the sequence to close to 10 per cent to compensate for the second dip in fuel pressure. This prevented the mixture from becoming oxygen-rich in the fuel preburner, which ignited at that time with a consequent slight acceleration in the speed of the high-pressure fuel turbopump.

At 1.25 seconds into the sequence, a check was made to confirm that the speed of the high-pressure fuel turbopump had reached the value required to start priming the main combustion chamber. This check was crucial because if the priming of the chamber were to occur with the fuel turbopump underperforming it would be unable to pump fuel against the rising pressure in the combustion chamber, risking engine burnout due to the resulting oxygen-rich combustion. Based on experience during tests, if the turbopump speed was below 4,600 rpm at 1.25 seconds then it would be too slow at combustion chamber priming for sustained pumping. If this was

All engines ignited.

detected, the engine would be immediately shut down because if such a low speed condition was allowed to persist beyond this point there would not be sufficient time to perform a safe shutdown.

At 1.4 seconds, the fuel preburner was finally primed with a rapid rise of pressure at the inlet of the high-pressure fuel turbopump that produced a high turbine pressure ratio and a significant acceleration in the pump speed. This was a desirable condition for a cool fuel-rich start and for the back pressure necessary to prevent a runaway condition in the pump.

Main combustion chamber priming was started by slowly opening the oxidizer valve to 60 per cent 0.20 seconds into the sequence to provide an oxygen flow rate to prime the chamber at 1.5 seconds. At this point, the rapid rise in chamber pressure acted as a brake on the high-pressure fuel turbopump, decelerating it.

Priming of the oxidizer preburner started at 0.12 seconds in the sequence, with the opening of the valve to provide just sufficient oxygen to ignite the ASI and to allow a small leakage flow into the preburner injector. At 0.84 seconds, the valve was opened to 46 per cent to allow a major flow. But this lasted only one-third of a second before restoring the valve so that the preburner again ran on leakage. The timing for this opening was scheduled to provide enough oxygen to the ASI to ignite the preburner prior to recovering from the second dip in the fuel system. On valve leakage flow, the oxidizer preburner priming finally occurred 1.6 seconds into the sequence, resulting in an increase in drive power to both high-pressure turbines. This power stabilized at about 2 seconds, with the combustion chamber pressure at approximately 25 per cent of the rated thrust level. Meanwhile, the chamber coolant valve was further closed in order to force additional coolant flow through the combustion chamber. The engine was allowed to run in this condition for additional 0.4 seconds in order to absorb the normal variations in propellant pressures and temperatures.

By using the sensors available on the engine, the engine controller verified proper ignition and operation of the preburners and main combustion chamber at 1.7 and 2.3 seconds. If both of these checks were positive, the controller started the closed-loop thrust control system at 2.4 seconds. From this point on, the controller compared the combustion chamber pressure with a preprogrammed profile and tried to nullify any difference by modulating the oxidizer valve of the oxidizer preburner, which had the side effect of regulating the oxidizer valve of the fuel preburner. At 3.8 seconds this valve started to be regulated in order to adjust the fuel flow rate to achieve the proper mixture ratio, thereby activating the closed-loop mixture ratio control system. By 5 seconds, the thrust was stabilized at the rated level with a mixture ratio of six and the engines were ready for the ride to orbit.

Ascent operations

At the precise moment of liftoff the SSME gimbal actuators, which were locked in their preignition position, were first commanded to their null position for solid rocket booster start and then allowed to control the thrust vector. From liftoff to main engine cutoff (MECO) all sequencing and control functions were executed automatically by the GPCs so long as the engines operated nominally, with the flight crew monitoring their performance and providing manual inputs in response to any malfunctions. As the vehicle accelerated in the lower atmosphere and the dynamic pressure increased, the GPCs throttled the engines down to a value in the range 67 to 72 per cent in order to minimize structural loading in the region of maximum aerodynamic pressure. At approximately 65 seconds after liftoff, the engines were once again throttle up to the maximum certified power level, typically 104 per cent.

As Thompson explained, "When we first started designing the SSME, everybody did their thing and decided what weight their element required in order to complete a mission. The engine requirements were based on those estimates. At that time, when we first started developing the engine, the requirement was 100 per cent power level. We called that 'rated power level'. Later on, the weight went up on the Orbiter and other elements and we really needed more than what we'd been planning for. We had enough margin in the engine – actually, as I recall, we certified it to fly 109 per cent. Then later 104, then 104.5 per cent were baselined for flight. Really, all that means is that flight power levels were 4.5 per cent higher than the early plans as to what the engine was required to do."

For a normal mission the power level remained at that setting until 3-g-throttling was initiated, approximately 7 minutes 30 seconds into the flight in order to abide by the 3 g operational limit imposed to prevent excessive physical stress on the "stack" and the flight crew. With 6 seconds remaining to MECO, all engines were throttled back to the minimum power level of 67 per cent in preparation for shutdown.[12]

Since the propellants had to be supplied to the main engines with adequate head pressure for proper engine operations, during ascent both propellants in the external tank were pressurized by an ullage pressure system.[13] This consisted of sensors, lines and valves to route gaseous propellants from the main engines to maintain the propellant tank pressures during engine operation. To pressurize the oxygen tank, each engine warmed liquid oxygen in a heat exchanger located inside the oxidizer preburner. The gaseous oxygen produced by the three engines was collected together and discharged into the top of the oxygen tank by way of an external manifold to maintain an ullage pressure in the range 20 and 25 psia. A similar arrangement was used to maintain an ullage pressure of 32 to 34 psia in the hydrogen tank using gaseous hydrogen from the turbine of the low-pressure turbopump. If the internal pressures exceeded the limits, a relief valve in the tank would open in order to reduce the pressure to an acceptable level. Ullage pressure had also to be maintained during prelaunch, but in this case it was by using helium routed from the T–0 umbilical.

During the entire ascent, valves were operated by the helium pneumatic system, while the engine helium system filled the cavity between the turbine and pump of the high-pressure turbopump. If this filling did not work properly, an immediate engine shut down would be commanded. In fact, this turbopump operated with the pump

---

[12] It is worth noting that the original specifications envisaged a minimum power level of 50 per cent. However, due to vibration problems encountered during tests the SSMEs were never flown at that power level. In practice, the minimum level was between 65 and 67 per cent.

[13] Ullage refers to the space in each tank not occupied by propellants.

122 Power to orbit: the main engines

Shockwaves in the main engine exhaust.

working on liquid oxygen and the turbine working on a fuel-rich mixture. Given that the pump and the turbine were on the same shaft, any leak of either fluid from one housing into the other would have caused a catastrophic explosion. (This happened once while testing a turbopump on the ground.) The best way to protect this piece of sophisticate turbo machinery was to keep the shaft penetrations in the pump and the turbine housing isolated by a complex labyrinth of seals, and filling the cavity with helium to act as a pressure barrier to prevent leakage.

Engine shutdown sequence

The engine shutdown sequence was much simpler than the start sequence, since it did not have to deal with all the complications of fuel system oscillations and priming. The first step was to power down the high-pressure oxidizer turbopump to adjust the mixing ratio for a cooler fuel-rich combustion. The oxidizer valve on the oxidizer preburner was commanded to close at a rate of 45 per cent per second. (A faster rate would cause a thrust decay of more than 700,000 pounds of force per second, and violate the structural limits.) Several milliseconds later the oxidizer valve on the fuel preburner was commanded to close, but with a closing rate such that the oxidizer side would power down first. During this process, the positioning of both valves was such as to maintain a low mixture ratio and a maximum oxidizer pressure decay sufficient to prevent hot gas in the oxidizer preburner from flowing back. In the meantime, the main oxidizer valve was commanded to close at a rate that would keep the pressure in the main combustion chamber high enough to maintain sufficient back pressure on the turbine inlets to prevent an overspeed condition. To accommodate the increased heat load due to the throttling, the chamber coolant valve was adjusted for a higher flow rate to cool the main combustion chamber and nozzle. The main fuel valve was then held open for more than 1 second in order to ensure a very fuel-rich shutdown. Then both valves were commanded to close as rapidly as possible, without causing any damage to the high-pressure fuel turbopump.

Post-MECO operations

Normally MECO was commanded by the GPCs once a specified velocity had been attained. Although it never happened, MECO could also have been prompted by a low-level cutoff condition in which a premature depletion of either propellant type was sensed. In this case, the task was to safely shut down the engines with sufficient propellant remaining to avoid cavitation in the turbopumps.

Once MECO was confirmed, some 8 minutes 30 seconds after liftoff, the GPCs executed the external tank separation sequence in which, amongst other things, the 17-inch disconnect valves of the liquid hydrogen and oxygen feedlines were closed and the gaseous hydrogen and oxygen feedlines were sealed at the umbilicals by self-sealing quick disconnects.

After external tank separation, typically 5,400 pounds of propellant remained in the Orbiter; 3,700 pounds in the feedlines and 1,700 pounds in the engines. There were several reasons for purging this from the system. Firstly, it shifted the center of gravity of the Orbiter by approximately seven inches, sufficient to cause guidance issues during atmospheric re-entry. Secondly, during re-entry the trapped hydrogen (liquid or gaseous) in the propellant lines could combine with atmospheric oxygen to create a potentially explosive mixture. Finally, if the trapped propellants were not dumped overboard, they would sporadically outgas through the feedline relief valves and cause slight vehicle accelerations that would complicate orbital navigation.

Because hydrogen evaporates quickly, prior to starting the propellant dumping sequence this was already passing out through the backup hydrogen vent line and the hydrogen relief feedline to prevent excessive pressure. In preparation for the nominal dump, all engine helium systems, including the pneumatic one, were commanded to interconnect to ensure that sufficient helium would be available to perform the dump. The dump started automatically at MECO plus 2 minutes.[14] To dump the oxygen, the GPCs opened the three liquid oxygen prevalves, commanded each engine controller to open its main oxidizer valve and to open the valves on the pressurization lines so that the helium could drive the oxygen trapped in the feedline

---

[14] In the event of abnormal pressure buildup in the feedlines, a manual dump could be attempted earlier than this time.

124 Power to orbit: the main engines

manifolds out through the nozzles of the engines. This dump was propulsive and provided a typical change in velocity of between 9 and 11 feet per second. The GPCs automatically terminated the dump after 90 seconds, closing first the valve of the pressurization line and, after another 30 seconds, closing the main oxidizer valves. The hydrogen dump occurred in parallel, starting with the opening of the fill and drain valves to allow the liquid to flow out, in this case without pressure from the helium subsystem. After 2 minutes the valves were closed. Dumping terminated, all valves of the helium system were closed and the engines gimbaled to positions with the nozzles all canted inwards in order to minimize aerodynamic heating during re-entry.

Despite the propellant dump, some liquid oxygen and hydrogen could still remain in the feedlines. Fifteen minutes after the propellant dump described above, a second cleansing sequence called vacuum inerting was automatically initiated and carried out for a duration of 2 minutes in order to get rid of these final traces of propellants. This consisted of opening the oxygen inboard and outboard fill and drain valves and the backup hydrogen dump valves in order to let the propellants vent to space. If the automatic sequence did not occur, the crew could initiate the procedure manually.

A second vacuum inerting was performed after the orbital maneuvering engines had been fired for orbital insertion. The vibration induced in the aft fuselage by such a firing could force any hydrogen ice still inside the manifold to quickly sublimate. Again, to avoid a pressure buildup, a second vacuum inerting lasting 3 minutes was carried out, although in this case only for the hydrogen side.

While on-orbit, the SSMEs required no action unless there was a malfunction such as a pressure buildup in the manifolds because of propellant remaining in the feedlines. In this case, after ruling out a sensor malfunction, another vacuum inerting would be performed in order to return the pressure to an acceptable value.

Re-entry operations

During deorbit preparations, the GPCs commanded the thrust vectoring system to gimbal the engines to their re-entry positions; they had been stowed after MECO but could have drifted, so this procedure was to make certain.

Starting at an altitude in the range 130,000 to 110,000 feet depending upon the re-entry trajectory, the aft compartment was continuously purged by helium to vent any hydrogen that might have accumulated during orbital flight. At the same time, helium pressurized the propellant feedlines to prevent contaminants entering the manifolds. (Post-flight removal of contamination from the manifolds and feedlines was a long and costly process that required disassembly of the affected parts.) This manifold pressurization continued until the ground service crew installed throat plugs in the main engine nozzles.

4

Power to orbit: solid rocket booster

THE ROCKET EQUATION

The "rocket equation" is without doubt the most important mathematical expression in rocketry, as otherwise it would not be possible to design even the smallest rocket. Its derivation is not difficult, but for this book it is sufficient simply to state its final expression as follows:

$$\Delta V = I_{sp}\, g_0\, \ln(m_{initial}/m_{final})$$

This deceptively simple equation offers important understanding into the nature of rocket propulsion. The term ΔV represents the change in velocity to which the rocket must be subjected in order to accomplish its mission. If a rocket has to leave a launch pad and attain a given orbit, then it will need to increase its velocity from that on the ground (which is zero if the rotation of the Earth is ignored) to that required to achieve and maintain the altitude of that orbit. I_{sp} is the specific impulse, a measure of how efficiently a rocket develops thrust. The term g_0 is the acceleration of gravity. The ratio in brackets is the mass of the rocket at the time of liftoff and its mass upon achieving the desired orbit. The fact that this mass ratio is a logarithmic function has an important effect on the way that modern rockets are designed.

Simply put, this equation says there are only two terms that a designer can vary in designing a rocket to achieve a desired change in velocity. Firstly, for a given mass ratio, the higher the specific impulse the greater is the change in velocity. The only engines currently capable of lifting a vehicle off the surface of a planet into orbit is a chemical rocket, and this technology has already reached its theoretical limit in terms of specific impulse at about 455 seconds. As for the mass ratio, it is evident that the higher its value the greater is the change in velocity. Although this may appear to be the way to achieve a high velocity, in reality the higher the mass ratio, the higher the initial mass and the greater the mass of propellant that must be carried. Furthermore, the nature of the logarithmic function causes the change in velocity to increase more slowly than the mass ratio. For example, to triple the maximum change in velocity it will be necessary to switch from a design with a mass ratio of 2.5

in which propellant represents 60 per cent of the initial mass to a design with a mass ratio of 20 in which the propellant accounts for 95 per cent! And bearing in mind that more propellant involves larger tanks and more plumbing, it becomes evident that the mass available to the payload rapidly approaches zero.

Rockets that are capable of leaving the Earth's surface and achieving orbit in just one piece are known as single-stage-to-orbit (SSTO), and represent the dreams of all rocket scientists since the inception of astronautics. The technological challenges are so formidable that for the foreseeable future such machines will remain in the realm of science fiction.

To enable a rocket to achieve orbit, designers developed a trick known as staging. This divides the launch vehicle into segments, or stages, so that it is not necessary for the entire mass of the rocket to achieve orbit. Each stage is a rocket incorporating its own engines, tanks, propellant and any other equipment necessary for its functioning. On depleting its propellant, a stage is discarded. The next stage will have less mass to propel. The result is that less propellant will be necessary to place the payload into orbit. What happens is that the final velocity achieved by a second stage will be equal to the velocity at the moment of its burnout plus the velocity that it inherited from the first stage. The final velocity reached by summing the changes in velocity produced by two stages exceeds that which could be achieved by a large single stage using today's technology. Of course, in adding upper stages, a designer must take care not to arrive at the point where the first stage is no longer able to lift off the ground owing to the mass of the stack above it. And the greater the number of stages the more complex is the vehicle, and greater complexity increases development costs and tends to reduce reliability. For these reasons modern rockets have between two and four stages, with each stage being successively smaller. There have been rockets with a greater number of stages, but these were the exception rather than the rule. Staged rockets provided the key to space exploration, and will probably remain in use for decades to come. To summarize, staging offers the following advantages:

1. Reduced total vehicle mass for a given payload and ΔV requirement.
2. Greater total payload mass delivered to space than a similarly sized SSTO vehicle.
3. Increased total velocity achieved for a similarly sized SSTO vehicle.
4. Lower engine efficiency required to deliver payload compared to the same payload carried by a SSTO vehicle.

But these advantages come at a price, in particular:

1. Increased complexity due to the extra sets of engines and plumbing for each stage.
2. Decreased reliability due to the extra engines and plumbing added to each stage.
3. Increased overall cost because a more complex vehicle is more expensive to build and launch.

THE SEARCH FOR A BOOSTER

The introductory section has provided insight into a fundamental aspect of the design of a rocket, and it explains why the Shuttle was designed as a two-stage system.[1] The selection of the first stage of the space transportation system was tricky, and there were several bizarre proposals intended to achieve the most important requirement of complete reusability.

When NASA began to study the Shuttle, it was still sending astronauts to walk on the Moon and, full of confidence, did not shy away from concepts that bordered on science fiction. As Royce Mitchell, who joined the Space Shuttle Task Team, recalls, "Many trade studies, many configurations were looked at. Operationally, the best solution was a liquid fuel booster carrying a liquid fuel Orbiter. The Orbiter had its own internal tanks, like any aircraft, but the resulting vehicle was monstrous. The fly-back-and-land-like-an-airplane-type booster was immense. It makes the Airbus 380 look like a popgun. It was a large booster." In May 1971 the wild dreams of the engineers suffered a setback when the Office of Management and Budget decided to limit spending for developing the Shuttle. This left the agency with enough money to design an Orbiter but not a booster as complex as that envisaged, and so the idea of a two-stage system that was fully reusable was put aside.

This cut in funding forced everyone back to the drawing board, this time with the task of designing an expendable booster. It was felt that an expendable booster could be developed fairly rapidly and cheaply in order to get the program running, and that when funding became available later on the expendable booster could be replaced by a fully reusable one. Boeing proposed a Saturn V first stage (S-IC) reconfigured to carry the Orbiter on top. Such a solution was of great interest for the company, which was eager to keep its S-IC production facility open as the Apollo program drew to a close. Likewise, Martin Marietta proposed a larger version of its Titan III, named the Titan III-L, as a possible booster for the Shuttle. This would use a new liquid fueled core with a diameter of 16 feet, as compared to 10 feet for the standard Titan III, and add as many as six strap-on solid-fuel rockets, each 10 feet in diameter. A third proposal was a joint venture between Thiokol, Aerojet, United Technology and the Lockheed Propulsion Company, all of which were solid rocket motor makers. They proposed a mixed cluster of their various rockets. Finally, NASA's Marshall Space Flight Center suggested a proposal consisting of a pressure-fed expendable booster. The attraction of this design was that it eliminated the requirement for complex turbo machinery to pressurize the propellant for feeding to the engines, thereby saving on developmental costs. The drawback was that the structure had to be thicker in order to withstand the internal pressures. The increase in weight and consequent reduction in performance was deemed a worthwhile

---

[1] Strictly speaking, the Shuttle was a one-and-a-half-stage launcher, because the first stage comprised both solid rocket boosters and the liquid-fuel main engines. After having jettisoned the boosters, the main engines kept running until orbit insertion. In this sense, the Shuttle was not a true two-stage system.

penalty to minimize the cost of developing the booster in order to allocate more funding to cutting-edge technologies such as the reusable main engines of the Orbiter, especially if the expendable booster was to be only an interim configuration.

Along with the expendable booster idea came external tankage for the Orbiter's propellants. When it was realized that with such a configuration the staging velocity could be cut to 5,000 (or perhaps even 4,000) feet per second, NASA did an about-face and restored the idea of a fully reusable booster because for that staging velocity the booster would be much smaller than originally envisaged and it would not require sophisticated thermal protection. With renewed hopes, Marshall kept proposing its pressure-fed booster, saying it could be made reusable even without providing it with wings. Its thick walls would create a structure strong enough to survive a parachute-assisted landing in the ocean and the ensuing perils of the sea. Furthermore, the thick walls would act as a heat sink against the heat generated during the re-entry phase, without the additional weight of a thermal protection system. Due to its inelegance in comparison to a winged booster, this concept was unofficially called the "big dumb booster".

Boeing also stuck to its initial proposal of a converted S-IC stage, this time fitted with wings, a tail, a nose with a flight deck, and ten jet engines for returning to the launch site. NASA was very interested in this configuration because it represented the agency's aspiration for a fully recoverable flyback booster. However, there were some practical issues. Charles Donlan, the Shuttle Program Director, put it this way, "We ran into a problem of pilot escape from the booster in the event of an abort. We never could quite figure out what to do about it. And then, as we looked at the development problems, they became pretty expensive." Another headache was the booster's thermal protection system. "We learned also that the metallic heat shield, of which the wings were to be made, was by no means ready for use. The slightest scratch and you're in trouble. It became increasingly evident that you want to have the cheapest possible configuration, but you put all this time and effort on a vehicle, the biggest part of which (the first stage), its only role was to get the Orbiter up high enough for it to fly itself. So you're spending all this effort on a part of the system that had no basic payoff. The important thing is the Orbiter – that's the payoff."

A fresh idea for a booster for the space transportation system came in the form of a thrust-assisted Orbiter Shuttle (TAOS). This would use a standard Orbiter with an external tank that was sufficiently large to permit the main engines to operate with "parallel staging" at liftoff. In this configuration, two liquid propellant boosters would flank the tank to provide additional thrust at liftoff and then be discarded once the staging velocity was attained. Interestingly, similar proposals were on the drawing boards of the many companies investigating designs for the new space transportation system but they had been rejected because NASA desired every part of the system to be piloted and fully reusable. And now the concept was promising to be as good as, if not better than, the two-stage configuration. But a question mark remained because solid-fuel rockets were a possibility. The performances for both types of booster were basically the same, but a solid booster was judged to be at least $1 billion cheaper to develop than its liquid counterpart. McDonnell Douglas was familiar with solids, and embraced a TAOS configuration using a solid rocket first stage. They also prepared an extensive report in which they analyzed all the failures that had occurred in solid boosters, proposing possible changes in design or in quality control procedures that would prevent a recurrence. The report noted in particular that in the event of such a mishap, it would usually be possible to safely abort a Shuttle launch. However, there was an exception. If the hot, high-pressure gas within a solid motor were ever to burn through its casing, and do so adjacent to the external tank or Orbiter, "timely sensing may not be feasible and abort not possible". If these words had been viewed through a crystal ball, they would have shown *Challenger* exploding only 72 seconds after it lifted off.

From a structural point of view, the solid booster would consist of a heavy casing that could easily withstand falling by parachute into the ocean. In addition, due to the low staging velocity (perhaps 4,000 feet per second), the heavy casing could act as a heat sink to absorb the thermal loads generated during re-entry. Basically, these were the same benefits as offered by the "big dumb booster" but less expensive in terms of development. Another advantage was that losing a solid booster at sea would be less expensive than losing a liquid booster. Financial limitations in the funding available to design the new space transportation system led therefore to the configuration with which the Shuttle flew all of its 135 missions: an Orbiter on the side of the external tank which mounted a pair of 156-inch solid boosters.

SRB STRUCTURE

The solid rocket boosters (SRB) of the Shuttle were the largest solid rockets ever built, the only ones rated for use with a human crew, and the first to be specifically designed for recovery, refurbishment and reuse. From a performance point of view, the two SRBs provided most of the thrust needed to lift the Shuttle off the pad and climb to an altitude of around 150,000 feet or 24 nautical miles. Furthermore, the entire weight of the fully loaded external tank and the Orbiter and its payload was transmitted through the structure of the SRBs to the mobile launcher platform below. At liftoff, after the thrust of the three main engines had been verified, the SRBs were ignited at exactly $T=0$ seconds. Each booster had a sea level thrust of approximately 3,300,000 pounds. While they were burning, the pair of boosters provided more than 70 per cent of the overall thrust.

The SRB was 149.16 feet long and 12.17 feet in diameter, with an approximate weight of 1,300,000 pounds at launch, 1,100,000 pounds of which was the propellant and the remainder being the structure. Due to this massive size, it was not possible to build each booster in one monolithic piece. Instead its casing was divided into four major subassemblies called the forward casting segment, the center casting segment (of which there were two in each booster), and the aft casting segment. Each of these was made of a given number of smaller segments, for a total of eleven. In particular, the forward segment had a forward dome segment and two cylindrical segments. The two center castings segments each consisted of a pair of cylindrical segments. The aft casting segment consisted of an attachment segment, two stiffener segments and the aft segment. The structure responsible for generating thrust was known as the solid rocket motor and the four major subassemblies were referred to as solid rocket motor segments.

The different types of segment were all manufactured in the same manner using a process that is conceptually quite simple. Royce Mitchell describes it in the following way, "It is important to understand that the rocket case segments are single pieces of metal that Ladish Company [Inc.] punched a hole in a billet of metal, D6AC steel, and then shear-formed these billets into cylindrical shapes – I'm sure wearing very large earplugs, because it's got to be a noisy operation. Ladish was very advanced in that."

The solid rocket motor was combined with other elements to create a complete solid rocket booster. At the very top of the booster, a nose cone assembly housed the booster recovery system, that is to say the parachutes. This assembly was split into a nose cap section and a frustum, with the former housing both the pilot and the drogue parachutes while the latter held the three large main parachutes, flotation devices and handling hardware for water recovery. Attached to the external surface of the frustum were four solid motors employed to separate the booster from the external tank at the appropriate moment during the ascent. The nose cap and the frustum were both made of machined and formed aluminum alloys. A forward skirt was positioned below the nose cone assembly and on top of the forward solid rocket motor segment. This was an aluminum cylinder, and contained the SRB/ET attachment fitting that transferred the thrust loads to the external tank, on which the Orbiter was mounted. The front of the skirt was sealed by a bulkhead. From a structural point of view, the forward skirt provided the structure to bear the parachute loads during deployment, descent and towing, and the internal mounting for elements of the electrical and instrumentation subsystem, the rate gyro assembly and range safety panels. Between the forward skirt and the nose assembly, there was a forward ordnance ring. This provided a plane of separation to allow the frustum to be severed in order to release the main parachutes.

On the aft casting segment at the bottom of the solid rocket motor, was a conical aluminum and steel aft skirt structure which provided aerodynamic and thermal protection and mountings for the thrust vectoring subsystem, other components of the electronic subsystem, and four separation motors. This skirt also had four attachment points to the launch platform structure to support the Shuttle on the launch pad in all conditions prior to booster ignition. These attachment points were called hold-down posts, and each had bolts holding the SRB and launcher posts together. A bolt had a nut at each end, but only one nut was frangible. The top one had two NASA standard igniters (NSI), and these were triggered by the solid rocket motor ignition command. At that time, the bolts traveled downward due to a combination of forces: the release of tension in the bolt (pretensioned before launch), NSI detonator gas pressure, and gravity. The bolt was halted by the stud deceleration stand that contained sand, and at the same time the nut was captured in a blast container. Each SRB bolt was 28 inches long and 3.5 inches in diameter. A second element attached to the aft casting segment was the SRB/ET attachment ring and fixture. A three-segment ring made of steel with three struts physically attached the SRB to the ET, and could react to loads of the ring and

SRB structure

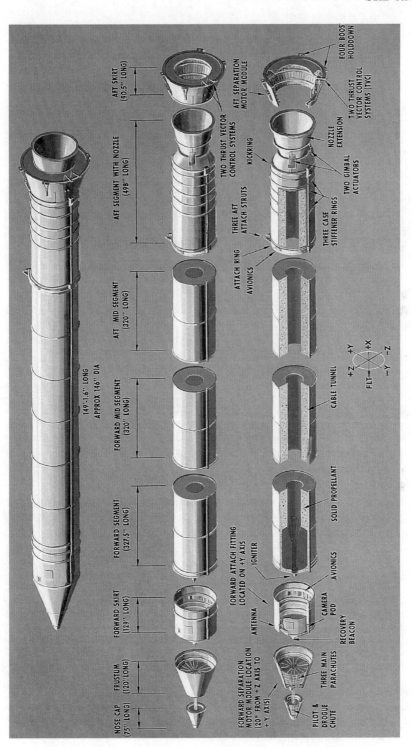

SRB technical view.

Power to orbit: solid rocket booster

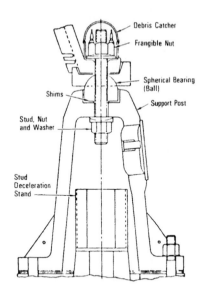

SRB hold-down post.

permit unrestrained contraction/expansion of the SRB and ET in the longitudinal direction.

Running outboard for the full length of the rocket motor was an aluminum tunnel housing the electrical cables associated with the electronic and instrument subsystem that provided lightning, thermal and aerodynamic protections. The tunnel also housed the linear shaped charge for the range safety system. Launching a rocket is always a risky business and despite the care taken in designing an ascent trajectory that would steer clear of populated areas there is always a risk of the vehicle veering off course and in uncontrolled flight threatening the civilian population. If this had occurred during a Shuttle launch, the range safety officer would have issued a radio signal to arm and then fire the self-destruct charge on each booster, with the resultant destruction of the entire vehicle and the loss of its crew. Fortunately, this situation never arose during the Shuttle program.[2]

Segment joint

As noted, each booster motor resembled a gigantic Lego construction with all of the pieces joined together. Owing to the immense pressures of combustion in an SRB, it was realized from the beginning that welding would not guarantee a flawless joint. It was therefore decided to use a more complex type known as a tang and clevis. There was a clevis at the top edge of each segment and a tang at the bottom edge. The joints that connected the segments to create one of the four castings were known as factory joints, while those between the four castings were field joints. Conceptually, they worked in the same way. As Royce Mitchell explains, "The joint is called a tang and clevis. You can think of it as a tongue and groove. If you are familiar with flooring, you have a board with a groove and then a board with a tongue that sticks out. The tongue goes into the groove. That's the way the motor was put together. There was a clevis, which is the groove part, and there's a tang, which is the tongue part. The tang drops into the clevis and then you pin the joint together with 177 one-inch steel pins. Keep in mind that the tang and clevis are pieces of the motor case, because it is all a

---

[2] The *Challenger* accident happened because the external tank exploded as a result of a leak of hot exhaust gas from the right-hand booster casing. The boosters separated from the stack and kept flying. They were blown up by the range safety officer. This is the only case in which the self-destruct system was used.

SRB structure   133

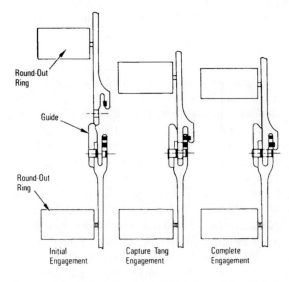

The working principle of the field joint of the SRB.

continuous piece of metal. There's no welding allowed on the motor case."

The most important difference between a factory joint and a field joint, apart from some changes in the configuration, involved the structural loads to which they were subjected at launch. "When the motor pressurizes," says Mitchell, "there's something like, depending on which joint you're talking about, anywhere from 13 million to 17 million pounds trying to pull those case segments apart. That's a hell of force for those pins and the tang and clevis to try to react against. That's mainly what caused the clevis to move away from the tang a little bit; only a few thousandths of an inch, but a very critical few thousandths." To understand why this seemingly tiny gap was so important, it must be remembered that when two of the four major segments were connected together there was a small gap left between the two blocks of propellant. During combustion, high pressure gas would fill this cavity, being prevented from reaching the field joint only by the casing insulation. Prior to the *Challenger* accident the small gap in the insulation between the two

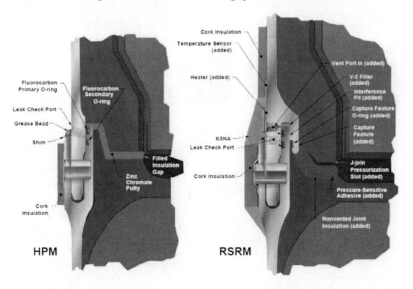

Pre-*Challenger* (left) and post-*Challenger* (right) SRB field joint schematics.

134 Power to orbit: solid rocket booster

segment casings was filled with zinc chromate putty. If the hot gas managed to reach the joint, the few thousandths of an inch gap between the tang and the clevis would have provided an easy escape route. In order to prevent this catastrophic scenario, a pair of O-rings were placed between the tang and the inner side of the clevis to guarantee a sealing action against escaping gases. That was the theory. In practice it did not work out that way, as the *Challenger* accident demonstrated.

"As the two segments to be mated were brought together, a layer of putty was placed on their surfaces, and then compressed as they came together," Mitchell says. "Unfortunately, it was impossible to avoid trapping air between the joints as you brought those two segments together, so there was air trapped in amongst and behind the putty. Over time, this air would work its way to surface, creating what was called a blowhole." *Challenger* was doomed by a blowhole in the putty that let hot gas impinge on the inner O-ring. As Mitchell continues, "When the flame had pushed its way through that blowhole, through the putty, as the motor continued to supply pressure, hot gas started to fill up the annulus, the circular tunnel in that joint, and the jet of hot gas that was hitting the O-ring was sustained ... When the motor finally pressurized and equalized, most of the problem was over because that hot gas was stagnated, the pressure was stable, and the transient was over. But you had a hot gas jet, and as the annulus filled around the whole circumference of the motor it sustained that jet, more and more gas tried to fill the circular tunnel, and that led to burnthrough of the O-ring." The chilly temperature sustained during the night prior to launch had caused the rubber O-rings to lose their flexibility, which in turn allowed the hot gas to easily escape from the field joint of the booster casing.

One possible solution was to prevent the O-ring from ever seeing any trace of flame. Another was to employ the immense pressure of the flames to push the joint components together and thereby seal the O-rings. Although this second option

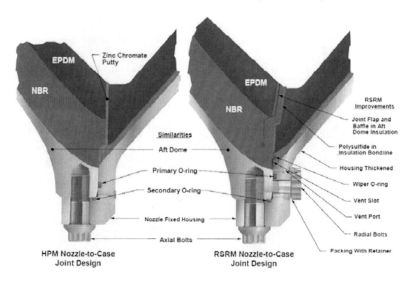

Pre-*Challenger* (left) and post-*Challenger* (right) SRB nozzle joint schematics.

would rule out any chance of a sustained hot spot burning through the O-rings, it was soon rejected because, as Mitchell recalls, "We could not be sure that the motor would not develop circumferential flow due to several factors which could produce that. If you get hot gas swirling around inside the motor, and it's traveling circumferentially over the O-rings, it might cause devastation." Since it was very difficult to model this, and there was no instrumentation to experimentally characterize it, the only option was to prevent flame from reaching the O-rings.

John Thomas got involved in redesigning the field joints, and as he recalls, "Our fundamental job was to stop the joint from opening when the motor was pressurized at ignition; if it didn't open, then the O-rings would remain in contact with the metal sealing surfaces and provide the necessary sealing function. The O-rings then didn't have to be so resilient to prevent hot combustion gas blowing past them."

The solution was to incorporate a capture feature on one side of the joint in order to prevent it from opening excessively. As Mitchell says, "It was like putting two clevises together, where the upper one gripped the lower one and the lower one, of course, gripped the upper one. That so-called capture feature, which was like another leg of a clevis on the tang side, gripped the inboard side of the clevis and kept it from moving." The new joint was then provided with a third O-ring, called the leak-check O-ring, in the metal capture feature. Additional protection was added. As Thomas recalls, "The second and most challenging part of it was to determine how to close the gap between the internal motor insulation at the joint... After evaluating several concepts, we selected one that produced an interference fit between the insulation mating surfaces of the adjoining motor segments." This configuration was named the J-seal, and it was to be the first line of defense against hot combustion gases reaching the O-ring seals.

As Mitchell explains, "The J-seal was a curved flap of internal insulation. It was called 'J' because a hook on the end was a part of the casing wall insulation that hung down. As you brought segments together, that circumferential flap interfaced with the next segment and sealed the joint. If the J-seal worked properly, the joint never would see the flame front." As an enhancement, an adhesive similar to that used on "Post It" notes was added between the J-seal mating surfaces. One final concern was the temperature of the joint. As Thomas recalls, "They wanted the temperature to remain relatively constant at the joint, so the O-rings would not be subjected to cold temperatures again if the temperatures at launch would fall below 50 degrees. Their concern was that low temperature would cause the O-rings to lose their resiliency. Our teams were not particularly concerned with this design feature, because tests showed that the O-rings would track joint opening under the specified temperatures at launch." But heaters were added anyway, to make the design of the new joint even more robust. The revised field joint was introduced with the return-to-flight mission, STS-26 in September 1988, and there were no further problems with the solid rocket boosters.

Subsequently, the field joint mating pins were lengthened and a retaining band added to improve the shear strength of the pins and increase metal part's joint margin of safety. Minor modifications to the factory joints increased the insulation thickness and layup to increase the margin of safety on the internal insulation. Longer pins were also added, along with a reconfigured retaining band and a new weather seal to improve factory joint performance and increase the margin of safety. Additionally, the O-ring and O-ring groove size was changed to be consistent with the field joint.

Nozzle

To generate thrust, the exhaust gas produced inside the solid rocket booster had to be channeled out through a convergent-divergent nozzle consisting of an ablative carbon cloth phenolic liner over a steel and aluminum support structure. The nozzle was a modular-type construction with parts grouped into assemblies to facilitate maximum reuse and refurbishment of structural members. The nozzle was partially submerged in the aft segment in order to minimize erosive conditions in the aft end of the motor and also to accommodate envelope length limitations.

The nozzle could also be gimbaled by dual hydraulic rock and tilt actuators that transmitted nozzle rotations around a flexible bearing that consisted of a flexible core contained between a pair of D6AC end rings. The core was a laminated structure of 10 spherical D6AC shims and 11 natural rudder pads. The end rings and shims would absorb the applied loads whilst simultaneously controlling the bearing motion during vectoring. Elastomeric pads were placed to transmit the loads whilst also permitting relative motion to occur between the structural members. Following each mission the flexible bearing was removed from the nozzle, placed in a test fixture, and subjected to extensive tests to verify its integrity prior to reuse in another

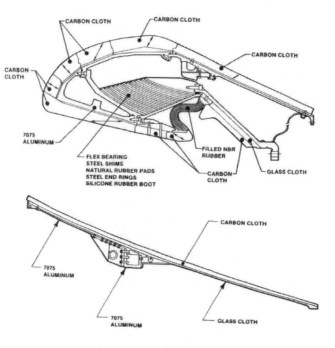

Main elements of the SRB nozzle.

nozzle. Based on calculations and subscale tests performed during development, it was determined that the elastomeric pads could be used up to ten times prior to replacement. Once this threshold had been reached, the bearing was cut apart, refurbished and the metal used to mold a new nozzle with new elastomeric pads. As some gas would leak anyway, thermal protection was placed on the bottom part of the bearing. This protection, known as the "boot", was a barrier of laminated rubber that eroded or burned away at a calculated rate and was thick enough to hold out until the motor expended all of its propellant.

Assembling the SRBs

Each booster segment was manufactured and/or refurbished by ATK in Utah. They were shipped by rail to the Kennedy Space Center on a schedule to suit the different milestones for the various missions of the Shuttle program. As engineer Anthony P. Bartolone says, the segments "are received at what we call the Rotational Processing and Surge Facilities. The railcar covers are removed, and the segments are taken off their railcars and rotated to vertical." This operation allowed the SRBs to be placed on purpose-designed steel pallets, one for each segment, so that when it was time to stack the SRBs they could be retrieved from storage in the desired order and in the correct orientation.

As Bartolone explains, "The first part of the SRB process is that we mate the aft motor segment to the aft skirt, and that happens in the [facility]. And then once that's done we call that the aft booster assembly, because it's now a motor segment." This first motor segment was then moved into the VAB and bolted to the mobile launch platform for that mission. "We go through the rest of the stacking of the Lego pieces, essentially all the different motor segments that make up the remainder of the SRBs." After stacking the pair of SRBs on the platform, the external tank could be mated to the boosters and finally the Orbiter added.

THERMAL PROTECTION

Early in the SRB design process there was little thermal design environment data. The low heating rates initially predicted suggested that a "heat sink" design would be feasible, with no thermal protection material being required. However, as the design trajectories were revised and wind tunnel tests yielded data on aerodynamic heating, the thermal design environment became much more severe, significantly increasing the predicted structural temperature levels. In particular, it was found that with the exception of the steel casing, the allowable maximum temperatures for all elements of the SRB were exceeded. It would therefore be necessary to develop an external thermal protection system (TPS) to maintain acceptable temperatures.

The first material studied was bonded cork, due to its availability and extensive prior application as an ablative insulation. But it was soon realized that cork would impose serious drawbacks in terms of the time and expense of its initial application and subsequent refurbishment. The cork bonding process on major flight hardware was extremely labor intensive, and would result in significant cost penalties over a large number of vehicles. Furthermore, the high density cork with its discrete layer of adhesive on the metallic substrate proved to be extremely difficult to remove during studies of refurbishment processing. The initial challenge was the development and qualification of a primary TPS system suitable for an area as large as the nose cap, frustum and forward skirt. Following an intensive formulation screening and spray development phase, a system was chosen that employed an aromatic, amide-cured, urethane-modified epoxy binder (or matrix resin) filled with a mixture of glass and phenolic microballoons as well as glass reinforcing fibers. Devised by the Marshall Space Flight Center, this ablator composition was named Marshall Sprayable Ablator or MSA-1. Various thicknesses were analyzed at each location until one was found that would prevent the structure from exceeding its design temperature limits. However, to ensure consistent material characteristics the maximum sprayable thickness had to be limited to one quarter of an inch. The high heating rates on the aft attachment ring, aft portion of the system tunnel, and the aft skirt, precluded utilization of MSA-1 on these structures due to thickness limitations and/or low tolerance to airstream shear forces. Hence cork insulation was employed on the aft skirt and areas of the system tunnel. For the aft attachment ring it was decided to use phenolic glass fairings. This latter choice was driven by the requirement for easy refurbishment, the ring structural configurations, and the high plume impingement heating of the main engines of the Orbiter during SRB separation.

A second ablator TPS with improved performance was later formulated with the aim of extending its use to more hardware components. Known as MSA-2, this was first used on the forward assembly by STS-29 in March 1989 and on the aft skirt by STS-28 in August 1989. The change in formulation reduced the susceptibility of the TPS to stress cracking during curing, enabling the maximum application thickness to be increased to half an inch. Other improvements of MSA-2 included the elimination of a carcinogenic curing agent, increased strength, improved adhesion, and shortened processing time. But MSA-2, like MSA-1, was still considered hazardous due to the solvents used in the application process.

To remedy to this issue, a third compound called Marshall Convergent Coating or MCC-1 was developed. This was not only more environmentally friendly and less hazardous to humans, but also had increased strength, better thermal characteristics, and was easier to apply. It reduced the time and expense of refurbishment.

RECOVERY SYSTEM

In the initial design definition phase of the SRB recovery system development, two primary issues were of major concern. Firstly, how to provide high altitude booster deceleration to achieve conditions suitable for parachute deployment and, secondly, how to optimize the recovery system for water impact. The solution had to minimize cost and complexity. This ruled out an active stabilization or attitude control system. Other options included various drag generators such as extendable flaps, drag petals, and inflatable ballutes, but a drawback to such devices was that they tended to orient the booster centerline in the stream-wise direction, causing a tremendous decrease in SRB drag. A high angle of attack or "broadside" re-entry mode would be preferable. In fact, the natural aerodynamic drag of the booster alone could potentially provide the deceleration required to yield the desired conditions for parachute deployment. Although the best method of achieving a near-broadside re-entry mode had not been determined, this re-entry concept was adopted because of the numerous advantages it offered.

Meanwhile, studies were made to optimize the final deceleration system. One key trade-off was an all-parachute versus a hybrid parachute-braking-rocket system. The results indicated that for water impact at velocities above 65 feet per second a pure parachute system would have been lighter. Studies of water impact had found that a tail-first impact at 80 to 100 feet per second was a good compromise between initial impact and slap down loads. The pure parachute system was therefore baselined as being both lighter and less complex.

During the program, two different type of main parachute were developed for the solid rocket boosters. The first one was the small main parachute (SMP). It had a canopy diameter of 115 feet and resulted in water impact velocities of approximately 90 feet per second. Starting with STS-41D in August 1984 a larger parachute called the large main parachute (LMP) was introduced. It had a diameter of 136 feet and a water impact velocity of about 76 feet per second. This was fully incorporated with STS-51D in April 1985. The larger canopy parachutes provided a two-fold benefit. Firstly, they reduced the water impact velocity and therefore the impact loads and the potential for structural damage. Secondly, they increased the SRB survivability in the event of a single parachute failure, where the water impact velocity under two LMP would be comparable to that of three SMP, thus reducing the structural damage from a hard water impact. However, the system failed several times.

The first failure was on STS-3 in March 1982 where a main parachute became entangled with floats installed to assist in retrieval of the parachutes. The floats were removed for STS-4 in June 1982 but another failure occurred where the parachute lines released from the deck fittings prior to deployment. The mechanism was meant to sever the lines upon water impact by sensing the g load, minimizing the possibility of a swimmer becoming entangled during recovery. Evidently during the separation of the frustum, the g load switch triggered the deck fitting pyrotechnics, prematurely releasing one of the two deck fittings on each main parachute. In this configuration, the parachutes gave no drag during the ensuing descent, resulting in a water impact at six times the intended velocity and the boosters sinking. As a result of this failure, the switch was deactivate until a timing system was incorporated that would release the deck fittings only when water impact was imminent. But as will be explained later, this system gave rise to a different problem.

STS-33 in January 1990 introduced ripstops in the canopy. These consisted of six horizontal bands of high-strength nylon added to the LMP in order to prevent a tear from propagating completely through the canopy, and thus enable a torn canopy to remain inflated. The spacing of the ripstops was defined by analytical models of the canopy loading. Ripstops evidently prevented total failure of a parachute system on seven occasions.

140 **Power to orbit: solid rocket booster**

Sequential stacking design (left) versus Hengel Weave (right).

A third improvement was in the design of the top of the canopy, otherwise known as the parachute vent. Starting with STS-51F in July 1985, a vent cap in the form of additional horizontal canopy members was added to the parachute vent to reduce its diameter to 5.6 feet. The vent cap consisted of 80 vent lines stacked in a clockwise sequence to produce large openings and a large mass at the center. This allowed the canopy to inflate more reliably to the first reefed condition but the openings allowed air to escape, which slowed inflation of the main parachutes and tempted the mass of lines to tangle, damaging the canopy. So starting with STS-95 in October 1998 the vent lines were woven together in an alternative stacking method called the Hengel Weave, with eight groups of ten vent lines to organize and constrain the vent lines. This eliminated two large slots in the vent and prevented the concentration of a vent line mass, thereby further helping the parachute to inflate more reliably to its first reefed condition.

PROPELLANT

Solid rockets were the first kind of rockets developed by mankind. Several thousand years ago, the Chinese developed rudimentary small rockets filled with gunpowder to be used as fireworks to celebrate important events and possibly even as weapons in battle. Since then, technology in the field of solid rocket propulsion has undertaken giant leaps forward, with the aim of providing more and more powerful and reliable rockets. Although several configurations are now available, every solid rocket motor uses the same working principle. Propellants are stored in a solid form, called grain, inside a pressure vessel (or case) made of metal or fiber reinforced plastic metal. An igniter spreads a flame over the exposed surface of the grain to initiate combustion, and the resulting hot gas is passed through the nozzle to generate thrust. The grain is consumed by a burning surface that "moves back" at a given regression (combustion) rate determined by the propellant composition, temperature, pressure in the chamber and velocity of the gas in the case.

For the solid rocket boosters of the Shuttle, a composite-type solid propellant was chosen, a formulation in which oxidizer crystals and powdered fuel were mixed and bound together by a matrix of synthetic rubber (or plastic) binder. Due to its quality,

SRB segment hoisted for assembly. It is possible to see the hollowed propellant grain inside the segment.

good performance, uniformity and availability, ammonium perchlorate was chosen as the oxidizer (69.8 per cent total solid composition) with aluminum powder as the fuel (16.0 per cent total solid composition). They were bonded with polybutadiene acrylic acid acrylonitrile terpolymer (PBAN or HB polymer) and an epoxy curing agent. The remaining 0.2 per cent was iron oxide as an additive to achieve the desired propellant burning rate.

A rocket using solid propellants must include a method for regulating the thrust level during burnout. Clearly this cannot be achieved using valves, since there is no propellant feed system. The only efficient solution is to vary the burning surface area by modifying its shape. This is quite intuitive, because the larger the burning surface the greater the amount of exhaust gas produced and the higher the thrust delivered.

The grain loaded into the SRB had an 11-point star central hollow cavity in the forward segment, and a double-truncated-cone central hollow cavity in the remaining segments. The forward segment configuration facilitated the high thrust needed to lift off from the pad at ignition. Although a solid rocket motor cannot be throttled in the manner of a motor with liquid propellants, the progressive burning of the star and its resultant widening, together with the burning of the propellant in the other segments, provided a reduction of the thrust by approximately one-third some 50 seconds after liftoff in order to prevent overstressing the vehicle during the period of maximum aerodynamic pressure.

Prior to mixing and casting the propellant into the casting segments, a number of operations had to be performed. Firstly, the raw material was inspected to verify its viscosity, moisture, particle size distribution, and other properties. Once the material was accepted, the standardization process could begin. This determined the amount of iron oxide required to give the desired burn rate and the amount of HB polymer to give the required mechanical properties. Next was a verification process to confirm the standardization values. Once it was decided to proceed with casting the segments, the propellant was mixed together. This was undertaken in two steps. In the so-called premix building, HB polymer, aluminum powder and iron oxide were combined and premixed in a 600-gallon mixer bowl for five minutes. Then an epoxy curing agent was "puddled" on top of the premix. This was normally done at the very last minute, because the propellant mix had to be cast into the segment within six hours of the epoxy curing agent being added. The premix was taken to the mixer building for the incremental addition of the oxidizer. A bimodal (ground and unground) mixture of oxidizer was used for optimum propellant processing and physical characteristics, and for a burning rate in the desired range. For the SRB, a 70/30 unground/ground ratio was optimal. All of this processing had to be done very carefully, as ammonium perchlorate is extremely unstable and can decompose explosively if subjected to heat or shock. The mixing cycle lasted around 50 minutes in total, with the addition of the oxidizer taking 25 minutes. A minimum of 15 minutes of mixing was required after finishing the oxidizer addition. Samples of the mix were analyzed in a laboratory to verify its properties and burn rate. If there were any anomalies, the entire mix would be scrapped. Otherwise the mixer bowl was moved to the casting pit. Each segment casting consisted of approximately forty 600-gallon mixes, accomplished by casting pairs of identical aft segments, central segments, etc., in sequence to ensure that the two boosters of a flight set would have burn rates as equally matched as possible. For each mix, samples were taken for subsequent analysis. On finishing the casting, the propellant was cured at 135°F for around 96 hours. The data from the final tests was used to predict the performance of that particular pair of flight motors.

It is worth mentioning that before casting, each segment was internally insulated to protect the casing structure from the heat of combustion. This included the primary insulation, forward-facing full web propellant grain inhibitors, and propellant grain stress relief flaps. The primary chamber insulation propellant stress relief flaps at the ends of each of the forward and center casting segments were to minimize insulation-liner-propellant bondline loads created at the grain termination surfaces after curing as a result of thermal shrinkage and during pressurization of the solid rocket motor. All insulation components were fabricated as integral assemblies within each casting segment as part of the refurbishing process. They were laid up in the SRM casting segment, bonded, and autoclave-cured in one operation prior to propellant casting. The inhibitors were to provide thermal protection to the propellant grain and prevent it from igniting and burning perpendicular to the surface of the inhibitor. For a solid rocket to function correctly, it is essential that the propellant burns only over the designed burning area. For the SRB, this meant that the top and bottom faces of the propellant castings had to be protected to prevent hot gas from penetrating the thin gap between the propellant castings of adjoining booster segments. To achieve this, a web of propellant inhibitor was provided on the forward facing side of each casting segment integral with the insulation. The aft-facing partial web inhibitor was cast and troweled on during the propellant curing process. A liner was then applied to bond the insulation and propellant in a manner that would fail cohesively in the propellant.

Igniter

Igniting the grain of a solid propellant rocket is much easier than starting a controlled combustion in a liquid fueled rocket engine. There is no need for complex plumbing and sequencing of valves. A solid rocket simply needs an igniter. In general terms, the process begins upon the receipt of a signal (usually electric), and involves heat generation, transfer of the heat from the igniter to the motor grain surface, spreading the flame over the entire burning surface area, filling the free volume (cavity) with gas and raising the pressure without serious abnormalities.

The igniter chosen for the SRB was of the pyrogen type and it was installed in the forward end, internally mounted. The ignition system had several components:

- The safe and arm (S&A) device to prevent inadvertent ignition or misfiring. It had a reusable electromechanical actuation and monitoring assembly with an electric motor, a manual safing and locking mechanism, a visual position indicator, and an electrical circuit switch deck. It also had a non-reusable barrier booster assembly with the motor pressure seal, the safety barrier that would be rotated to "ARM" by the actuation and monitoring assembly, two NASA standard igniters with a pyrotechnic booster charge and an "armed" and a "safe" indicator electrical switch.
- An igniter initiator, which was a small multi-nozzle, steel cased solid-fuel igniter containing a case bonded 30-point star propellant grain.
- A rocket motor igniter which was a single nozzle, steel cased, internally and externally insulated solid-fuel igniter containing a base bonded 40-point star propellant grain.

The igniter initiator and the motor igniter were arranged in a Russian doll fashion, with the initiator contained within the igniter. They were cast at the same time using the same batch of propellant.

Several days prior to launch, the manual lock pin was removed from the S&A device to enable it to be armed remotely five minutes before ignition. A number of conditions had to be satisfied before the ignition command could be issued. Firstly, within three seconds of booster ignition all three main engines on the Orbiter had to be delivering at least at 90 per cent of their rated thrust without any failure alarms.

Secondly, none of the circuits for activating the hold-down posts could be indicating a malfunction. When the ignition command was issued, the NASA standard igniters in the S&A device would fire through a thin barrier seal into the pyrotechnic pellet charge in the device, located behind a perforated plate. The pellet charge ignited the propellant for the igniter initiator, whose combustion products ignited the propellant of the rocket motor igniter, which then spread its flame all over the internal surface of the booster grain to start that grain burning.

It is mind-boggling to realize that both boosters were required to develop a thrust of 1.9 million pounds within an interval of 0.15 to 0.45 seconds, with an increase in thrust not exceeding 150,000 pounds in the next ten milliseconds. During the ignition transient the maximum thrust imbalance between the two SRBs was 300,000 pounds and there could not be an ignition pressure overshoot. In 135 missions flown during the program, the SRBs never failed to ignite. Of course, if one of a pair had failed to start, this would have been catastrophic for the vehicle.

SRB THRUST VECTOR CONTROL

Almost every book about rocketry claims that one of the advantages of a solid rocket booster is its absence of moving parts. This is easy to understand, because there is no need for turbo machinery or a plumbing system with associated control valves. But the SRB has a thrust vectoring control system consisting of a pair of auxiliary power units for the hydraulic systems that drive the two nozzle actuators.

The solid rocket boosters of the Titan III, the largest built before the SRB, used a vectoring technique called "liquid injection thrust vector control". In this system, the booster carried a long tank that held a pressurized liquid which could be squirted into the exhaust from within the nozzle. On coming into contact with the exhaust gas, the liquid would flash to a gas. This would produce a localized shockwave within the supersonic flow of the exhaust and thereby alter its direction. This concept was well understood and proven, so the initial idea was to employ it for the SRB. Furthermore, because the Orbiter would steer itself into orbit by gimbaling its main engines after booster separation, it was intended to use these to steer the stack during the first stage of the ascent. The nozzle of the SRB would be fixed in a canted position and liquid injection steering would merely aid the main engines. But this approach proved to be impractical. Because the two boosters were quite far from the centerline of the stack, their nozzles would have to have been canted at eleven degrees, and this would have caused such a major loss of forward thrust as to require each booster to carry an extra 52,000 pounds of propellant. In addition, the cant angle would have produced a side load of 790,000 pounds to be borne by the external tank structure and also by the separation motors at booster jettison. It would have imposed a major weight penalty. Finally, it would have been necessary to perform several expensive tests in order to characterize the nozzle environment for such a high cant angle, due to the possibility of an instability forming inside the exhaust gas as it expanded in the nozzle. A canted nozzle was discarded. And so was liquid injection thrust vector control. Given the phenomenal thrust of the SRB, the amount

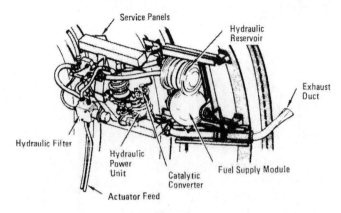

SRB HPU system.

of liquid to be carried to provide a decent steering capability would have been prohibitive.

The only remaining solution was to adopt a movable nozzle driven by hydraulic actuators. This proved to be very beneficial, because it provided a net increase in the performance of the Shuttle during ascent. Soon after leaving the pad, the Orbiter was required to roll in order to align itself on the azimuth to achieve the desired orbital inclination. With the enormous thrust provided by the boosters, this maneuver would have been impossible to make by steering only the main engines. The movable nozzle improved the maneuverability of the Shuttle during ascent.

Hydraulic power

Each SRB possessed two self-contained independent hydraulic power units (HPU) consisting of an auxiliary power unit (APU), a fuel supply module, a hydraulic pump, a hydraulic reservoir, and a hydraulic fluid manifold assembly.

During the technical studies to choose a SRB thrust vector control (TVC) system, several options for providing hydraulic power were evaluated. A key influence was the development program then underway at Sundstrand Aviation for the APUs of the Orbiter. Their requirements very closely approximated or exceeded those of the SRB, and they were physically compatible with the SRB concept of reusability for at least twenty missions. It was therefore decided that the Orbiter APU should serve as the basic power element for the SRB. This offered the prospect of reduced development costs and early hardware availability.

The APU was a monopropellant hydrazine fueled gas turbine engine that drove a variable displacement hydraulic pump. At 28 seconds prior to liftoff, the APUs were initiated by opening an isolation valve to allow fuel to flow to the APU fuel pump and control valves, and on to the gas generator. In the gas generator, a catalytic action decomposed the fuel, creating hot gases which expanded in a two-stage gas turbine. The hydraulic pump was set in rotation by a reduction gearbox in order to pressurize the hydraulic system. The turbine exhaust for each APU flowed over the

Power to orbit: solid rocket booster

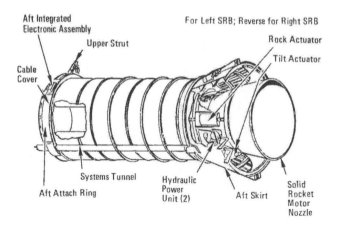

SRB hydraulic actuators location.

exterior of the gas generator to cool it prior to being directed overboard through a vent duct.

Although the APU of the Orbiter was the basic power element of the APU of the SRB, the two units were very different. The Orbiter APUs were started 5 minutes prior to liftoff, operated continuously through to orbit insertion and then shut down. In preparation for deorbit and re-entry, they were restarted and operated throughout atmospheric maneuvering to landing. They therefore had a total operating time of about 82 minutes. The SRB units were started less than 30 seconds prior to launch and then operated only for two and a half minutes through to SRB separation. Thereafter, the SRB's mission was one of survival. From the shock of splashing down at 91 feet per second to SRB removal from the ocean at dockside, the APUs were subjected to various combinations of seawater immersion and water pressures. Complicating the situation was the installation. Whereas the Orbiter APUs were installed in an aircraft-type compartment that offered protection against aerodynamic, vibration and thermal extremes during all phases of operation, the SRB APUs were located in the aft skirt, next to the motor where protection was available only during ascent. It was therefore obvious that the Orbiter APU would have to be redesigned to cope with the severe conditions endured by the SRB.

Two modifications were noteworthy. The first involved the development of the gearbox. For the Orbiter APU the gearbox was surrounded by an intricate externally attached lubrication oil cooling loop, and also by a pressurization system designed to support long duration and on-orbit operation in zero g and a vacuum environment. If these had been necessary for the SRB APU, they would have complicated the TVC system and added weight and volume to the aft skirt. And to redesign the externally mounted components to deal with the in-flight vibration and splashdown loads would have been expensive. The solution was obvious. An analysis of the operating times and heating loads showed that it would be feasible to tailor the operational profile of the SRB APU to stay within the 300°F thermal limit of the gearbox. This eliminated the coolant loop, accumulator, and pressurization system. The second modification was necessary because the vibration loads expected for the SRB APU were greater than those of the Orbiter APU, prompting concern about the operability and life of several components. To preclude having to redesign these components, a vibration isolation system was developed employing three individually tuned vibration damping mounts attached between the APU at its mounting lugs and the primary mounting structures. This attenuated the vibration loads on the SRB APU to levels well below those for a hard-mounted APU and, in some cases, below those experienced by the APUs on the Orbiter.

Each SRB had two hydraulic gimbal servoactuators, one for rock and one for tilt, to provide the force needed to gimbal the nozzle and provide thrust vectoring control. Each actuator was connected to both HPUs, with one HPU serving as the primary hydraulics source for the servoactuator and the other as the secondary hydraulics for the same actuator. A switching valve allowed the secondary hydraulics to power the actuator if the primary hydraulic pressure dropped below 2,050 psi.

SRB RECOVERY AND REFURBISHMENT

To make Shuttle launches as economical as possible, the reuse of flight hardware was essential. Unlike solid rocket boosters previously used in the space program, the SRB casings and associated flight hardware were recovered at sea, returned to the factory, disassembled, refurbished, and reloaded with solid propellant for reuse.

SRB separation

Separation was initiated when the head-end chamber pressure of both SRBs reached a pressure less than or equal to 50 psi. A small fraction of the propellant remained in the case but the thrust generated was so little that for all practical purposes the SRBs could be considered to be incapable of providing any useful propulsion contribution. They had then to be jettisoned, because the main engines of the Orbiter would not be able to fly the desired trajectory carrying their inert mass. To protect against chamber pressure sensor biases, SRB separation would in any case be triggered at a given time after booster ignition. A third line of defense against possible separation failure was a push-button on the flight deck of the Orbiter.

When the separation sequence was initiated, the

Right-hand SRB separating from the ET.

Forward (left) and aft (right) SRB separation motors. The aft separation motors were slightly canted outward and positioned asymmetrically in order to push the SRB away from the Orbiter at separation.

SRB nozzles were placed in the null thrust vectoring position and the Orbiter held its yaw and pitch for 4 seconds to allow a clean separation without risk of recontact. Each SRB was connected to the external tank by forward and aft attachment points, with the latter taking the form of three struts. An explosive bolt in each attachment point severed that mechanical link within an overall interval of 30 milliseconds. At the same time, two groups of four small solid motors, one on the nose cone assembly and the other on the aft skirt, were fired to rapidly displace the booster from the fast ascending Orbiter and eliminate the risk of collision.

Each booster separation motor was 31.1 inches long and 12.85 inches in diameter, and gave an average thrust of 22,505 pounds for a duration of 0.55 seconds. Despite seeming simple, their development caused some headaches for Gerald Smith, whose first assignment as program manager was to develop the SRB separation system. "We started out as a rather traditional program where we planned to use state-of-the-art propellants for small motors. Early on, we tested a motor against some Orbiter tiles. I think they were 20 feet away, which was the approximate distance we thought we'd have the motors separate from Orbiter – and it destroyed the tiles." This was a major problem and required a rapid solution. An investigation was promptly started to understand what could be done "to avoid both impinging on the

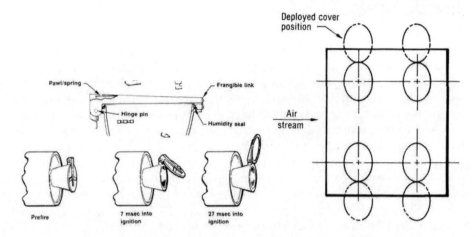

The working principle of the forward SRB separation motor covers.

Orbiter with the motor firing, and at the same time what kind of propellant we could use to mitigate the damage". Working alongside the Air Force, Navy, and industry, a propellant was developed using only 2 per cent of aluminum additive. As Smith recalls, "This was really pushing the state of the art at the time. It necessitated a lot of testing, because the aluminum is a stabilizer for combustion stability. We had to know whether or not we could produce this material and have the motor stable, because if it goes unstable it can explode since you get pressure oscillations inside the chamber. It was critical to us to establish a stable propellant design, so we did a lot of testing." However, there were other problems involving the forward group of motors. "We had planned to put the motors on the side of the forward skirt, and we realized we needed to move them further forward. We also had to terminate the burn quickly, so that we brought the pressure down before the Orbiter would fly through the plume. That drove us to move the motors forward and cant the nozzle away from the Orbiter. After about eight-tenths of a second the motor firing was terminated."

But the forward motor still presented a small problem. Smith and his team "were worried about the nozzle being canted upward. The ascent heating could prematurely light the propellant, so we put a cover over it. We examined a lot of different cover designs, and everything we looked at would generate debris. When you ignited the motor and blew the cover off, it could hit the Orbiter. Finally, we decided on a metal cover design that would swing open at motor ignition. It had a ratchet device that would keep the cover open and lock it in place. That solved the problem. It had never been done before. A first in the business. We did a lot of experimentation on different cover designs, and finally chose one that would ratchet open and at the same time not come off and generate debris." Thanks to very extensive testing, no Shuttle ever had any problem with these motors.

Power to orbit: solid rocket booster

SRB descent and recovery

After separation, the SRBs did not start falling towards the ocean immediately, since their momentum kept them going uphill for another 70 seconds to a peak altitude of 38.6 nautical miles. The recovery sequence began at an altitude of 15,700 feet, 188 seconds after separation, with a high-altitude baroswitch firing the pyrotechnic nose cap thrusters to eject the nose cap and deploy the pilot chute. The 11.5-foot-diameter conical ribbon pilot chute provided the force to pull the lanyard to activate the zero-second-cutter that severed the loop that secured the drogue retention straps. This allowed the pilot chute to pull the drogue pack from the SRB, causing the drogue suspension lines to deploy from their stored position. Upon full extension of the dozen 95-foot suspension lines, the drogue deployment bag was stripped away from the canopy and the 54-foot-diameter conical ribbon drogue chute inflated to its initial reefed state. It dereefed twice after specified delay intervals, reorienting and stabilizing the SRB in a tail first attitude ready for main parachute deployment.

At 5,500 feet, 243 seconds after separation, the low-altitude baroswitch fired the pyrotechnic to sever the frustum from the forward skirt. As the drogue chute carried the frustum away, the suspension lines of the main parachute were drawn out of the deployment bags located in the frustum. At full extension of these 204-foot lines, the three main parachutes were hauled from the deployment bags and inflated to their first reefed condition. After specified delays, the main parachute reefing lines were cut and the canopies inflated to their second reefed and finally fully open states. The parachute cluster could now decelerate the SRB to terminal conditions. The nozzle extension was severed by pyrotechnics 20 seconds after the low-altitude baroswitch triggered. Water impact occurred around 277 seconds after separation approximately 140 miles off the Atlantic coast of Florida at a velocity of 76 feet per second.

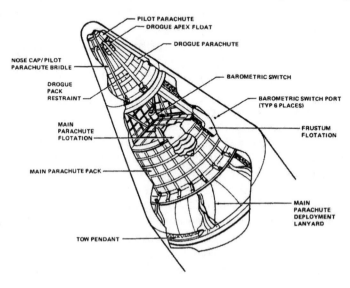

SRB parachute assembly.

SRB recovery and refurbishment   151

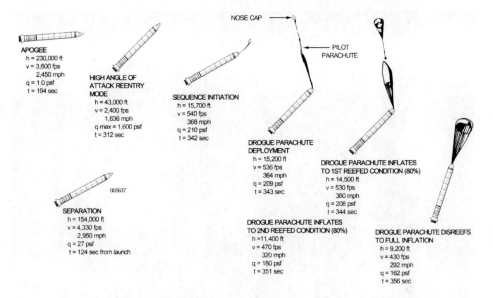

SRB recovery sequence from separation through to drogue chute dereefing.

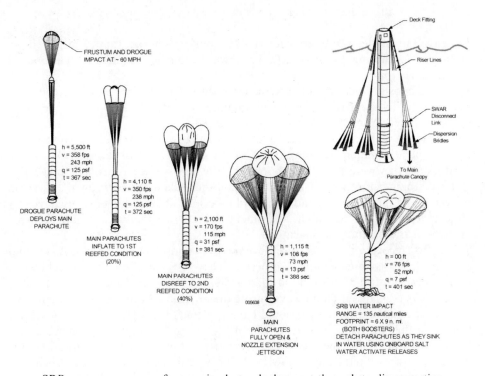

SRB recovery sequence from main chutes deployment through to disconnection.

152 Power to orbit: solid rocket booster

Left-hand SRB falling into the ocean after the launch of STS-5.

Since the parachutes produced a nozzle-first impact, air was trapped in the empty motor casing, causing it to float with the forward end projecting about 30 feet out of the water.

As the boosters were falling, the recovery ships *Liberty Star* and *Freedom Star*, built specifically to retrieve SRBs, were standing by some 8 to 10 nautical miles from the impact area. After the boosters were in the water and their command destruction systems were confirmed to have been safed, the ships moved in. Joe Chaput served as marine operations manager and skipper of the *Liberty Star* for several years. As he recalls, once they were close to the boosters, "We do a surface assessment, take video and everything before we touch anything, and then, likewise, we throw the divers in the water and they go down and video and take stills of everything because it is very important during the post-flight assessment to know whether an anomaly occurred in flight or is something we've caused."

The first recovery task was to retrieve the pilot and main parachutes. As Chaput explains, this was an evolutionary process. "They just about double in weight when they're wet, and I think it's right around 4,000 pounds when we recover them. Now things changed many different ways during the program. At the beginning they used to be connected and the divers would have to detach them. There was a link forty feet down that the divers would have to hang onto and disconnect. Later on, as a booster hit the water there was a sensor that would trigger a charge and the parachute would literally get blown off. They had their own floats. When you approached you'd have the booster, the frustum, and three parachutes floating on

SRB recovery ship. The conical shape visible at the aft end of the ship's deck is an SRB frustum retrieved by the crane.

their own. So you just had to go over and recover them. At that point we could do that whole recovery process in less than four hours. If the parachutes weren't fouled, it was really an easy thing." But on several occasion, especially in the presence of a crosswind in the splashdown area, the boosters could hit the water not exactly tail first but at a significant angle. If this was the case, on releasing the parachutes there was nothing to prevent the casing from making a strong impact with the water. This often caused the booster to develop a so-called "banana shape". The most severe damage was STS-63, when the aft skirts of both boosters were severely damaged. In the hope of reducing structural damage, starting with STS-86, sea-water-activated switches were incorporated to release the lines once the SRB had settled more gently into the water.

This influenced the way the retrieval crews worked. "Now it is a lot harder," says Chaput. "If everything goes well, the parachutes, when they hit the water, have things called SWARs, saltwater activated release, that cut the risers, which is the thing that connects the parachute to the booster, and then they are just retained by a couple of Kevlar lines. The divers just have to go in, tie floats to the parachutes, cut the Kevlar lines, and they float free." This solution worked quite well for the booster's structure, but could cause some problems for the divers. "I have to say, though, that there have been times, especially when it's calm, where the parachutes drape over the booster. Then the divers are really at risk because they're trying to free these parachutes and they are in an entanglement nightmare." The solution? As Chaput puts it, "We have a safety diver who goes in before everybody else. His only job is to watch everybody else, and he has underwater communications with the ship. Then the divers swim down in pairs or groups." The main parachutes were then reeled on board the ship by means of three reels. A fourth reel hauled on board the

Power to orbit: solid rocket booster

Main parachute reeled onboard.

drogue parachute, and then the attached frustum was hoisted onto the deck by a small crane.

With the parachutes and the frustum recovered, the effort switched to the booster casing. As Chaput explains, "We send the divers in again to put in a DOP, a diver-operated plug. It's 1,400 pounds, but just slightly buoyant in the water. That's been a whole evolutionary process in-house, the development and improvement so that they can swim it down about 110 feet, stick it into the aft skirt and lock it into place using hydraulic arms. We pump air to it, the air displaces the water and brings the booster back up to the surface. Then we connect the tow and come on home. It sounds like an easy operation, but it takes a lot longer than you might think." After the booster was horizontal in the water, like a log, air pumping continued until all the water had been expelled from the casing. The final step in the retrieval procedure was to connect the tow line. Once the two connections were made, the divers returned to the ship. Upon arrival at Cape Canaveral Air Force Station, the booster would be moved from the stern tow position to a position alongside the ship ("on the ship") to provide greater control during towing.

The only difference between the two SRBs was a stripe around the forward skirt of the left-hand one. Although there was no specific reason, *Liberty Star* always got the right-hand booster while *Freedom Star* got the left-hand one. The retrievals were conducted separately, without communications between the ships. What about night launches? Chaput remembers, "Early, early, early in the program we used to dive at night, but it's so hard to keep track of the divers. You would think if you put enough Cyalumes (chemiluminescent material) on the guys, you'll be able to see them. In a dynamic environment, it's just not worth the payoff, so we stopped diving at night.

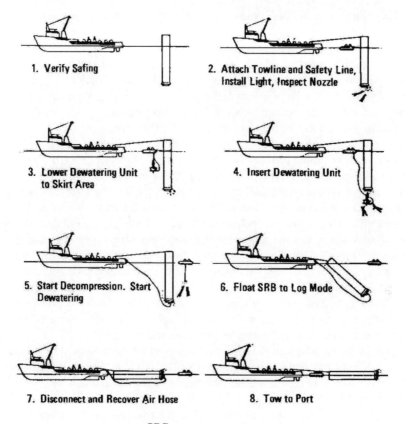

SRB recovery sequence.

We approach the hardware, we have good searchlights on the ship and we just hold station and keep an eye on the hardware until daylight, and then begin operations just like normal there."

Post-launch processing

Once in port, the boosters were lifted one at a time onshore by means of a straddle lift crane, while at the same time the frustum and parachutes were removed from the ships and taken to their refurbishing facilities. Everything was thoroughly cleaned to remove any saltwater contamination.

Then the booster was moved to a safing area and lowered onto a rail dolly, where technicians performed the initial safing. The booster was driven through a wash bay for a second cleaning and returned to the safing area for removal of ordnance from the forward skirt and depressurization of the thrust vectoring control system. Other technicians carried out an assessment of any damage that might have occurred either in flight or post-flight. Once it had a clean bill of health, the booster was returned to the wash bay for a cleaning process known as "hydrolasing", to remove the thermal protection system from the tunnel covers, separation motors, and aft skirt attachment points.

After hydrolasing, the booster was taken to a hangar for removal of the aft skirt. Since there was hydrazine inside the thrust vectoring control systems, non-essential personnel were banished from the hangar. The aft skirt was taken to a "deservicing" facility where the hydrazine was drained and then the auxiliary power units, thrust vectoring control system and aft booster separation motors were removed. It was then taken to the robotic hydrolase facility for further disassembly. The nozzle was removed, followed by the aft casting segment and the other segments. Because of the impact of a booster falling into the ocean, all internal segment debris would lodge in the forward segments. This type of debris was therefore rinsed out into special carts and taken away. After a final check by an assessment team, the segments were placed on flatbed trucks and transported to the Kennedy Space Center, where each segment was placed on a rail car and protected by a yellow cover. They were then shipped to the manufacturer in Utah for refurbishment and reloading with propellant.

In the meantime, refurbishment of the remaining flight hardware continued at the hangar at Cape Canaveral Air Force Station. The frustum and the forward/aft skirts underwent a hydrolasing process to remove the thermal protection system, a media blasting to remove paint, the application of alodine for corrosion prevention, and then the application of primer, paint and sealant. Finally they were taken to the assembly and refurbishment facility for buildup and testing.

The parachutes were returned to the parachute refurbishment facility for cleaning, repairs and repacking. A hanging monorail system was employed to transport each parachute into a 30,000-gallon washer and then a dryer heated with 140-degree air at 13,000 cubic feet per minute. Typically, each canopy and deployment bag required hundreds of repairs. This damage occurred for several reasons. The parachutes were deployed so rapidly that their fabric and lines could easily suffer friction burns. Sea conditions and hot debris from the SRB nozzle could also cause damage. Because the pilot chute/drogue chute deployment bag assemblies could not always be recovered, these often had to be manufactured afresh. After the parachutes had been cleaned and repaired, they were carefully packed into their bags ready for the next mission.

5

Shuttle propulsion: the external tank

INTRODUCTION

As with the design of the solid rocket boosters, the external tank underwent several changes in configuration before settling upon the shape that was used throughout the program. The initial Orbiter that NASA aspired to would have been fully reusable in all its components. A fully reusable booster would have carried the spaceplane to an altitude from which it would have been able to achieve orbit using its own engines. It was therefore natural to opt to place the propellant tanks for these engines within the structure. This configuration demonstrated that the human imagination is sometimes unconstrained. This pair of fully reusable vehicles suffered from a number of serious flaws.

Hydrogen and oxygen were to be the propellants due to their high performance in a rocket engine, but hydrogen has a drawback. As the most basic course in chemistry will readily show, hydrogen is the lightest element in the universe and as such it has a very low density, around one-fourteenth that of water. Although it would constitute only about one-seventh of an Orbiter's propellant load in terms of weight, with most of the rest being liquid oxygen, liquid hydrogen would occupy nearly three-fourths of the volume. Being low in density and hence light in weight, to carry the hydrogen in the structure of the Orbiter would have required incorporating very large tanks with two major consequences. Firstly, despite being made of aluminum, these tanks would add an incredible weight to the Orbiter at the expense of its payload capacity. And of course the resulting very large structure would have been difficult to protect from the heat of re-entry. Finally, even supposing that it would have been possible to fly such a monster, the development costs were well beyond the funding that even the most enthusiastic members of Congress were willing to allocate.

This led to consideration of partially reusable configurations in which the Orbiter would carry its hydrogen externally in an aluminum shell that would be expendable. The Orbiter would have much less volume to enclose within its structure and much less surface area to protect thermally. In addition, new studies showed that with this configuration it would be possible to reduce the staging velocity at which the booster was separated. This would ease the development of both vehicles and increase the payload capacity of the Orbiter. The next logical step was to stretch the external tank to carry liquid oxygen as well. This would have a two-fold benefit. Firstly, it would reduce the size of the Orbiter to that required to perform the mission. Secondly, with all the propellant carried externally, the spacecraft could achieve a standard design that was independent of the tank. The tank could grow to a particularly large size to further reduce the staging velocity of the booster. In turn, this would reduce the size of the booster, thereby cutting the cost of the program. This reasoning led directly to the familiar configuration of an Orbiter on the launch pad attached to its enormous expendable external tank, to which were attached a pair of reusable boosters.

The backbone of the Shuttle

Having a length of 153.8 feet, a diameter of 27.6 feet, and a capacity of 1,385,000 pounds of liquid oxygen and 231,000 pounds of liquid hydrogen, the external tank (ET) was the largest and heaviest (when loaded) component of the Shuttle stack. As the central, integrating structural element it was subjected to static loads from the Orbiter when on the pad and to thrust loads from the SRBs and the SSMEs in flight. During propellant loading, it had to accommodate flexure arising from cryogenic and ullage pressures. The attachments for the Orbiter and the SRBs, and the various supports for pipes and cable trays, introduced point loads with bending stresses resulting in stress discontinuities in addition to the pure membrane stresses of a pressure vessel. Indeed, the ET could reasonably be thought of as the backbone of the Shuttle.

Jim Odom joined the Shuttle program at the very start of the configuration studies and has nicely summarized the complex structural loads imposed on the ET, "The tank takes a lot of asymmetrical loads, which really affect the structural design and capability of the tank. It is between the two solid rockets. Their thrust is taken up at the intertank [of the ET], at the front end of the motors. The weight and thrust of the Orbiter is carried at the back end of the tank. So you have got the roughly 250,000-pound Orbiter hanging on the back end of the tank with all the SRB thrust after liftoff going principally in at the front end." Furthermore, "You've got the heavy weight of the liquid oxygen tank on the front end and the longer, lighter hydrogen tank at the back end, yet the Orbiter thrust goes in at the back end of the hydrogen tank. It makes the load paths very complicated, which makes it structurally a lot more complex than it looks on the surface."

The moment of launch is one of the most demanding in terms of dynamics and stresses. As Odom explains, "If you visualize the tank sitting on the pad, then it is bolted down by the two solid rocket motors. One unique thing was there's the tank in the middle with the Orbiter hanging off to the side. So here you've got 250,000 pounds hanging on the side of this fairly rigid tank and solid rocket motors. When you light the Orbiter's engines, it actually pushes the tank over. The top of the tank goes off-vertical about three or four feet. Then you light the solids. Once they pressurize, it wants to pop it back straight, and you time all of that such that it lifts off when it is again vertical." This motion of the external tank, called "twang", is

Introduction 159

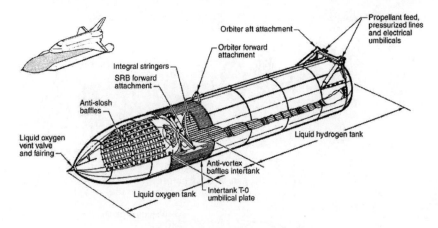

External Tank major components.

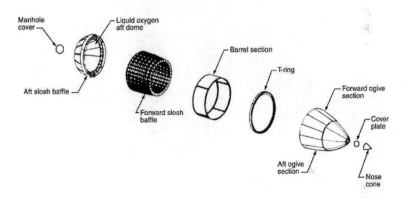

ET oxygen tank structure.

clearly visible in video footage of Shuttle launches and once again demonstrates the impressive loads and stresses to which the full stack was subjected.

As a result, the structural configuration chosen for this colossal piece of hardware was simple: two tanks, one for each of the two propellants used by the main engines, joined by a strong cylindrical structure known as the intertank.

Because the ET was the only expendable part of the entire system and was carried to orbital altitude, it was imperative to minimize its weight. The material selected for the three principal components was aluminum alloy, with some composite materials for fairings and access panels, and steel and titanium for the fittings and attachment points.

Despite there being more oxygen than hydrogen, the LO_2 tank was much smaller because the oxygen was denser. Vehicle controllability meant the oxygen tank had to be located at the front of the ET. It was an aluminum monocoque consisting of a fusion-welded assembly of preformed, chem-milled gores, panels, machined fittings

160 Shuttle propulsion: the external tank

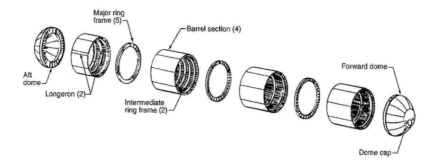

ET hydrogen tank structure.

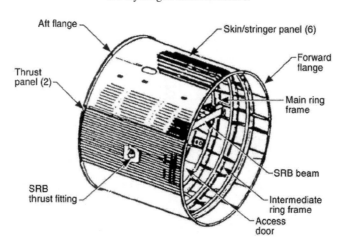

ET intertank structure.

and ring chords. Internally, anti-slosh and anti-vortex provisions in the bottom part minimized liquid residuals and damped out destabilizing oscillations of the fluid. To minimize aerodynamic drag and heating, the tank had a double-wedge nose cone that also served as a lightning rod.

In a similar fashion, the liquid hydrogen tank was an aluminum semi-monocoque of four fusion-welded barrel sections, capped by forward and aft ellipsoidal domes, all joined together by five major ring frames. Anti-slosh barriers were added at the propellant line outlet in order to minimize a fluid vortex that could otherwise induce instability in the delivery of fluid to the outlet.

When the ET was designed, it was decided to add a third structural component to prevent the two propellant tanks being in direct contact with each other. In particular, this eliminated the conceptual and operational complexity of a common bulkhead. The intertank was a 22.5 foot long hollow steel/aluminum semi-monocoque cylinder whose primary functions were to receive and distribute all the thrust loads from the SRBs and to transfer loads between the propellant tanks. Its major structural elements were two machined thrust panels connected by a

Inside the intertank, showing the SRB crossbeam, lower dome of the oxygen tank and upper dome of the hydrogen tank.

transverse beam and six stringer-stiffened panels joined mechanically. Its presence simplified the oxygen tank, which, being located above the attachment points, would not require to withstand the static and dynamic loads imposed by the attached Orbiter and solid rocket boosters.

The two thrust panels were so-named because they had external fittings (one per panel) for the SRBs and they distributed the concentrated axial SRB thrust loads to the LO_2 and LH_2 tanks and to the adjacent intertank skin panels. Each was machined from an aluminum block to create a panel with variable thickness having 26 integral longitudinal ribs and 7 integral circumferential ribs to provide flutter resistance and prevent buckling. Each panel also had two longerons mechanically fastened (one on each side of the SRB fittings) to provide added stability for the compressive loading of the forward section of each thrust panel. The two panels were connected by a box-section crossbeam (commonly referred to as "the SRB beam") to limit rotation at the SRB attachment points. Structural requirements in designing the beam were strength, stability and the maximum deflection at midspan, which was limited by the available space between the two tanks. The six stiffened-stringer panels were each made of two aluminum skins mechanically spliced longitudinally by internal and external butt straps. Doublers were added to provide the necessary reinforcement of areas where the skin was penetrated by feed lines or access doors. On each panel, 18 stringers were fastened on the external side to provide flutter resistance, prevent buckling, distribute loads to attachment flanges, and accommodate mounting brackets for propulsion and electrical subsystem lines and cable trays. All eight panels were mechanically joined together by means of splices. The final cylinder was further stiffened with five ring frames. Four ring frames stabilized the entire cylinder. The fifth ring, which adjoined the SRB thrust fittings, was intended to handle the ET ground-handling loads and to transmit the transverse SRB thrust load.

162 Shuttle propulsion: the external tank

Lateral views of the External Tank. On the right it is possible to see the ET/Orbiter interface structure. Running for the full length of the hydrogen tank is the LO$_2$ feed line paralleled by the lines for pressurizing the LO$_2$ and LH$_2$ tanks. Also on the right it is possible to observe the bottom of the ET where three access panels allowed installation of sensor to detect a condition of LH$_2$ tank almost empty.

Propellant was fed to the ET through two 17-inch-diameter insulated pipes made of aluminum and corrosion resistant steel. The oxygen feedline started at the aft dome of that tank, passed out through a slotted port in the intertank skin, and ran the exterior length of the hydrogen tank to the right-hand umbilical disconnect plate. The line assembly consisted of nine parts, using a combination of fixed and flexible joints in order to accommodate fabrication and installation tolerances, differential thermal expansion between the line and tank structure, and relative motion during liftoff and ascent. The full line was supported at eight places, including at the interface with the liquid hydrogen tank. Five of the remaining supports were of the pivoting type which incorporated pin-and-ball joints to accommodate misalignments caused by tolerance buildup, line movement resulting from thermal contraction, vehicle vibrations, and protuberance airloads. To prevent geysering on the oxygen

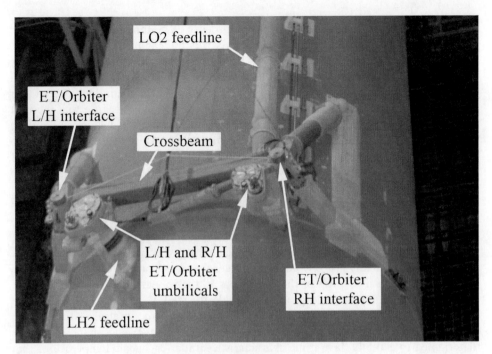

Detailed view of the ET/Orbiter interface structure.

feedline during propellant loading, gaseous helium provided by the launch facility was injected into the aft end of the line to provide evaporative cooling of the feedline and reduce the possibility of the LO_2 reaching the local saturation temperature and causing a geyser. In the event of a geyser, the refill of the feedline would have produced damaging water-hammer effects.

The hydrogen feedline was much simpler because it started inside the tank in the bottom part of the aft done, came out of a flanged port on the aft dome wall and went to the left-hand umbilical disconnect plate. This configuration was chosen because it minimized the length of the costly jacketed line that would have been required for an external exit (like that of the oxygen line). The internal configuration also reduced problems associated with base heating and acoustics effects on the ET originating from the SRBs and the main engines. The external part of this line was supported by two arms that were attached to the crossbeam assembly and bolted to pads on each side of the elbow section.

STS-133

On 5 November 2010, during the countdown for STS-133, which was to be the final mission of *Discovery*, a hydrogen leak from a hardware interface in the intertank was detected. Due to the magnitude of the mysterious leak, the launch was scrubbed. Then other issues developed. As Anthony Bartolone, the lead ET/SRB project manager at that time, recalls, "During the process of de-tanking and draining the external tank as a result of the hydrogen leak, our ET engineers known as the 'ice team', the people who do the final inspection of the external tank, saw from their camera view that the flange on the liquid oxygen tank at the top of the intertank had actually cracked. You could see that there was an offset in the foam that covered that flange." This was not something to be taken lightly, not only because that piece of foam could damage the Orbiter, as occurred when *Columbia* was lost, but because it implied damage to the underlying intertank structure. "We ended up having to go and remove the foam out at the launch pad. We discovered that there was a series of actual parallel cracks that went down from fastener hole to fastener hole on the top nine inches of two stringers, and that had caused them to almost buckle. They had come away from the skin of the external tank, and that was what caused the offset in the foam that we were able to visually detect."

In total, cracks were discovered on four stringers and as Bartolone recalls, "We ended up rolling back to the VAB and implementing what we called a radius block mod, which is basically just a structural doubler that was installed over the top of the first nine fasteners at the top of these stringers all the way around the external tank. A total 108 stringers were modified to include this radius block modification and that gave us the structural strength to allow the tank to be cleared for flight." Following an investigation to understand the causes, "The stringer material ... was determined to actually have some characteristics under cryogenic temperatures that were less than ideal," Bartolone explains. "Although they met procurement specification, there were fracture toughness qualities about them that made them a little less than what we had normally seen for external tank production, and that was ultimately deemed the root cause." *Discovery* was able to lift off on 24 February 2011. It is worth pointing out that cracks in the intertank stringers were not a novelty. But STS-133 was peculiar for being the first time that such damage was found with the stack already on the pad. Previously structural damage to the ET was found during production, and rectified long before the countdown.

ET structural evolution

The ET underwent two major evolutions during the 30 years of the Shuttle program, but these modifications were not visible from the outside. The tank flown on STS-1 was of the so-called standard-weight tank (SWT) or heavy-weight tank (HWT) type that was designed with a higher structural load margin owing to uncertainties in the loads to which it would actually be subjected in flight. Static and dynamic tests were carried out prior to the first launch in order to characterize the distribution of loads and stresses all over the structure, but this data needed to be confirmed by measurements in flight. To play safe, the first external tank design incorporated significant margins. But work on an improved and lighter version began even before the first flight, and the new tank, called the light-weight tank (LWT), flew for the first time on STS-6. Whilst the original tank weighed about 76,000 pounds empty, the new one was some 6,000 pounds lighter as a result of resizing the structural membranes, eliminating hardware elements and making material substitutions.

Considering that the ET was carried to orbital altitude and then discarded to re-enter the atmosphere and burn up, every single pound of weight saved contributed to improving the payload capacity of the Orbiter.

The need to carry into orbit the heavy elements of the International Space Station prompted another weight saving exercise, with the result that STS-91 introduced the super-light-weight tank (SLWT). The weight saving request was for at least 7,500 pounds, and this was achieved by changing the material used to make the structure. Both the SWT and the LWT were made of aluminum alloy 2219. In 1986 Lockheed Martin Laboratories in Baltimore, Maryland undertook the challenge of developing a high-strength low-density replacement for 2219, without losing its excellent welding characteristics and resistance to fractures when exposed to cryogenic temperatures. The result was a family of aluminum-lithium alloys called Weldalite, from which the 2195 alloy was selected. This is 30 per cent stronger and 5 per cent less dense than the 2219 alloy, can be welded and can withstand fractures at −423°F, the temperature at which the liquid hydrogen is stored on board. Owing to the significant challenges of welding elements with complex curves, such as the forward dome of the liquid oxygen tank, a productivity enhancement study was carried out to include a return to using 2219 for these elements. This resulted in the development of three versions of the SLWT, the last of which was flown for the final two missions of the program.

GROUND PROCESSING

After an ET was manufactured, assembled and passed the final acceptance testing at NASA's Michoud Assembly Facility (MAF) near New Orleans, the delivery process began with the tank being mounted on a wheeled transporter and towed by tractor to a 225-foot barge waiting at a dock one mile away. This involved an eight-man crew and took eight hours. Once on the barge, the transporter platform and the tank were secured using hydraulic locks to reduce load shifting and vibration during the 1,000-mile voyage to the Kennedy Space Center. The barge was towed by one of the two ships that NASA used to retrieve solid rocket boosters from the ocean. Once clear of the Mississippi River Delta, the ship steered across the Gulf of Mexico, heading for the southern tip of Florida. While in open waters, the barge was actually towed using a line about a quarter of a mile long in order to make it easier for the twelve-person crew to handle it in the ocean's strong currents.

Upon arrival at the barge turning basin at the Kennedy Space Center, the barge was secured and ballasted to dock level. The doors were then opened in preparation for transporting the ET to the VAB. Firstly, the four pedestals and tie-down chains used to secure the tank for the voyage were removed and a visual inspection made of the payload for any apparent damage. If no problems were found, the transporter was towed to the VAB and the tank prepared for transfer to a checkout or storage cell.

"Once it arrives in the Vertical Assembly Building, it's taken off of its transporter and rotated to the vertical. Then it's lifted up into one of four checkout cells that we

166 Shuttle propulsion: the external tank

Pegasus, one of the two barges used for transporting ETs from the assembly factory in Louisiana to KSC in Florida.

Transporting an ET to the VAB.

have in the west side of the VAB," Bartolone explains. "Once it's up and over the top of those checkout cells, a crane lowers it down past the initial set of platforms that are in the cell. We have a team of engineers, technicians and quality inspectors who inspect the tank for anything that's visually detectable in the foam or the structure as it is being lowered into the checkout cell where we do the first stage of processing on the ET, getting it ready to mate to the SRBs."

Ground support equipment and launch processing systems were connected as part of this process to be able to perform a checkout of the electric system simulating the presence of the Orbiter interfaces. The vital umbilical connections were checked by Boeing technicians from California, since these connections were designed and built there. As Bartolone explains, at KSC these technicians "do a series of what we call

Lowering an ET between the two SRBs.

angle and tip loads, which is where they balance out the valve plates that are on the 17-inch lines that feed liquid oxygen and liquid hydrogen to the Orbiter. They adjust them and make sure that there is no corrosion present. They do any repair necessary to get them back within compliance or specification for the vehicle that they're to be mated to, and they go about clearing those and getting them prepped for mating with the Orbiter."

Following these checkouts, the ET would be transferred by crane to the storage or integration cell. The process of mating to create a Shuttle started by lowering the ET onto the forward fittings of a stacked pair of solid rocket boosters, firstly using pins and guides for seating alignment and then installing the frangible bolt/nut assemblies. At this point, the ET was suspended entirely from the two forward attachment points. The next task was to attach the aft stabilizing struts. Firstly, the right and left-hand diagonal struts were attached to the ET upper fittings. The upper lateral struts were attached to each SRB, adjusted to fit, and bolted to the upper ET fittings. The lower struts were then added in the same manner. The operation was completed by mating the electrical pull-away connectors positioned on the two upper lateral struts, and by electrically connecting the frangible bolt of each forward attachment point. The next dish on the menu was the Orbiter!

168 Shuttle propulsion: the external tank

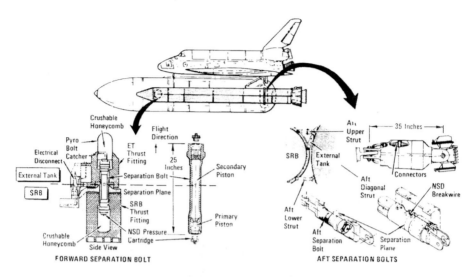

ET/SRB attachment points.

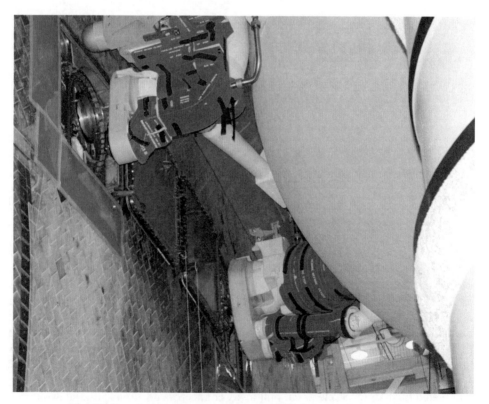

The Orbiter about to be mated with the ET lower attachment points.

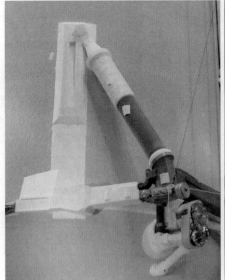 

Left-hand ET/Orbiter interface (left) and right-hand ET/Orbiter interface (right).

As Barlotone says, "The Orbiter is finally brought in somewhere in the four-to-six week timeframe from when SRB stacking started. It is rolled in on a transporter into the VAB and then lifted off this transporter, rotated to vertical, and then lowered into the integration cell where we go through soft mate and a hard mate to get the various different bolts and fasteners installed to secure the Orbiter to the side of the external tank." The two aft ET/Orbiter structural interface points were attached first. During mating, the right-hand interface was taken as a fixed reference point and the left-hand attachment point was left free-floating laterally once the Orbiter socket had captured the ET hemisphere. The reason for this was that the right interface was designed as a tripod structure of structural elements to constrain any movement of the ET/Orbiter interface, whereas the left one was designed as a bipod formed by only two structural elements that allowed pivotal movements in the X/Z plane.

Next was attachment to the forward ET/Orbiter interface, which was achieved by drawing the Orbiter onto the pivotal pod, which was a bipodal structure consisting of a pair of identical hollow aluminum tubes approximately five feet long with flanges on each end. Both tubes were attached to a yoke fitting installed on the Orbiter. This yoke was a large machined titanium casting with a 74-degree apex angle between the two tube attachments. On the other two ends, the tubes were flanged onto two spindle assemblies mounted on the flange of the intertank with the liquid hydrogen tank. The spindle assemblies also had springs that rotated the bipod forward when the ET was jettisoned as a precaution to prevent contact with the belly of the Orbiter. Mating this assembly was a two phase operation. The first phase required the ET bipod struts to be installed at the two tank fittings. The apex yoke fitting would have been installed on the Orbiter using a frangible bolt/nut assembly

Forward ET/Orbiter interface.

prior to the hoisting operation. Final assembly occurred by drawing the Orbiter onto the ET pivotal bipod and connecting first the left strut to the yoke and then the right one, which had an adjustable sleeve to provide mechanical linkage with the yoke. The structural attachment created in this manner formed a locked kinematic chain, constrained to carry loads in the Y/Z plane. For each of the ET/Orbiter interface structures, numerous pinned and spherical joints allowed for multi-directional motion, whilst also minimizing the magnitude of any bending induced in the support structures as a result of thermal and structural loads. Motion and adjustment for alignment, tolerance, strain, and thermal distortions were provided by shims in the attachment points on the aft ET/Orbiter interfaces and by an adjustable threaded sleeve on the right-hand strut of the forward bipod.

With all the mechanical interconnections completed, the ET/Orbiter aft umbilical disconnects containing the electrical, pneumatic, and fluid interfaces could be mated. Finally, the protective covers were removed from the disconnect halves. The Orbiter halves of the umbilical plates were extended from their retracted positions, aligned to the ET halves, and secured using three frangible bolt/nut combinations per assembly. A leak check was then performed on the pneumatic and fluid systems, and a system integrated test was performed. The complete stack of Orbiter, ET and SRBs was now ready to be rolled out to the launch pad.

LAUNCH PAD OPERATIONS

While a Shuttle was on the pad, both propellant tanks of the ET were maintained at a nominal positive pressure of 3.7 psig in order to preclude structural damage from any negative delta-pressure caused by thermal and atmospheric pressure changes.[1] Prior to starting propellant loading, the tanks were purged with gaseous helium in order to remove residual air and gaseous nitrogen and to ensure a dry state.

Both propellants were supplied from storage facilities via the Orbiter through the LO_2 and LH_2 feed systems. Propellant loading started at about T-5 hours 15 minutes with the fast-fill part of the liquid oxygen tank. This was concluded at the 98 per cent level at T-3 hours 30 minutes. Meanwhile, the fast-fill of the liquid hydrogen tank to the 98 per cent level was completed at T-3 hours 45 minutes. A topping off process then slowly raised the levels to 100 per cent. From this point through to immediately prior to lift off, a continuous process of replenishment was carried out to compensate for boil-off losses. This stepped propellant-loading process is standard for a launcher that uses a cryogenic fluid, in order to provide thermal conditioning of the tank structures. At the start of the process, the tanks were essentially at room temperature and when supercold liquids came into contact with the warm tank walls the liquids boiled. By starting with a slow fill, the fury of the boiling, and therefore stress on the structure, could be minimized. Once a tank had been chilled to a temperature better suited for maintaining the propellant, the fast filling could proceed.

The ET could vent excess LH_2 and LO_2 by converting them into gases during the loading operation, and thereafter automatically relieve any subsequent pressure buildup within the tanks over predetermined limits. Both the vent and relief operations could be accomplished using a single dual-purpose valve for each propellant tank. Venting operations started at T-6 hours in preparation for the propellant loading sequence. Gaseous hydrogen was routed via an intertank vent line and out through the so-called ground umbilical carrier assembly (GUCA) where a vent line carried the gas to a "burn stack" located far from the pad. The vent line was attached to the GUCA using a 48-foot arm that also permitted contingency access to the intertank compartment. Also known as the external gaseous hydrogen vent arm, this would be retracted after the umbilical/vent line mating, typically at about T-5 days, leaving only the umbilical vent line connected to the external tank to support propellant loading and launch. The line would disconnect from the vehicle at first motion of liftoff, retracting vertically downward to a stowed position. The gaseous oxygen was vented through two louvers on opposite sides of the nose cone of the external tank for non-propulsive venting. To prevent the venting oxygen vapor from condensing to form potentially damaging ice, a so-called ET gaseous oxygen vent arm was attached to the ET just over the nose cone. This retractable arm had at its end a 13-foot-wide vent hood, also known as the "beanie cap", into which heated gaseous nitrogen was pumped to warm the liquid oxygen vent system. At about T-2

---

[1] A psig value does not include the ambient pressure, so at sea level 14.7 psi must be added to the "gauge pressure" to obtain the absolute pressure.

minutes 30 seconds the vent hood would be lifted clear of the external tank, and at T-1 minute 45 seconds it would swing in close to the pad structure without being latched. In the event of a hold in the countdown the arm would return and the beanie cap reinstalled on the external tank. As the vehicle lifted off, the arm would finally latch into position to prevent vibrations from causing it to come into contact with the stack with potentially catastrophic consequences.

THERMAL PROTECTION SYSTEM

The ET did not receive much attention from the public until *Columbia* disintegrated 15 minutes from its scheduled time of landing on 1 February 2003. The investigation into this tragedy determined that a suitcase-size piece of thermal protection had fallen off the tank and struck the leading edge of the left wing of the Orbiter, opening up a breach in the thermal protection system that would prove fatal during re-entry at the end of the mission.

The ET had not only to contain propellants and deliver them to the main engines of the Orbiter, it also had to deliver them as liquids. Ground equipment liquefied the oxygen and hydrogen at extremely low temperatures for loading into the tanks. It was the task of the ET to ensure that the propellants remained in liquid form throughout the countdown and the ascent to orbit. This required the tanks to be shielded from the temperatures of the external environment, particularly from aerodynamic heating in flight. Every rocket with cryogenic propellants is provided with some kind of thermal protection, transforming its tanks into thermos flasks. Several requirements must be taken into consideration. For example, the thermal protection system of the ET was required to maintain the LO_2 and LH_2 boil-off rates within the capabilities of the vent valves, contribute to loading accuracy and increased propellant densities, ensure that the LO_2 and LH_2 at the Orbiter interface were at specific temperatures, minimize air liquefaction on the LH_2 tank, and minimize ice formation on the ET surface. These were quite demanding requirements, and it is interesting to note that in the beginning they were not fully understood.

Mike Pessin followed the development of the ET from the beginning. "The initial ET design only had foam on the side walls and forward dome of the hydrogen tank,

The oxygen vent line and "bennie cap" just elevated from the ET.

and an ablator that Martin [Company] had developed for the Mars Viking landers on the aft dome." But after having tested the resilience of the newly developed tiles for the Orbiter a new requirement was added. "They reported that an ice cube dropped four inches would crack a tile. We were not required to eliminate foam from coming off, but needed to eliminate ice. That was the big driver. So we covered the liquid oxygen tank with foam." However, as Pessin recalls, "Then the program said, 'On all of your protuberances, all the pieces that stick out that have thermal shorts where they could get cold enough to freeze moisture out of the air, we want those insulated. But you've got 500 pounds of weight that you can use. Start at the front end. Work your way as far back as you can.' That was a rather strong challenge." It resulted in a quite complex thermal design configuration.

Two materials were selected for building this thermal barrier against the external environment. For the majority of the outer ET surface, which was almost one-third of an acre, it was decided to use a polyurethane type of foam using five primary ingredients: polymeric isocyanate, a flame retardant, a surfactant, a blowing agent, and a catalyst. A surfactant controls the surface tension of the liquid and hence cell formation, and a blowing agent makes millions of tiny bubbles to produce the cellular structure of the foam. The foam had not only to keep the propellants at optimum temperatures, it had also to be sufficiently durable to endure a 180-day stay at the launch pad, withstand temperatures up to 115°F, humidity as high as 100 per cent, and resist sand, salt, fog, rain, solar radiation, and even fungus. During launch, it had to tolerate temperatures as high as 2,200°F generated by aerodynamic friction and radiant heating from the 3,000°F solid rocket boosters and 6,000°F main engine plumes. Finally, when the external tank began re-entry into the atmosphere, about 30 minutes after launch, the foam had to maintain the structural temperatures so that it could safely disintegrate over a remote ocean location.

Application of the foam, whether hand-sprayed or by an automated process, had to meet NASA's requirements for finish, thickness, roughness, density, strength and adhesion. To achieve these requirements, the foam was applied in specially designed environmentally controlled spray cells and applied in several phases, often over a period of several weeks. Prior to spraying, the foam's raw material and mechanical properties were tested to confirm that it would meet NASA specifications. After the spraying was done, multiple visual inspections of all foam surfaces were performed. Most of the foam was applied by the ET manufacturer in New Orleans, including most of the "closeout" areas where the foam was manually applied. Some additional closeout areas were manually completed at the Kennedy Space Center. Although the foam insulation on the majority of the tank was only 1 inch thick, and between 1.5 to 3 inches in the most thermally stressed areas, it added 4,823 pounds to the weight of the structure.

The second material developed was an ablator, that is to say a material designed specifically to withstand high temperatures through controlled energy absorption by way of sacrificial degradation. The need for such material arose when analyses of the thermal loads that would be experienced during the ascent showed that heating from aerodynamic and plume sources required the thermal protection in certain areas to act as an ablator to prevent structural overheating. Since prohibitive thicknesses of foam would have been required, it was necessary to employ an ablator capable of withstanding significantly higher heating rates. But this would be at the cost of more weight and insufficient cryogenic insulation properties.

Prior to the development of the ET, whenever protection of cryogenic surfaces was required, ablators were generally applied to separately attached heat shields or standoff panels. This was due in part to the problems of strain compatibility between the substrate and the ablator material when subjected to cryogenic temperatures. But for the ET the Martin Marietta Corporation recognized the cost and weight benefits of applying the ablator directly to the cryogenic substrate. After a feasibility study, a material called SLA561 was baselined and maintained through the entire program.

It is worth noting that because the ET was designed to be economical to produce as the only expendable part of the Shuttle, the construction technique chosen made it impractical to use an internal insulation. For this reason, the thermal protection was applied externally. This plagued the program from the beginning, and ultimately led to the *Columbia* accident.

Redesigning the thermal protection

The Columbia Accident Investigation Board (CAIB) reported that the physical cause of the tragedy was a breach in the Orbiter thermal protection system "initiated by a piece of insulating foam that separated from the left bipod ramp of the External Tank and struck the wing." Consequently, the board recommended that NASA "initiate an aggressive program to eliminate all External Tank Thermal Protection System debris-shedding at the source with particular emphasis on the region where the bipod struts attach to the External Tank."

However, this area presented serious problems as far as the application of thermal protection was concerned. As Pessin explains, during propellant loading, "The tank is chilled to –423°F. It shrinks several inches in relation to the Orbiter, so the forward attach, which is a bipod spindle, has to move. Then as you fly and you start putting warm [ullage] gas in the front end of the hydrogen tank, it begins to grow. The bipod has to move in the other direction."[2] This presented a challenge in applying foam at the point where the bipod strut was connected to the ET. As recognized by the CAIB, when the ET was designed, particular aspects of the design (structural, thermal, etc.) were designed one after the other by different teams of engineers. The bipod design was optimal from a structural point of view, but its complexity posed problems in the application of foam and ablator. To thermally protect the fitting, the entire structure apart from its end point was covered in a ramp of manually applied foam insulation to prevent and/or eliminate ice formation and provide protection from ascent heating. It was this foam ramp that detached and struck the wing of *Columbia* when launched as STS-107.

---

[2] The bipod was canted 0.5 degrees forward when mated to the Orbiter, but cryogenic shrinkage of the ET resulted in an additional 5.5 degrees of forward cant.

Afterwards, engineering analysis and dissection of existing bipod ramps indicated that hand-spraying over the complex geometry of the fittings was apt to create voids and defects that contributed to foam loss during ascent. Consequently, the main goal of the bipod redesign was to eliminate this large piece of foam as a potential source of debris. In fact, the new design completely eliminated the foam block. Instead, the fitting was mounted on a copper plate that was threaded with four rod heaters placed to prevent ice formation during propellant loading. The copper plate with heaters was sandwiched between the fitting and an existing thermal isolating pad, which reduced the transfer of heat from the copper plate into the cryogenic hydrogen tank.

The newly designed ramp was introduced with STS-114 in August 2005, which was the first mission after the *Columbia* accident, but it did not work as expected.

Forward bipod fitting thermal protection pre-*Columbia* (top) and post-*Columbia* (bottom).

176 Shuttle propulsion: the external tank

PAL ramps.

Imagery obtained during the ascent and on-orbit showed a 7 by 8 inch divot of foam was missing close to the left-hand bipod attachment fitting. This was attributed to a process known as cryoingestion. As part of propellant loading, the intertank had to be filled with gaseous nitrogen to prevent condensation and to prevent liquid hydrogen and oxygen from combining. Any nitrogen gas that penetrated into regions under the foam at cryogenic temperatures would first condense as liquid and later boil off by aerodynamic heating during the ascent. This rapid expansion would cause an increase in pressure beneath the foam and potentially break off material. During the STS-114 ascent, the nitrogen could have penetrated to the tank structure through the electrical harnesses servicing the bipod heaters and temperature sensors. The harnesses were modified to reduce the potential for nitrogen leakage from the intertank through the cables into the cryogenic region near the bipod fittings, and voids beneath the cables were eliminated by using an improved bonding procedure to ensure full adhesive coverage.

The two so-called protuberance air load (PAL) ramps were deleted from the ET for the next mission, STS-121 in July 2006. These thick, manually sprayed layers of foam had been designed as a safety precaution to protect the tank's cable tray and pressurization lines from an air flow that could potentially cause instability in these attached elements. One PAL ramp was near the aft end of the liquid oxygen tank, just above the intertank, and the other was below the intertank, along the upper end of the liquid hydrogen tank. During the ascent of STS-114 it was observed that a piece of foam was lost from near the hydrogen PAL. For this reason it was decided to further investigate the requirement for these protuberances. As a result, three design options were studied as possible improvements. After tests of actual flight hardware revealed that the cable tray did not suffer instabilities it was decided to eliminate the PAL ramps. As a result, starting with STS-121, external tanks were flown

Thermal protection system

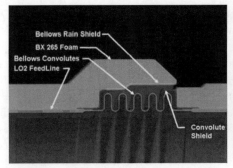 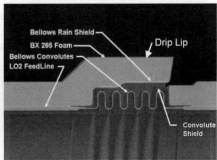

Liquid oxygen feedline bellows thermal protection pre-*Columbia* (left) and post-*Columbia* (right).

without these ramps. Interestingly, the intertank had to be slightly modified to reduce the risk of shedding debris. As explained above, the intertank was the structural connection between the two propellant tanks. The connection took the form of two flanges affixed at the top and bottom of the intertank, and functioned like the seam of a skirt. After the flange was bolted to the LH_2 and LO_2 tanks, the area was insulated with foam. To minimize the cryoingestion problem of the gaseous nitrogen used to purge the intertank coming into contact with the structure of the cryogenic hydrogen tank, the flange bolts were reversed and a sealant applied to the threads.

Cryopumping, a process similar to cryoingestion, was believed to have caused foam loss from larger areas of coverage. If there are cracks in the foam, and if these penetrate through to voids at or close to the surface of the propellant tanks, then air chilled by the cryogenic temperatures could liquefy in the voids. As propellant levels fell in flight and aerodynamic heating of the exterior increased, the temperature of the trapped air would rise and the increasing gas pressure inside the void would break off pieces of foam. Since the thermal protection on the intertank was applied manually, a new mold injection foam closeout process for the intertank stringers was introduced in order to provide a technician with a less complex base and thereby reduce spray defects. In addition, an enhanced finishing procedure was added that improved foam application to the intertank ribbing area as well as to the upper and lower area of the flange. This enabled technicians to apply a higher quality product by way of process verification and more stringent controls.

Other areas subjected to scrutiny were the bellows of the liquid oxygen feedline. These joints allowed for fabrication and installation tolerances, differences in thermal expansion between the line and the tanks, and relative motions in flight. There were two bellows internal to the intertank and three outside of the ET. The entire feedline was insulated with foam, but the three exterior bellows had to be free to move. With a surface temperature of $-297°C$, it is easy to see how this could be another source of ice during propellant loading. Although there were no reported losses of foam on this feedline, there was concern that if ice on the bellows became dislodged during liftoff it might damage the Orbiter. To play safe in the wake of *Columbia*, it was decided to revise the thermal protection system of the bellows by adding a strip heater to reduce the amount of ice and frost formed and a "drip-lip" to allow air moisture to run off. Procedures called for activating the heater as soon as the liquid oxygen fast-fill began and turning it off at about T-2 minutes.

Hail damage

On 26 February 2007 *Atlantis* was on the pad with launch scheduled several days hence. A severe thunderstorm caused more than 4,000 points of damage on the foam of the ET, mostly on the upper part. NASA decided to return the Shuttle to the VAB to assess the full extent of the damage and come up with a repair plan. As a hands-on inspection was required to determine the depth of the damage, workers assembled an intricate network of scaffolding around the vehicle to gain access to the hail-damaged areas. The damage was "mapped" in terms of its location, size and depth. This initial assessment showed damage extending from the liquid oxygen tank ogive to the aft interface where the Orbiter was attached. A thermal and aerothermal assessment was performed for each case of damage to determine the required extent of repair. About 1,400 to 1,500 sites were near the top of the tank in a location known as the "pencil sharpener". To avoid numerous individual repairs, it was decided to remove from this area a layer of foam half an inch thick and replaced it with a single, large manually applied spray using a type of material referred to as "BX" foam. This same technique was used to repair another large area on the side of the oxygen tank which had almost 500 damage sites.

While BX material had been used for numerous applications on return-to-flight hardware, these spray repairs required a demonstration in order to ensure that the application process was repeatable and the performance would satisfy or even exceed the thermal protection and debris minimization requirements. Engineers and technicians devised a high-fidelity mockup of the upper part of an ET at the Michoud Assembly Facility and duplicated the exact conditions in which they would be working on the spray and final foam machining process in Florida. This even included replicating the precise conditions in which technicians would carry out the work. The team also tried the manual spray. Afterwards they dissected the finished "repair". With this success, BX foam was applied for real, and technicians used a new pneumatic tool that they had improvised at Michoud to trim and machine the foam to the precise dimensions of the "pencil sharpener". This one-off tool fitted over the nose cone spike, then sanded the foam using a sandpaper covered roller.

Thousands of other damage sites were repaired by using a mechanical grinding tool to remove the foam and then reapplying specialized pourable foam known as "PDL". This process was familiar to Michoud, since it had been used several times to repair incidental damage to the foam material during tank construction. Another 900 sites were sufficiently shallow to be repaired by a technique called "sand and blend". In this case the damaged foam was first sanded away by hand using a coarse sandpaper and smoothly "blended" into the surrounding foam. Such a repair was only feasible where analysis and testing showed adequate foam thickness would remain to protect against ice formation prior to launch and from aerodynamic heating during ascent. Finally, the remaining 400 damage sites were left alone because analysis established they would not have ill effects. *Atlantis* was successfully launched on 8 June 2007 for the STS-117 mission.

6

Maneuvering in space: the orbital maneuvering system and reaction control system

INTRODUCTION

When the first American astronauts were sent into space, their Mercury capsules had very limited capabilities for performing maneuvers. A rudimentary manual reaction control system could be used for minor attitude adjustments and adopting the proper orientation to return home by firing solid-fuel retro rockets. Gemini and Apollo had propulsion systems to enable them to carry out rendezvous and docking operations.

The Shuttle had even greater need for maneuverability, both due to its complex shape (no longer a ballistic capsule) and due to the large variety of missions it was to conduct. The final configuration had a separate Orbital Maneuvering System (OMS) and a Reaction Control System (RCS), although they were fully integrated.

The OMS was to provide propulsion during the orbital phase of the flight and in particular orbit insertion, orbit circularization, orbit transfer, rendezvous, and deorbit. Sometimes the OMS was even used during the ascent to augment the main engines in order to permit heavier payloads to be carried into orbit. After several iterations in the developmental phase, the final configuration saw two independent pods on each side of the Orbiter's aft fuselage, each housing one OMS engine with thrust vectoring capability and the systems to pressurize, store and distribute the propellants. On the other hand, the RCS used a collection of jet thrusters, each of which was fixed to fire in the general directions of up/down, left/right and forward/aft. By selectively firing individual jets or combinations of jets, the RCS controlled the attitude of the Orbiter in terms of pitch, roll and yaw (rotational maneuvers) and small changes in velocity along the principal axes (translational maneuvers). With a developmental phase very similar to that for the OMS, the final configuration saw the RCS divided into three modules, one on either side of the aft fuselage and the third in the nose section. Each module had a cluster of thrusters and all the hardware for their proper functioning, like helium storage tanks, pressure regulators, valves, propellant distribution system, etc.

From an operative point of view, the RCS was used much more extensively than the OMS. The first time the RCS was employed during a mission was at around two and a half minutes into the ascent, when the solid rocket boosters were jettisoned. At that time the forward facing jets on the nose of the Orbiter were briefly fired in order to disperse the particulates from the small solid motors that were used to separate the boosters from the external tank. This firing was not a flight control requirement, but a countermeasure to prevent the particulate matter from hitting the forward windows of the cockpit and thus reduce the time and cost of window inspection and maintenance during turnaround between missions. The first time that the RCS was used for flight control was at main engine cut off, firstly to maintain the attitude of the vehicle until external tank separation, and then to maneuver clear to preclude a collision with the spent tank. Following a clean separation, the RCS was used again to adopt the correct attitude in preparation for an OMS burn to complete the orbit insertion process. Any minor velocity adjustments after this would be made by the RCS. The RCS would remain active during most of the re-entry in order to provide attitude control until the aerosurfaces became effective in the dense lower atmosphere. The RCS could also be used in off-nominal situations. For example, in the event of losing two main engines on ascent the RCS would provide single-engine roll control. During an abort, it could assist with the dumping of ascent propellant to decrease the weight of the Orbiter, improve its performance, and control its center of gravity. If the OMS gimbaling system did not perform adequately to control vehicle attitude during a burn, the RCS would be used to help to maintain attitude in what was called "RCS wraparound".

DEVELOPMENT

Although the OMS and RCS were developed independently and were different from an operational standpoint, they shared many of the same design requirements. Three common requirements were that they had to be a fail-operational/fail-safe design, and have a 100-mission reuse capability over a period of a decade. In addition, the RCS was to be able to operate both on-orbit and during re-entry. The fail-operational/fail-safe design not only introduced additional hardware but, perhaps more importantly, also a complex redundancy management (RM) system. The reusability and calendar life requirements posed problems in material selection and material compatibility, in ground handling and turnaround procedures, as well as wear-out problems. Finally, the requirement to operate both on-orbit and during re-entry complicated propellant-tank acquisition system design.

An early issue to be resolved was the propellant to be used, because this was not specified by the requirements. The selection would have a significant impact on the design. Early in the program definition phase, oxygen and hydrogen were baselined as the reactants for all propulsion and power systems. This choice was made for a number of reasons. This propellant combination delivers a high specific impulse, the liquids are relatively clean and nontoxic, and the exhaust products are noncorrosive – all desirable attributes for a reusable system. Furthermore, having a single propellant combination would minimize costs. As the component technology and systems study programs progressed, the expected advantages in terms of weight diminished. The heavy accumulators in combination with redundant turbopumps and heat exchangers offset the weight advantage afforded by the better performance in the total impulse. And because the dry weight of the oxygen-hydrogen system was high, this imposed a vehicle landing weight penalty. A further factor that became increasingly apparent as technology work progressed was that such a system would be costly to develop and build, and its extreme complexity raised concerns about reliability. When it became clear that the weight of an oxygen-hydrogen RCS system would be no better than a monopropellant system, the baseline was changed to a hydrazine system to reduce cost and complexity. As the design of the Orbiter evolved, performance requirements dictated a switch to a bipropellant system using monomethyl hydrazine (MMH) and nitrogen tetroxide (NTO). Although less powerful than oxygen-hydrogen, the MMH-NTO combination was lighter, simpler and cheaper. This propellant combination was independently chosen for the OMS, allowing an easier interconnection of the two maneuvering systems. The major perceived disadvantages of using these propellants were the higher maintenance requirements resulting from their corrosive nature and the risk to personnel during handling due to their toxicity. These considerations were addressed by fitting the OMS and the RCS into modular pods that could be readily removed from the Orbiter, and thereby decouple maintenance or refurbishment from other Orbiter turnaround activities.

In parallel with the studies conducted to determine the best propellant choice for both OMS and RCS, it was necessary to decide the number of engines required and the amount of propellant required for on-orbit and deorbit maneuvers. Based on Apollo experience, a pair of OMS engines in conjunction with other components would have represented an acceptable level of safety consistent with the fail-operational/fail-safe philosophy. This prompted the rule that the system must be designed for full mission capability after an engine failure, which in turn defined the thrust level and the total impulse based on performing all mission phases using a single engine. At this point, two different configurations were studied: a common propellant supply system and a modular-type (pod) system. The first, having only one propellant supply, offered the capability of using all the system impulse through either OMS engine. In the event of failure of the propellant supply system, however, maneuverability would be totally lost for both engines. From a standpoint of fail operational/fail safe, the modular configuration offered more redundancy. On the other hand, carrying propellant and pressurant supplies in separate pods would effectively halve the system impulse capability in the event of an engine failure unless each pod had full system capacity. But to provide both pods with full capacity would have imposed an excessive weight penalty. When the modular system became the preferred choice, it was decided that the OMS be designed such that it could expend all of the propellant through either engine by way of crossfeeds. This allowed the propellant supply components to be only doubly redundant and still make the Orbiter fail-operational/fail-safe in terms of the OMS.

The structural configuration of the OMS evolved along with the Orbiter. When in

182 Maneuvering in space

View of the two OMS/RCS pods.

1972 Rockwell International was awarded the Orbiter contract, the OMS changed from an internal installation to separate modules mounted on the aft sides, projecting into the fuselage. And the configuration changed again when the McDonnell Douglas Astronautics Company was hired to make the pods, becoming shoulder mounted for aerodynamic reasons. This configuration had the pods extending to the payload bay, a fairing on the bay doors, and a kit in the bay that had the same components as the pods with as many as six propellant tanks on an all-aluminum structure increasing the total capability of the system to undertake large maneuvers. Propellants would be provided to the pods by transfer lines which joined the engine feedlines upstream of the engine isolation valves. Closed valves in the pods would have kept these lines dry until they were needed to feed the pods from the tanks in the cargo bay. But after the OMS interconnect lines were incorporated, this third OMS kit was deleted. As a result, the final configuration featured shoulder mounted pods on the top part of the aft fuselage without an extension into the payload bay. From a structural standpoint, the OMS pods were initially intended to be made of conventional aluminum, with the emphasis on low cost and simple field inspection and maintenance.

The pods were attached to the Orbiter at four points with shear pins and threaded fasteners for a rapid mate/demate capability. But when it was realized that a large weight saving could be achieved by using a graphite epoxy skin similar to that of the payload bay doors, some 250 pounds was trimmed off each pod. In their final form, the pods were built using skin panels made of graphite epoxy honeycomb sandwich

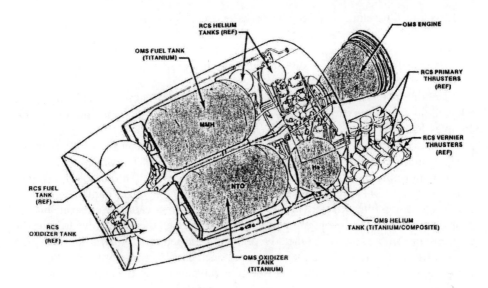

OMS/RCS pod schematics.

and internal structures of aluminum, graphite epoxy, titanium, and corrosion-resistant steel as appropriate. A small box constructed of aluminum sheet metal and panels of graphite epoxy honeycomb sandwich was installed on the aft part of each OMS pod to accommodate the aft RCS modules and their plumbing. The exposed areas of the OMS pods and the RCS were covered with a reusable thermal protection system, and a pressure and thermal seal was installed at the interface between the OMS pod and the RCS seal. Other thermal barriers were added to interface with the RCS thrusters and the reusable thermal protection system. Finally, 24 doors in the surface provided access to the OMS and RCS and the attachment points.

OMS AND RCS CONFIGURATION

Since the OMS and RCS used the same propellants and incorporated crossfeeds, their system layouts became virtually identical. In both cases the propellants were fed to the engines by pressurized helium gas stored in appropriate tanks. For the OMS pods, the decision to use a single helium supply was to ensure that both propellants would remain at the same pressure and thereby avoid an improper mixture ratio. However, the RCS had two helium supplies, one for each propellant line.[1] In both

---

[1] Below a certain propellant quantity, there is enough residual helium pressure in the propellant tank to effectively use all the propellant in that tank. This quantity is referred to as "max blowdown". For the OMS it is about 39 per cent propellant quantity remaining, while it is 24 per cent for the aft RCS and 22 per cent for the forward RCS.

cases, before the helium reached the tanks it had to pass through different sets of valves meant to isolate the helium manifolds from the tanks and to regulate the pressure at the inlets to the propellant tanks. These latter valves were used to reduce the pressure from the high value in the helium storage tank to a more suitable pressure for the propellant lines.

A complex of redundant valves was located just upstream the propellant tank inlets to prevent a reverse flow of liquid propellant or vapor in the helium lines. This complex had four poppets in a series-parallel arrangement. The series arrangement restricted the backflow of propellant vapor and maintained propellant tank pressure integrity in the event of an upstream helium leak. The parallel arrangement ensured that helium would reach the tanks if a series check valve failed in the closed position. A pressure relief valve regulated excessive pressure. These valves contained a burst diaphragm and a filter. In the event of excessive pressure created by either helium or propellant vapor, the diaphragm would rupture and open the relief valve to vent the excessive pressure overboard. Once the pressure returned to the operating range, the relief valve would close. Since there was only one helium supply for each OMS pod, the helium system had an additional isolation valve placed in the pressurization line for the oxidizer tank to prevent vapors that might diffuse into the helium line from migrating upstream into the fuel system and causing a hypergolic reaction.[2] Actually, there were two such valves in parallel so that if one vapor isolation valve were to get stuck in the closed position, one path would still remain open. As explained above, it was decided to use MMH as the fuel and NTO as the oxidizer for both the OMS and RCS. One advantage of these propellants was that they are hypergolic, meaning that they spontaneously ignite upon coming into contact with each other. Eliminating the need for an ignition system increased the overall reliability of the engines and saved some weight.

Another important characteristic of hypergolics is that they remain liquid at room temperatures, simplifying the ground handling of these toxic substances. It is for this reason that hypergolics are defined as "storable", meaning they can be stored without particular concern for the temperature of the ambient environment. Once forced from their tanks by helium pressure, so long as the downstream isolation valves were open the propellants would reach the feedlines of the OMS engine or RCS thruster, where so-called "bipropellant valves" would regulate their flow to deliver the desired thrust for the specified duration.

The supply line distribution of the RCS was more complex than that of the OMS. Downstream of each tank isolation valve the feedlines branched to supply individual manifolds, where isolation valves controlled the flow of propellants to the thrusters associated with that particular manifold. Common to both the OMS and aft RCS was the crossfeed capability. In the event of an imbalance of the propellants in a given pod or the failure of one engine or propellant tank, an "OMS crossfeed" could be set up to transfer propellants from one pod to the other.

---

[2] The oxidizer was chosen because it has a higher vapor pressure than the fuel.

OMS and RCS configuration 185

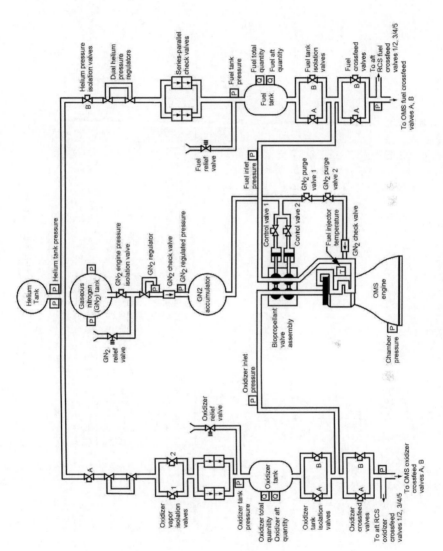

OMS pressurization and propellant feed system of one engine.

186 **Maneuvering in space**

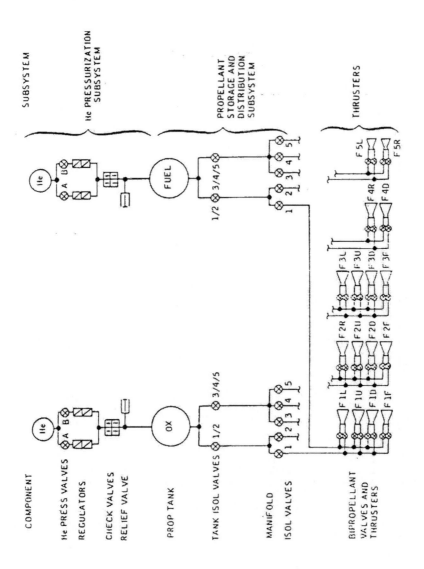

Forward RCS jet thrusters scheme.

OMS and RCS configuration 187

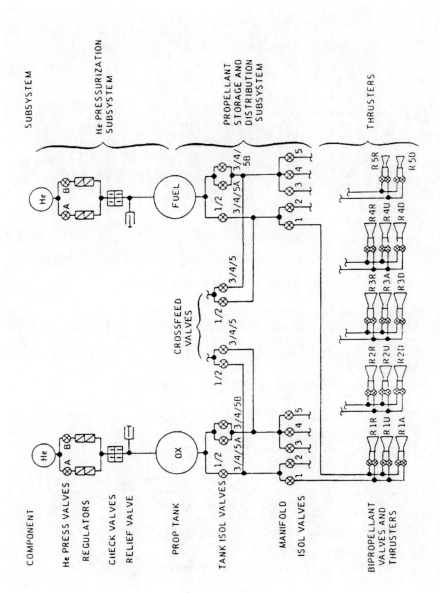

Right-hand aft RCS jet thrusters scheme.

188 Maneuvering in space

Internal view of an OMS/RCS pod.

Internal view of the forward RCS pod.

The crossfeed lines connected the left and right OMS propellant lines at a point between the tank isolation valves and the bipropellant valves. Each crossfeed line had two crossfeed valves configured in parallel for redundant paths for propellant flow. If necessary, OMS propellant could also be fed to the aft RCS from either pod, in what was known as "OMS-to-RCS interconnect". Normally, an interconnect involved one OMS pod feeding the RCS on both sides. This type of interconnect would have been set up manually and used while on-orbit. The most important use of an OMS-to-RCS interconnect would have been during an aborted ascent, when the interconnect setup was automatic. As with the OMS, the aft RCS was also capable of crossfeeding. This would be used if, for some reason, the propellant system of one of the aft RCS had to be isolated from its thrusters. In this event the valves that tied the crossfeed manifold into the propellant distribution lines below the tank isolation valves would be set so that one aft RCS propellant system could feed both left and right thrusters. No matter their type, all of the possible crossfeed interconnections were set up in the same way. The crossfeed valves on the feeding lines and on the receiving lines had to be open, while the propellant tank isolation valves on the receiving side had to be kept closed. It was possible for the crew to manually set up any required crossfeed or to let the on board computer operate the crossfeeds automatically anytime that it was required.

OMS gaseous nitrogen system

An additional system peculiar to the OMS provided pressurized gaseous nitrogen to regulate propellant flow to the combustion chamber of an engine. The layout of this system was very simple. A spherical storage tank formed by welding two titanium halves together was mounted next to the combustion chamber and provided nitrogen to the bipropellant valve that had the all-important function of regulating the flow of propellants to start and stop an OMS engine. The origin of this valve traced back to the quad-redundant valves used by the main engine of the Apollo service module. In designing a new valve for the OMS, the basic requirement was for a modular type of valve that would be cost-effective, not only for maintenance and servicing but also during fabrication and testing. This would have enabled preassembled subassemblies to be installed and removed without disturbing other parts of the valve, minimizing developmental time and fabrication issues and allowing servicing and maintenance goals to be met with a smaller inventory. Although the quad-redundant concept was advantageous because parallel flow was provided in the case of a failure to open, it introduced the risk of additional leak paths.

On revaluation, it was decided that series redundancy was more than sufficient for the OMS since, unlike the single engine of the Apollo spacecraft, the OMS engines were redundant to each other and their tanks and feedlines could be configured to provide complete functional redundancy in the event of a valve failing in the closed position. The design selected for the valve assembly therefore consisted of two fuel valves in series and two oxidizer valves in series. Having two valves in series for each propellant provided redundant protection against leakage, but it also meant that both valves had to be open to enable propellant to flow to the engine. Each fuel valve was mechanically linked to an oxidizer valve so that they opened and closed as one. The name "bipropellant valve" derived from the fact that each linked pair controlled the flow of both the fuel and oxidizer. Two solenoid-operated control valves on each engine allowed gaseous nitrogen to control the bipropellant control valve actuators and ball actuators.

When the solenoid control valves were energized open, the nitrogen acted against the piston in each actuator, overcoming the spring force on the opposite side. Each actuator had a rack-and-pinion gear. The linear motion of the actuator's connecting arm was converted into rotary motion and this simultaneously opened a pair of ball valves, one for fuel and the other for oxidizer, to allow propellants to flow to the combustion chamber. To enable the engine to function properly, the ball valves had to rotate from the fully closed (0 per cent) to full open position (100 per cent). If the position was less than 70 per cent for either valve, ignition would either not occur because there would not be sufficient propellant flowing, or it would occur with the possibility of a hard start or some other combustion instability that would result in structural failure and/or chamber burnthrough.

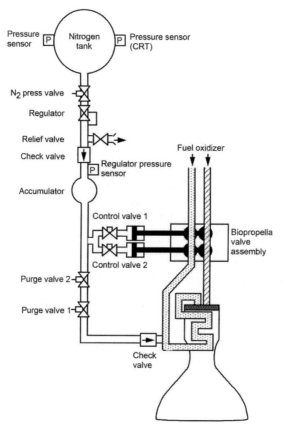

Gaseous nitrogen system.

When established thrust was commanded off, the solenoid control valves were de-energized, removing gaseous nitrogen pressure from the actuators, and venting the nitrogen left in the actuators. The springs forced the piston of the actuator to move in the opposite direction, driving closed the fuel and oxidizer valves simultaneously.

In the event of loss of pressure from the upstream side of the nitrogen supply line, a small spherical accumulator would provide sufficient pressure to operate the engine bipropellant valves at least once. Protection from upstream leaks of the accumulator was provided by the one-way check valve just upstream of the accumulator. Pressure sensors in the fuel and oxidizer lines were located just above the bipropellant valves. They provided an indirect indication of propellant flow rates. Once the engine was firing, both propellants had to be supplied at the right pressure otherwise an incorrect reaction mixture in the combustion chamber could result. For example, with a fuel-rich mixture the chamber pressure and temperature would both decline as combustion choked off. If instead the mixture was fuel-lean, the chamber temperature would rise and the engine would be damaged. If the fuel-lean condition was the result of a low fuel flow rate, the situation was serious because fuel was used to cool the outside of the combustion chamber. In this case a breach in the chamber wall would be highly likely.

The nitrogen supply line had the secondary function of purging the fuel lines after an engine burn. Although the feedlines were exposed to the freezing temperatures of vacuum in space, this was not in itself a problem because any fuel and oxidizer left in the lines that froze would eventually sublimate. However, if a second burn were to be ordered within ten minutes then the frozen fuel in the cooling lines of the combustion chamber could cause damage by trying to force fuel into clogged lines. This situation was avoided by forcing nitrogen through the fuel lines immediately after shutdown. With the lines clear, it would be possible to perform another burn immediately. Each OMS engine had a pair of gaseous nitrogen purge valve in series, solenoid-operated open and spring-loaded closed. Opening the purge valves allowed nitrogen to flow through the valve and check valve into the fuel lines downstream of the ball valves. This purged residual fuel from the combustion chamber and injectors, to enable the engine to restart safely. Afterwards, the purge valves were de-energized and spring-loaded closed. Check valves downstream of the purge valves prevented any fuel from flowing to the engine purge valves during an engine burn. The purge sequence was automatically started 0.36 seconds after the engine thrust was commanded off and it lasted for two seconds.

PROPELLANT STORAGE AND GAUGING

Whereas the SSME system used turbopumps to deliver pressurized propellant to the engines, the OMS and RCS relied on a pressurized gas propellant feed system. In principle this is straightforward, in that the propellants are forced out of the tanks by displacing them with high-pressure gas. Propellant discharge is controlled by feeding in the gas at a controlled pressure. This type of propellant feeding is favored for low thrust or short-duration applications such as for a spacecraft reaction control system. It is reliable, and does not require any heavy and complex hardware. Nevertheless, its implementation can be far from trivial. Prior to the Shuttle, one of the favorite gas-pressure systems involved placing a Teflon membrane inside the tank to serve as a separation barrier between the propellant and the pressurizing gas (normally an inert one, such as helium). Although well suited for a satellite or space probe that slowly consumes a single load of propellant, a tank with a membrane was not practicable for a propulsive system that was to have a 100-mission lifetime because the membrane would rupture after several expulsive cycles. At that time, no elastomeric membranes had been developed that were sufficiently compatible with the corrosive propellants used by the Shuttle to assure a decade of life.

Another common method was to insert a bellows into a tank, but this was heavy. For these reasons, the Shuttle engineers opted for a so-called propellant acquisition device (PAD). This uses the surface tension of a liquid to create a barrier as effective as a Teflon membrane. The primary design objective for a surface tension PAD is to keep the tank outlet covered with liquid whenever outflow is required, ensuring that only propellant, and not the pressurant gas, leaves the tank until it is nearly depleted. Secondary design objectives for a PAD can be to minimize the effects of propellant sloshing in the tank, and to hold all the liquid at a specific position so that its center of gravity is known. Furthermore, if it is desirable to vent the pressurization gas in a low acceleration environment without venting liquid, the device can be configured to hold the liquid away from the vent.

As surface tension forces are very weak, a PAD for rocket engine applications has somehow to amplify them so that a liquid can be expelled from a tank without there being bubbles of gas that could damage the engine. After studying several options, it was decided that the RCS tanks should employ channels covered with a very finely woven stainless steel mesh or screen that would be strategically located on the inner wall of the tank and fully connected to the outlet manifold.

It is important to understand why such devices were necessary. In a low-gravity environment, pressurant gas in a tank of propellant will be freely intermixed unless there is some provision to separate it. Otherwise it is unlikely there will be propellant at the outlet manifold. Nevertheless, for a rocket engine to function properly it must be fed propellant which contains no gas. A meshed channel can achieve this. When liquid wets the screen, it starts to fill the channels by capillary action. Since all the channels are connected to the outlet manifold, the result is to fill that with liquid. Thanks to the amplification of the surface tension force that occurs on the surface of the screens, pressurant gas is prevented from penetrating the channels. The surface tension on the screen will therefore create a natural barrier for the pressurant gas to act upon and force only liquid into the channels. The strength of the liquid barrier is finite, and the pressure differential at which gas will be forced through the wetted screen is known as the "bubble point". When this is exceeded, the screen "breaks down" and allows gas to penetrate the propellant lines. So long as the differential remains less than the bubble point, gas is unable to penetrate the liquid barrier and only liquid is drawn into the channels.

Developing a PAD for the RCS of the Shuttle was complicated by the fact that a system that was primarily to work in a low-gravity environment could be tested only

Propellant storage and gauging 193

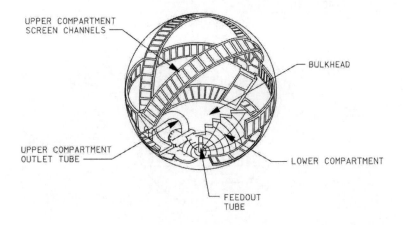

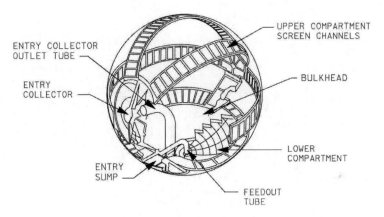

Internal schematics of the forward RCS tank (top) and aft RCS tank (bottom). (Courtesy of www.nasaspaceflight.com)

in a one-gravity environment, giving biased results. This was the greatest engineering challenge encountered in the tank development program. Sophisticated mathematical models were developed to characterize the on-orbit performance of the tanks. One of the most important results was discovering that when combinations of thrusters were commanded to fire simultaneously, the decrease in pressure associated with opening the thruster valves was transmitted through the supply lines to the tank. In particular, when more than three thrusters were pulsing, the pressure drops were sufficient for a momentary breakdown of the screen. If gas were to cause a thruster to misfire, then it might be "deselected" at a critical phase of the mission. To prevent this, the number of thrusters that could be simultaneously commanded to fire per aft and forward RCS module was constrained to only three for all mission phases except for re-entry and a Return To Launch Site abort. For re-entry the acceleration vector caused propellant to cover the outlet, thereby eliminating the

194 Maneuvering in space

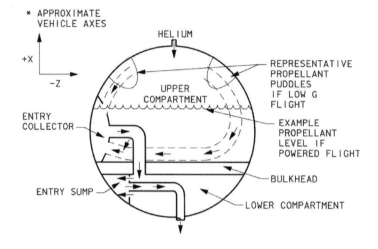

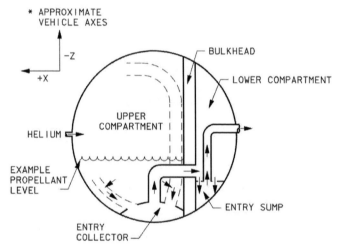

Propellant levels during powered and low-g flight (top) and re-entry (bottom) in an RCS tank. (Courtesy of www.nasaspaceflight.com)

possibility of gas ingestion. For a Return To Launch Site abort, owing to the need to get the Orbiter and its crew home as quickly as possible, the flight rules made an exception which allowed more than three RCS thrusters per pod to be commanded to fire simultaneously. Of the many criticalities involved in attempting such an abort, the risk of having RCS jets deselected was the least worrisome.

As regards the propellant tanks, those for both the aft and forward RCS modules were spherical and internally divided into two compartments of unequal sizes. The larger "upper" compartment contained the pressurization inlet and all the propellant, while the smaller "lower" compartment had the propellant outlet and did not contain any liquid at all. Although identical in conceptual terms, the propellant tanks of the forward and aft RCS were different in construction, since the aft propellant tanks had an entry collector in the upper compartment and an entry sump in the lower one. This configuration allowed propellant acquisition to occur with the change in propellant level orientation during the acceleration loads of re-entering the atmosphere. The forward tanks lacked the entry collector and sump hardware because this module was not used for control during re-entry. An outlet manifold in the upper compartment collected the liquid and directed it out of the tank. The forward RCS was not used during re-entry in order to preclude the thruster plumes and exhaust from damaging the vehicle and disrupting the boundary layer over the wings. This could have caused control issues due to aerodynamic instabilities. In fact the propellants of the forward RCS had to dumped overboard prior to re-entering the atmosphere in order to improve the center of gravity of the Orbiter, so there was very little, if any, liquid available in these tanks for control purposes.

The same problems had to be solved for the OMS tanks. These had a cylindrical domed shape and were made of titanium with their interior divided into forward and aft compartments by a mesh screen PAD called a "trap". The solid-walled vessel had one or more windows consisting of a porous capillary element that allowed filling and emptying of the aft compartment to hold a liquid static during relatively large accelerations. There were four stubby galleries in the aft compartment that acted in a similar manner to the channels in the RCS tanks. The capillary and surface tension properties of the liquid were exploited to deliver propellants to the engines free of gas bubbles. The galleries were tubular sections with rows of windows on one side which allowed liquid to pass by capillary action. The galleries were linked to the collector manifold at the bottom of the tank that supplied propellant to the tank outlet. As long as the four galleries were full there was only liquid on the outlet manifold, allowing a safe engine start. Once a burn was initiated, the acceleration caused the fluid to move from the forward to the aft compartment, passing through the mesh screen and then into the galleries. Helium had to be confined to the forward compartment until the fluid was below the one-third level. The aft compartment included about one-third of the volume of the tank and was completely full at the time of launch. As long as the screen remained wet, the surface tension of the liquid meant it could flow into the aft compartment but was unable to return to the forward compartment during periods of OMS inactivity. At low propellant level, the trap would not be wet, so the propellant would disperse back into the forward compartment. If an OMS burn was required at a low propellant level, then it had to be preceded by a short RCS burn to apply an acceleration to drive the OMS propellants to the aft end of their tanks and fill the galleries with liquid. In the low level condition, helium could enter into the aft compartment because the trap was no longer wet.

Such a propellant acquisition system is called a refillable compartmentalized trap, with the compartmentalization referring to the four galleries. This configuration was chosen after it was decided that the OMS and RCS should use the same propellants. A new requirement was added, that the OMS must be capable of supplying 500 kg of

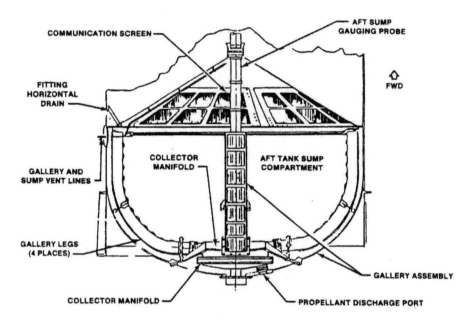

OMS tank propellant acquisition system.

propellant to the aft RCS without compromising its capability to fire its own engines ten times. The trap plus galleries combination in the aft compartment of the OMS tanks guaranteed there would always be sufficient propellant available to the OMS engines in the event that a crossfeed to the RCS was necessary.

The commonality of propellants also caused a change in the gauging system used to determine the amount of propellant remaining in the tanks. The gauging of OMS propellant quantities was accomplished by a combination of physical measurement using a probe inside the tank and computations performed by the gauge totalizer unit. The probe was a rod along the center axis of the tank and was therefore divided into a forward and an aft section, and consisted of two concentric capacitance probes with associated electronics.

ORBITAL MANEUVERING SYSTEM ENGINE

In 1971 a number of Orbiter configurations using external main propulsion system propellant tanks were studied. The results showed that smaller Orbiters with external, expendable tankage would be more cost effective than the larger vehicles envisaged when drawing up the initial baseline requirements. The reduction in the weight of the Orbiter enabled a significant reduction of OMS impulse requirements, and this, along with the decision to allow scheduled OMS refurbishment, stimulated further consideration of using storable propellants. For the smaller, lighter Orbiter with external tankage, the internal volume needed for an oxygen-hydrogen OMS was a significant penalty. The higher density storable propellants were very attractive from this standpoint. To be consistent with the philosophy of that time, only existing engines were considered and therefore the early studies indicated that an OMS using the Apollo lunar module ascent engine would have offered the lightest system weight of a storable propellant configuration. In terms of burn-time considerations, the lunar module descent engine and the engine of the Agena rocket stage were also good candidates. However, when thrust levels, specific impulse, burn time, refurbishment costs, and the low margin for vehicle weight growth were taken into account, it was decided to reject all three of these engines and to develop a brand new one instead.

At this point, two separate contracts were awarded with the primary objective of determining the viability of a reusable thrust chamber with storable propellants. This would provide basic engineering data to potential vehicle contractors to assist them in evaluating and selecting between various engine configurations. One contract was to study an engine that would be regeneratively cooled by circulating fuel through the wall of the chamber. The second study was of a chamber that would be insulated by columbium. It was concluded that regenerative cooling could provide a lightweight, stable, high performance propulsion system that would be reusable. Furthermore, the studies indicated that MMH-NTO propellants were the best in terms of performance, weight, development risk, cost, safety, maintainability, operating life and reliability.

Acoustic cavities, used either independently or in conjunction with baffles, had been shown to be an effective method of suppressing acoustic modes of combustion instability in rocket engines. In applications with requirements for both long duration firings and reusability, cavities were preferred over baffles because they were easier to cool, which made them attractive for use in the orbital maneuvering engines. Yet there were concerns regarding the effect of the high fuel temperature associated with regenerative cooling. Therefore tests were undertaken to evaluate the effectiveness of acoustic cavities in conditions closer to those of the OMS engines. The results proved that stable operation could be maintained with a variety of cavity configurations, an indication that a moderate margin of stability would be possible.

In 1974 the California-based Aerojet Liquid Rocket Company was awarded the contract to develop and build the OMS engines. Aerojet already had a track record in the development of liquid storable propellant engines, having built the engine of the Apollo service module. That engine was a marvel of engineering, because if it failed to ignite when the time came to head home to Earth, it would have meant sure death for the astronauts stranded in lunar orbit. Drawing upon their Apollo experience, Aerojet developed an engine that was as simple as possible, whilst also offering the maximum possible degree of reliability. The basic design was a direct application of predevelopment technologies and of the various design drivers for the OMS engine. The main design drivers were operating life, envelope, applied environment, specific impulse, combustion stability, reusability, and propellant inlet feed pressure. The key requirement of a long life influenced the design of the acoustically stabilized, flat-face, photo-etched injectors, the regeneratively-cooled slotted combustion chamber, and the redundant ball valve. The nozzle area ratio of

198 Maneuvering in space

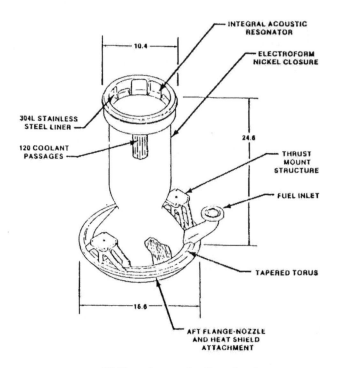

OMS engine combustion chamber.

55:1 efficiently exploited the allocated envelope length, but only 44 inches of the 50-inch envelope diameter. The result was a smaller nozzle skirt having a lighter configuration. The propellant valves were side mounted to reduce feedline length and engine length, and to permit shock mounting to modulate engine input. For ease of access the filters were placed at the inlets of the feedlines. Static leakage was controlled by redundant seals at all of the flanges and interconnecting fluid joints. Inlet line routing to the engine was through propellant lines located in the same plane as the gimbal ring (for thrust vectoring) and the chamber throat. Installation, maintenance and servicing were simplified by designing the engine as a line replaceable unit and mounting it on the inlet manifold-mounted gimbal ring. Large tolerance stackups were eliminated and thrust alignment was simplified.

To demonstrate the robustness of the design, the performance of the engine was tested under operational parameters well outside its nominal operating conditions. For example, one important test was to establish the operability of the engine after having been exposed to the harsh vibration environment caused by the main engines located alongside the OMS pods. As a result of extensive testing at the component level, the various problems were resolved long before testing at engine level began. Qualification testing continued until 1979, and in May of that year the first flight-worthy OMS pods were mounted on *Columbia*.

The combustion chamber had a similar configuration to that of the SSME, being a stainless steel liner with rectangular coolant channels enclosed in an electroformed nickel shell. It contained 120 longitudinal, milled, rectangular-shaped passages, each with a constant width but a varying depth to provide the optimum configuration for cooling efficiency, engine performance and chamber lifetime. The remaining portion of the regenerative cooling system was the fuel inlet-distribution ring welded to the outer wall. The liner was fabricated from 304L stainless steel, selected because of its adequate strength properties at the operating temperature of the engine, its chemical compatibility with the combustion environment, its thermal fatigue properties for the applied temperature gradient and channels pressure, and its superior machining and electron beam welding characteristics. The nickel outer shell thickness was governed principally by the strain along the length of the chamber resulting from aerodynamic loading on the nozzle.

After passing through the bipropellant valves, the oxidizer line ran directly to the engine injector. The fuel was routed through a cooling jacket around the combustion chamber on its way to the injector. The injector system consisted of an oxidizer/fuel manifold, a core billet, a fuel distribution ring, a platelet injector, and the manifolds. The distribution ring was mated with the combustion chamber regenerative cooling passages to deliver the fuel to the manifold, where oxidizer and fuel passages were machined into the stainless steel core billet separated by parental metal or redundant metallurgical joints. Six thin platelet disks provided a pattern of sixteen concentric alternating rings of oxidizer and fuel orifices, with the outermost ring spraying fuel onto the wall of the chamber for film cooling. The manifold covers included bosses on which to mount pressure and temperature sensors.

Integral to the inner wall of the chamber was the converging-throat-diverging (initial) section of the nozzle. The converging section blended into the throat area and diverged to produce the initial section of the bell-shaped exhaust nozzle. The nozzle extension was made entirely of columbium, was attached by a split retainer ring with a graphite gasket and 36 bolts, and expanded from the regeneratively cooled interface to an area ratio of 55:1. It consisted of a flange, a forward section, and an aft section. The mounting flange consisted of a bolting ring made from a forging and a tapered section that could either be made from a forging or be spun and provided a smooth transition from the 0.100-inch-thick flange to the 0.030-inch-thick forward section. The forward and aft nozzle sections were each made from two panels butt welded to form two cones that were welded circumferentially to each other to form the nozzle. This assembly was then bulge formed into the final configuration and coated with a silicide compound to prevent corrosion. The original design included three stiffening rings approximately at the midpoint of the nozzle, but the final design with a single flange at the nozzle exit was dictated by changes in the magnitude and the location of aerodynamic loading and by changes to the expected aerodynamic noise level.

Whilst the propellants would be liquid at the temperatures normally experienced during a mission, care was taken to prevent them from freezing during long periods on-orbit when the OMS was not in use. This involved insulating the propellant lines and walls that enclosed hardware components, and installing line-wraparound heaters and blanket-type heaters. The heater system was split into two areas: the OMS/RCS pods and the aft fuselage crossfeed and bleed lines. Heater patches were used in the pods. Each patch consisting of a redundant set of wires formed into a flat

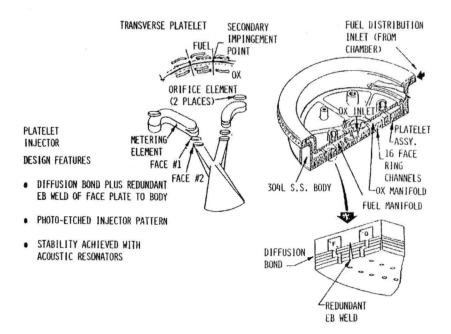

OMS engine injectors.

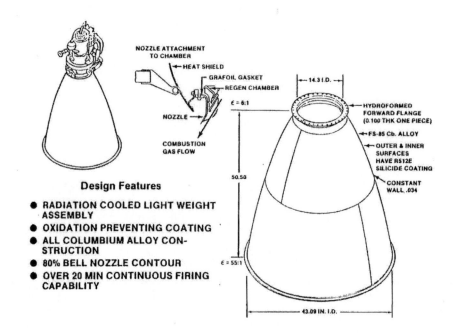

OMS engine nozzle.

and tightly spiraled pattern. As electricity flowed through the highly resistive wires, the heat they generated served both to directly warm the hardware to which the patch was attached and to radiate into the surrounding volume. Sensors were located throughout the pods to supply temperature information to the telemetry.

REACTION CONTROL SYSTEM THRUSTERS

Unlike the OMS, which had only one engine in each pod, the RCS was a collection of 44 small thrusters distributed between the forward and two aft modules, with each aimed in a given direction. By selectively firing individual thrusters or combinations of thrusters, the Orbiter could execute rotational maneuvers and small translational maneuvers. Of these 44 thrusters, 38 were primary jets able to deliver thrust levels up to 870 pounds of force in a vacuum. The other six thrusters had a maximum thrust level of 24 pounds of force in a vacuum. For this reason they were called vernier jets and used only for fine attitude control. Each thruster was identified by the propellant manifold that fed it and by the direction of its plume. The first identifier designated a thruster as forward (F), left aft (L) or right aft (R). A second identifier of digits one through five designated the propellant manifold. Finally, a third identifier designated the direction of the plume: aft (A), forward (F), left (L), right (R), up (U) and down (D). The forward module had 14 primary jets and 2 side-firing verniers, while the aft modules each had 12 primary jets and 2 verniers, with one set of verniers side-firing and the other one down-firing. In the event of losing a down-firing vernier, the entire control mode would be denied due to loss of control authority. If a side-firing vernier was lost, control would remain for everything except some robotic arm operations.

Although much weaker in terms of thrust, the RCS engines were similar to their OMS counterparts. Each primary jet and vernier had a combustion chamber made of columbium and had a columbium disilicide coating to prevent corrosion. Propellants reacted in the chamber after having passed through an injector plate that consisted of a collection of so-called doublets. Each doublet was two injector holes (one for each propellant) canted towards each other to promote stream impingement and therefore faster reaction time. The chamber for the primary jets had 84 doublets arranged in a circular "showerhead" pattern with two concentric rings. Additional fuel holes were provided near the outer edge of the injector assembly to cool the interior wall of the chamber. There were acoustic cavities to dampen out any combustion instability that might arise. The injector plate of the smaller vernier engine had just one doublet, and slightly more fuel was injected for film cooling of the chamber wall.

After combustion, the hot gas expanded through a nozzle to complete the process. As with the combustion chamber, the nozzle was made of columbium with a profile tailored to match the external shape of the RCS module into which it was to be fitted. Insulation was placed around the combustion chamber and the nozzle to prevent heat from radiating into the Orbiter's structure.

Another major component was the reaction jet driver, a black box in which the

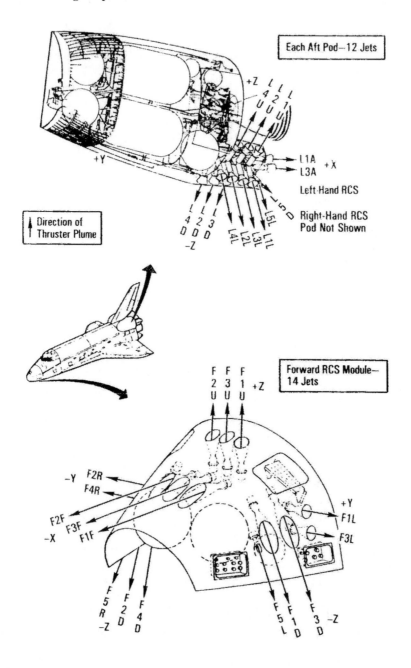

RCS jet location and identification.

Reaction control system thrusters 203

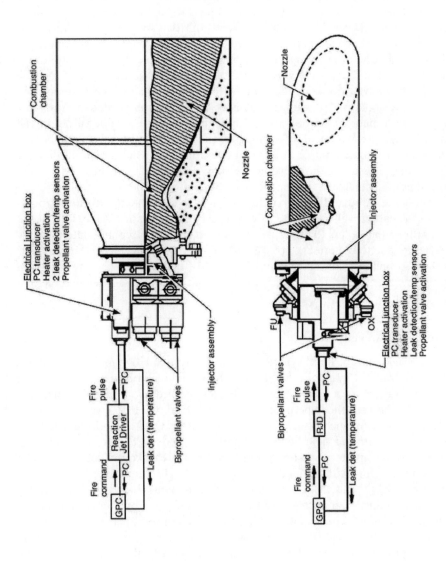

Schematics of primary RCS jets (top) and vernier RCS jets (bottom).

firing commands issued by the GPCs were converted into the voltage needed to open the bipropellant valve and thereby initiate the combustion process. Like the OMS, the RCS bipropellant valve had the function of regulating the flow of propellants into the combustion chamber. Because early technology work had shown that a hydraulically operated valve would be lighter than a solenoid valve for such an engine, this concept was chosen when designing the bipropellant valve. But this design introduced a new problem. Each bipropellant valve contained two injector solenoid pilot poppet valves, one for each propellant, mechanically linked. These poppet valves were operated by coaxially wound coils that were energized open by a firing command and thereafter spring-loaded closed. As the pilot valves opened, the propellant's hydraulic pressure opened the main poppet valves to enable propellant to flow to the injector. Thus the valve relied on pressure imbalances established by the pilot poppet valve to provide the opening force. It was subsequently discovered that these valve-actuating pressure imbalances could also be created by transient pressure waves generated by operating other thrusters or isolation valves in the system. Although these momentary openings of the valves were established to be safe in flight, they were unacceptable for ground operation. It was also realized that gas trapped in the recesses of the valve could slow down both opening and closing response times and increase the tendency of the valve to "bounce" with pressure transients. To minimize the possibility of a valve bouncing during ground operations, isolation valve operation was procedurally limited to cases with a pressure differential of less than 25 psi across the valve. To accommodate the slow valve response due to the presence of gas, the minimum thruster firing duration was doubled to 80 milliseconds, satisfactory for control purposes. When the thrust-on command was terminated, the valves were de-energized and closed by the force of a spring.

RCS ground operations

The basic design concept for a hypergolic system with toxic, flammable propellants was to incorporate the systems in modules or pods that could be detached from the Orbiter and taken to a dedicated facility for hazardous repair or checkout operations. In this case the hypergolic maintenance facility (HMF) was constructed to perform hazardous operations in parallel with other Orbiter work, and thereby save valuable turnaround time between flights. The pods were also provided with access panels and plumbing connections to enable checkout functions to be performed in the OPF with the pods installed on the vehicle.

Developing the ground checkout philosophy was a major challenge for a reusable system that contained highly corrosive propellants, was used on a continuous basis in flight, and included considerable redundancy. A full electrical and mechanical checkout of each individual RCS module was performed by the manufacturer using a simulator, prior to shipment. Upon arrival, the most critical components and the integrity of the plumbing were checked out again in the HMF. Once a given module was installed on an Orbiter for the first time, the electrical components were checked for proper end-to-end channelization, where possible by actual physical response.

A view of the protective covers applied to the RCS jets.

Once the Shuttle was on the launch pad the modules were loaded with propellants and helium. During actual flight the system pressures, temperatures, quantities and valve positions were monitored closely for any indication of a malfunction. Special procedures were used to obtain functional data by changing from an operational component to a redundant component during each mission. Special hot-fire tests were also performed in flight to check out jet thrusters that might not normally be used. In this way, the functional capability of as many component as possible was confirmed as a means of reducing ground checkout requirements. After each landing, a further component checkout was performed on a very limited number of components based on their criticality and whether they were tested in flight. Most components were only checked every five or ten missions in order to screen for unanticipated deterioration. The plumbing was leak checked by monitoring the pressure decay in normal pressure checks after every flight. The engine chamber and nozzle coating were also inspected for defects after every flight. This enabled the turnaround time to be minimized without significantly undermining reliability.

One of the major challenges encountered in actual operations was the requirement to provide rain protection for some of the engines of the Orbiter after the launch pad protective structure was moved away and the ground covers were removed from the RCS engines. It was essential to protect the three upward-facing engines and eight of the left-side engines from rainwater accumulation. In particular, the up-firing engine covers had to prevent a water accumulation that could freeze in the injector passages during ascent, blocking the sensing port and causing the engine to be declared "failed off" when it was first commanded to fire. For the side-firing engines the covers had to prevent water from accumulating in the bottom of the chamber and protect the chamber pressure sensing ports. The original design concept was to install Teflon plugs in the engine throats (side-firing) and a combination of a Teflon plug tied to a Teflon plate that covered the nozzle exit (up-firing). This added vehicle weight, necessitated special procedures to eject the plugs in flight, and posed the risk of accidental ejection during the ascent causing damage to the thermal protection system tiles. The second concept studied involved Teflon sheets that were glued to the nozzle exits and pulled off by lanyards when the crew access structure was retracted. However, this would not have provided protection all the way to launch, or for all engines, and was unnecessarily complex. The final solution was a novel one of using ordinary plastic-coated freezer paper shaped to fit the exit plane of the nozzle and glued into place. Tests proved this would provide a reliable seal under all expected rain and wind conditions. The covers would blow off during ascent prior to the vehicle going supersonic. The solution was ideal, because the covers were cheap, simple, safe, and added no significant weight.

7

Heart and lung of the Orbiter: the environmental control life support system and electrical control system

ENVIRONMENTAL CONTROL LIFE AND SUPPORT SYSTEM

Earth provides all the necessary conditions for a wide variety of life forms to survive, and life is present in even the least hospitable parts of the planet. But human life can survive only in conditions that allow our bodies to function properly. Space is one of the most inhospitable places for life, so much so that thus far no life has been found there. A spaceship for a human crew must duplicate the intricate and interdependent functions the human body relies upon in the terrestrial environment. This is the task of the environmental control life and support system (ECLSS). It must provide for atmosphere revitalization, atmosphere control and supply, temperature and humidity control, water recovery and management, waste management, and fire detection and suppression.

The Shuttle's ECLSS functionality was divided into four main subsystems:

1. <u>Pressure control system</u>. This maintained the crew cabin at 14.7 psia with a breathable mixture of oxygen and nitrogen.
2. <u>Atmosphere revitalization system</u>. This used air circulation and water coolant loops to remove heat, control humidity and clean the air.
3. <u>Active thermal control system</u>. This used two Freon loops to collect waste heat from the Orbiter systems and transfer it overboard.
4. <u>Supply and waste water system</u>. The supply system stored the water that was created by the fuel cells and made it available for drinking, personal hygiene and Orbiter cooling. The waste water system collected and stored water from the humidity separator and crew fluid waste.

Pressure control system

NASA lost three astronauts on the evening of Friday, 27 January 1967. While Virgil "Gus" Grissom, Edward H. White and Roger B. Chaffee were performing a dress rehearsal for the first launch of an Apollo capsule, a violent fire swept the cabin. In a matter of seconds the entire cabin was aflame. Because of the lengthy procedure for opening the hatch the three men were doomed. The commission that investigated the accident determined that the composition of the atmosphere inside the cabin was one of the principal contributing factors to the tragedy. As with the Mercury and Gemini programs, NASA intended that while the Apollo spacecraft was in space it should be pressurized with pure oxygen at 5 psi. But to perform a pressure check that fateful evening it was necessary to raise the cabin pressure slightly above ambient, in this case to 16 psi. If oxygen is harmless at low pressure, at high pressure it becomes one of the most toxic and dangerous of substances because in that environment virtually anything will burn. With an atmosphere of high pressure pure oxygen, the spark that occurred somewhere in the electrical circuitry readily triggered an explosive and deadly fire. As a safety precaution, the spacecraft was modified to employ a nitrogen-oxygen atmosphere until in space, when it was changed to the planned pure oxygen at 5 psi.

The Shuttle's ECLSS was designed for a cabin atmosphere of 20 per cent oxygen and 80 per cent nitrogen, a composition very similar to that on Earth. That oxygen is essential to human life is obvious. Less well known is that for us to breathe properly, it must be at the right pressure, stated in the jargon as the partial pressure of oxygen (PPO_2).[1] At too low a PPO_2, the oxygen molecules do not have sufficient strength to fix themselves to the hemoglobin in red blood cells, thereby creating a condition called hypoxia. Yet oxygen at too high a PPO_2 becomes toxic to the human body as well as increasing the fire hazard. To preclude both conditions in the Shuttle cabin, it was decided to regulate the PPO_2 within the range 2.95 to 3.45 psi. The nitrogen that brought the cabin pressure up to that of sea level on Earth is an inert gas that acts as a fire suppression agent. For a comfortable environment, it is also necessary to hold constant the total pressure, because the reduced heat capacity of the air at pressures below the minimum of this range impairs the ability of the temperature and humidity control subsystems to properly cool the habitat. In order to match terrestrial sea level, a total pressure of 14.7 psi was chosen as the design point for the Shuttle.

It is also important to note that for any pressurized spacecraft there are always tiny leaks that lower the pressure over time. For a crew cabin, is necessary to have a system that periodically replenishes the atmosphere to maintain the total and partial pressures at their required values. For the Shuttle these tasks were assigned to the pressure control system (PCS). For redundancy, this was designed as two identical separate systems (PCS 1 and PCS 2), each capable of sustaining the crew on-orbit for the full duration of the mission. The design was straightforward but effective. Each PCS consisted of a liquid oxygen tank (which was shared with the electrical power generation system) and two gaseous nitrogen tanks. Pressure regulators, filters, sensors, check valves, interconnections with the Orbiter coolant loop for warming up

---

[1] Partial pressure refers to the fraction of the total pressure that is accounted for by a particular gas in a gaseous mix.

Environmental control life and support system 209

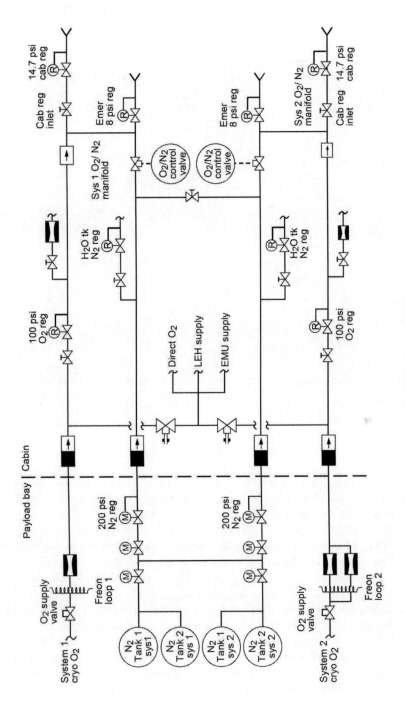

ECLSS pressure control system.

210 Heart and lung of the Orbiter

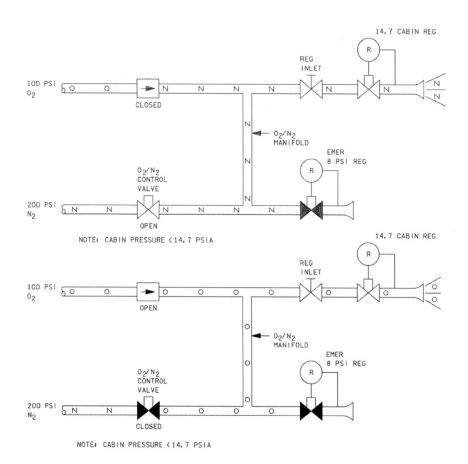

O_2/N_2 control valve open (top) and closed (bottom).

the oxygen, and a set of crossover valves between PCS 1 and PCS 2 completed the configuration. The tanks were beneath the payload bay, but the pressure regulating hardware was inside the cabin so that it could be operated manually if necessary.

The heart of each PCS was its O_2/N_2 manifold. This was connected to the 14.7 psi regulator valve, PPO_2 sensor, and O_2/N_2 control valve. If the total pressure fell below 14.7 psi while within the design range for PPO_2, then the O_2/N_2 control valve would open to permit nitrogen to flow into the O_2/N_2 manifold and from there into the cabin through the 14.7 psi regulator valve. If the PPO_2 fell below the minimum value, the O_2/N_2 control valve was closed, and oxygen would start to flow into the manifold as soon as the total pressure sensed became less than 14.7 psi. If the PPO_2 sensed was above the maximum allowed value, the O_2/N_2 control valve was opened to flood the manifold with gaseous nitrogen, and as soon as the sensed total pressure fell below 14.7 psi, the 14.7 psi control valve would allow the nitrogen in the manifold to enter the cabin in order to restore the proper total pressure. Because the gaseous nitrogen pressure was greater than the oxygen pressure, when the O_2/N_2

control was open the higher pressure would close the check valve of the oxygen manifold, blocking the flow of oxygen into the manifold. In the event of rapid loss of air, an 8 psi emergency control valve would guarantee a flow of gas to prevent the total pressure of the cabin from falling below 8 psi.

Two positive pressure relief valves ensured that if the PCS regulator failed, the pressure would not increase sufficiently to compromise the structural integrity of the cabin. These were located behind the back wall of the toilet area and vented into the payload bay. Conditions in which the external pressure exceeded the cabin pressure would also threaten structural integrity. For this reason, two negative pressure relief valves would open when the ambient pressure was 0.2 psi greater than the internal pressure. These valves were located below the side hatch and caps were provided as a redundant seal to prevent leakage overboard. When the pressure outside the cabin increased above the cabin pressure, the relief valve would crack, the caps would pop off and air would flow into the cabin to equalize the pressure. Finally, a cabin vent valve and cabin isolation vent valve would vent the cabin into the payload bay while the Orbiter was on the ground. These were used prelaunch during cabin leak checks, when the cabin pressure was raised to 16.7 psi and held for 35 minutes to detect any decay that would indicate a leak. Once this test was finished, the valves were opened to allow the cabin pressure to equalize with the external one. Since these valves had very high flow rates, the post-insertion checklist called for their power to be disabled in order to prevent any inadvertent opening that could rapidly reduce the pressure of the cabin.

Atmosphere revitalization system

A spacecraft crew cabin must recreate within its enclosed environment conditions as similar as possible to those that humans require on Earth. In order to create a livable environment several functions have to be considered. For example, while people can survive in a relatively wide range of temperatures and humidity conditions, the range in which we are comfortably able to live and work is fairly narrow and a function of the activity being undertaken. The ideal temperature range is 18°C to 27°C, and the ideal humidity is 25 to 70 per cent. Experience during long duration space missions has shown that temperatures below 19°C with relative humidity above 70 per cent are considered by the astronauts to be either "cool" or "cold". A temperature in the range 22°C to 24°C is optimal for crew comfort. Control of humidity is equally important. To avoid adverse physiological effects it is necessary to keep the water vapor content at appropriate levels (a relative humidity or RH of between 25 and 70 per cent). For instance, if the RH is too low then throat and nasal tissues become dry; if it is too high, perspiration does not provide adequate body cooling. And proper atmospheric water content is necessary to protect avionics and other electronic equipment, and also to prevent the growth of fungi and bacteria. A high RH can cause condensation on windows, walls and optical equipment. Compared to dehumidification systems for aircraft, the design of a humidity control system for a spacecraft is more challenging owing to the absence of gravitational forces. The removal and storage of atmospheric condensate requires capillary devices and/or rotating machinery to produce an artificial sense of gravity.

It is also essential to create forced ventilation inside a cabin. In weightlessness this is the only method of ensuring good mixing of the atmospheric constituents for adequate removal of CO_2, water and trace contaminants, and to provide sufficient O_2 for metabolic requirements and the removal of body heat. Ventilation flow rates are determined by medical requirements to preclude stagnant regions in which either the O_2 level may get too low or the CO_2 level too high, and by the requirements to deal with the waste heat generated by crew, equipment and experiments. Another factor in selecting the ventilation rate is the cabin air total pressure, since lower total pressures require higher ventilation rates for the same amount of cooling capacity. Cooling the equipment and experiments creates another problem for an atmosphere revitalization system. As the natural convection of air on Earth due to the buoyancy of warm air is absent or reduced in weightlessness, the cooling of electronic equipment and payload requires either forced ventilation or a "cold plate". In the latter, heat is drawn off by conduction to a liquid contained in the plate on which the apparatus is mounted. This is used when forced convection is insufficient to provide effecting cooling.

The fairly small size of a spacecraft cabin does not provide the volume relative to contaminant sources of the Earth's atmosphere, and consequently does not offer much buffering capacity to dilute contaminants. Even a small amount of contaminant may result in a hazardous condition. It is therefore essential to monitor the amounts of contaminants and have an efficient means of removing excess contaminants. Trace contaminants derive from outgassing by materials, metabolic byproducts of the crew (perspiration, urine, feces, etc.), food preparation, housekeeping cleaners, and scientific experiments. Passive contamination control, such as careful selection of materials to minimize outgassing and dust particles can significantly reduce the amount of contaminants requiring to be removed, and is the first necessary step in designing an ECLSS. However, active contamination control is necessary for long missions.

On Earth dust is removed by gravity and rainfall, and trace gases are transformed by physical, chemical and biological processes. In the weightlessness of space, other means of removing dust particles are needed. The standard methods are screens and high efficiency particulate atmosphere (HEPA) filters in the ventilation system. Trace gases that are potentially harmful to the crew must be actively removed by means of either absorption, adsorption, or catalytic oxidation. Forced ventilation inside a small cabin in reduced gravity can very rapidly spread microorganisms. HEPA filters can be used to remove them from the atmosphere, depending on the pore size of the filter and the size of the microorganisms. The filters are placed at strategic locations in the ventilation system to ensure that microorganisms and particulate contaminants will be held to acceptable levels.

Another key factor is the concentration of CO_2, which must be held at a very low level in order to prevent adverse physiological effects. Since CO_2 is generated during normal metabolic respiration at an average rate of 1 kg per person per day, it is easy to understand how the CO_2 would quickly increase to unacceptable levels in a closed volume without some means of removing it. For a spacecraft, the preferred methods employ absorption (a chemical or electrochemical reaction with a sorbent material), adsorption (a physical attraction to a sorbent material), a process called membrane separation, or biological consumption. Historically, manned US space programs used absorption with canisters of lithium hydroxide (LiOH). But in the case of the Shuttle the atmosphere revitalization system (ARS) utilized most of these functions. From when the ingress/egress hatch was closed prior to launch to when it was opened again after landing, the ARS drew the heated cabin air through a duct installed beneath the middeck floor. After passing through the cabin fan and a trap for debris removal, the air was sent to a canister of LiOH where CO_2 was removed by means of a reaction that created lithium carbonate. Because this reaction is irreversible the canisters had to be changed periodically, generally once or twice per day. In replacing a canister, it was necessary to switch off the fan to prevent problems of eye and nose irritation caused by canister dust stirred up by the fan. The LiOH canisters also provided for removal of acidic contaminants by chemical absorption. Next the air was exposed to active charcoal to remove odors and contaminants, and passed through a heat exchanger to shed excess heat to a coolant water loop. The cold temperature of the heat exchanger enabled the humidity in the air to condense and be drawn into one of two separators. A fan in each separator created the suction that drew water-laden air away from the heat exchanger. The condensate was isolated by centrifugal force of a rotating drum and stored in the waste water tank.

A bypass valve just upstream of the heat exchanger regulated the amount of air to be cooled, so that by mixing the flow of air exiting the heat exchanger with the one that bypassed the exchanger it was possible to achieve the desired temperature in the

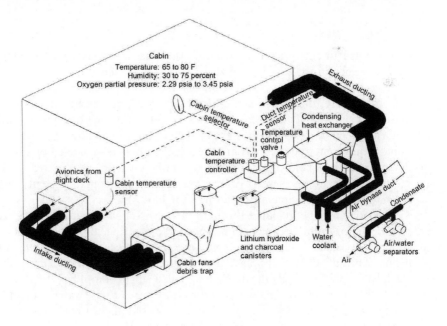

Cabin air system.

214 Heart and lung of the Orbiter

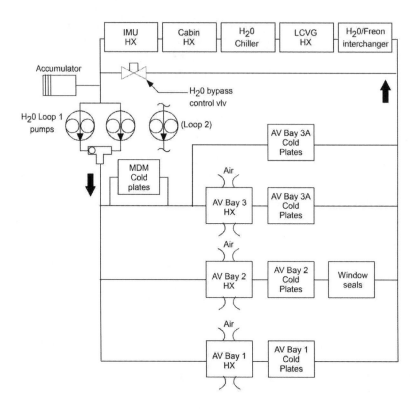

Water coolant loop 1.

cabin. The cooled air emerging from the heat exchanger was then further purified by the ambient temperature catalytic oxidizer (ATCO), which removed the carbon monoxide generated by the crew and from the outgassing of non-metallic materials in the cabin. Finally, the air from the heat exchanger and the bypassed air came together in the supply duct and was fed into the cabin through the consoles of the commander and pilot and through ducts at the various workstations.

Avionics equipment and the inertial measurement units were cooled by means of forced ventilation and cold plates. A closed ventilation loop was created for each bay, resembling an atmosphere revitalization system but containing only a heat exchanger to cool the air circulating through the bays.

A water coolant loop circulated through the crew compartment to collect excess heat and transfer it to the Freon coolant loops. Two separate, almost identical water coolant loops ran side by side although only one was active at any given time during the mission. The only difference was that loop 1 (normally used as backup) had two pumps whilst loop 2 (normally the active one) had a single pump. For items on cold plates, heat transferred to the plate was in turn transferred to the water coolant loop and carried away. For maximum cooling action the cold plates were connected in a series-parallel arrangement with respect to the water coolant loop flow. Downstream

Environmental control life and support system

of each water pump the water flow split into three parallel paths, each of which was passed through one of the three avionics bays. The three parallel paths in each loop then rejoined upstream of the Freon/water heat interchanger, which was outside the cabin in order to protect the crew from the toxic Freon 21 in the event of a leak. The flow path then split, with one parallel path in each water coolant loop going through the Freon/water interchanger to cool the water loop. The water then passed through the potable water chiller, cabin heat exchanger, and heat exchangers of the inertial measurement units on its way to the appropriate pump package. The other parallel path in each water loop carried warm water that was directly returned just upstream of the pump package. This bypass permitted temperature control of water leaving the pump package for each coolant loop. Running both coolant water loops for lengthy periods was avoided, since together they would pass too much water through the water/Freon interchanger and produce a significant increase in the cabin temperature. That is, two active water loops would pick up more heat than the heat exchangers could transfer to the Freon loops and, over time, the water loops would accumulate heat that would decrease their cooling efficiency.

Active thermal control system

The active thermal control system (ATCS) of the ECLSS was designed to cool/heat other subsystems using interface heat exchangers, to transport heat from sources to sinks using coolant loops, and then to reject heat by a variety of means according to the mission phase. The backbone on this system comprised two identical Freon loops that received heat from a network of cold plates and heat exchangers.

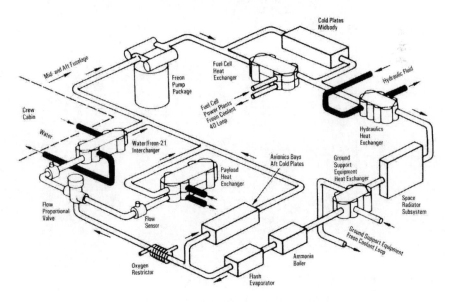

Freon coolant loop.

216 Heart and lung of the Orbiter

Extensive trade-off studies were made during the definition phase of the Shuttle in order to arrive at the optimum mode of operation. In particular, it was found that the best option was to operate both Freon loops at the same time in normal operation and a single loop only in the event of a failure, thereby providing on-line rather than standby redundancy. This saved considerable fixed weight and power relative to an approach in which a single loop would provide total cooling in normal operation and the second redundant loop was on standby. The dual-loop design needed redundant pumps in each loop and a reduction in heat load (power usage) from normal when in one-loop mode, but this was deemed acceptable because it would not adversely affect the safety of the crew and vehicle. Studies identified the need for four different heat sinks for heat rejection during each mission phase. Both prior to launch and for some time after landing, cooling would be provided by a ground-based refrigerator system that worked with an on board ground support equipment (GSE) heat exchanger in the Freon loop. In order to minimize the size of this GSE, cold Freon was supplied to the vehicle at high pressure by fly-away disconnects. The increased safety and reliability that this configuration provided were deemed more important than the small increase in weight from having to carry the GSE heat exchanger into orbit.

Heat rejection on-orbit was by radiation to space. It was preferable to expendable evaporation. If water was used as an evaporant, approximately 16,400 pounds would be needed for the maximum payload heat load over a seven-day mission, but the fuel cells would produce only 2,400 pounds. The remaining 14,000 pounds would have to

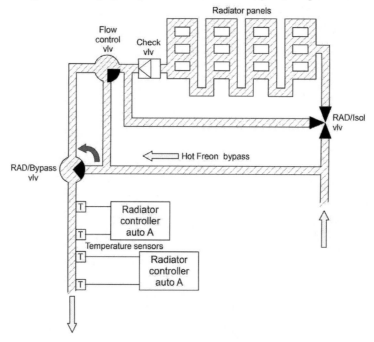

Radiator/flow control assembly overview.

be carried into orbit and would be a tremendous weight penalty. The radiator system of the Orbiter consisted of four radiator panels attached to the inside of each payload bay door. Based upon the heat rejection requirements for a specific mission, the two forward radiators could be deployed once the doors were open. The aft radiators were not deployable.

The radiator panels on the left side serviced Freon loop 1, while those on the right serviced Freon loop 2. For both loops, coolant flowed in series from panel to panel, and in parallel within each panel through an array of tubes that were connected by an inlet and outlet connector manifold. Each loop had a pair of electronic controllers (referred to as Auto A and Auto B) which commanded the flow control valve and the RAD/BYPASS valve. In order to keep the temperature of the coolant emerging from the radiators at one of two temperatures (38°F or 57°F, plus or minus two degrees in each case), the flow control valve was operated to permit some of the "hot" coolant to mix with the "cold" coolant emerging from the radiator. The RAD/BYPASS valve served two roles. In the BYPASS position it bypassed the radiators, preventing them from being used as a means of cooling. In the RAD position it allowed Freon to flow into the radiators and into the Freon loop. It could be controlled automatically by the radiator controller or manually by the crew. The temperature of the radiator output was constantly monitored. If it dropped below 33°F (plus or minus half a degree) the "undertemp" logic would command the valve to the BYPASS position to prevent the water in the stagnant water loop from freezing in the water/Freon heat exchanger. In the BYPASS state, a small amount of Freon continued to flow through the radiator to prevent the chill of space freezing the coolant in the panel. The large exposed area of the radiator panels posed a high likelihood of a micrometeoroid impact puncturing a Freon loop. A leak would be detected by a drop in the coolant accumulator below the preset lower limit of 12 per cent that lasted more than five seconds. In the event of a leak, the radiator isolation valve would redirect the coolant upstream of the flow control valve without passing through the radiator, and a check valve would prevent it from passing back through the flow control valve into the radiator and leaking out.

For mission phases such as launch and re-entry, when the radiators could not be used because the payload bay doors were closed, heat rejection was accomplished by water evaporation. Although this secondary heat rejection system added weight to the Orbiter, the mission phases in which it was necessary were brief and the penalty was small. Heat rejection by water evaporation was chosen for three main reasons:

1. Water has the largest latent heat of vaporization of any candidate fluid, thus minimizing the evaporant weight penalty for the cooling capacity needed for launch and an abort following launch.
2. The fuel cells could not produce sufficient water for total heat rejection over the entire mission, but their water would be convenient for supplementary heat rejection on-orbit at times when the attitude of the Orbiter reduced the radiator efficiency.
3. A water evaporator could also be used to expel excess water as steam when the water storage tanks reached maximum capacity.

NASA therefore installed a flash evaporator system (FES) in the aft fuselage of the Orbiter. It had two separate evaporators, known as the high-load evaporator and the topping evaporator. The only major difference was that the high-load evaporator had larger spray water nozzles for a greater cooling capacity. Each was a cylindrical shell similar in size to an office wastebasket and had dual water spray nozzles at one end and a steam exhaust duct at the other end. The shell consisted of two separated finned packages, one for each Freon loop. The hot Freon flowed around the finned shell as water from the supply system was sprayed onto the shell by the nozzles. The water would immediately vaporize, thereby cooling the Freon. To work properly, the spray chamber had to be maintained at a saturation pressure low enough for the water to evaporate at a temperature below the desired Freon outlet temperature. Since the maximum desired Freon outlet temperature was 40°F, the chamber pressure had to be maintained at or below 0.1 psia. This also allowed the water to instantly evaporate on coming into contact with the hot chamber walls, preventing flooding and subsequent freezing from causing the device to fail. Vent lines carried the steam to nozzles on the sides of the aft fuselage to be expelled at high speed. In particular, while the vent lines for the topping evaporator were placed on either side of the aft fuselage so that they would cancel out their thrusts, the vent line of the high-load evaporator was on the –Y axis to make it a propulsive nozzle. This would minimize the degree to which the thrust of venting disturbed the guidance and navigation system and also reduce contamination of space in the immediate vicinity of the vehicle.

Another heat rejection system used ammonia evaporation. The addition of a third system was required by the fact that the FES could be used effectively only when the external pressure was low enough for water to quickly evaporate and cool the Freon loops to 40°F, which meant above an altitude of about 140,000 feet. This meant that during re-entry the FES could not be used at altitudes below this threshold. For this reason, an ammonia evaporation system was used during this mission phase until the cooling GSE was connected after landing. Ammonia was selected for being the most efficient evaporant after water, having a latent heat of vaporization about half that of water. Despite its toxicity, the brief period of operation required a very small amount, minimizing both handling and contamination hazards. The system consisted of two individual ammonia storage and control systems, and a common boiler that contained ammonia passages and the two Freon coolant loops. The boiler was a shell-and-tube configuration with a single pass of ammonia on the ammonia side and two passes of each Freon coolant loop. As in the water boiler of the FES, upon coming into contact with the hot loops the ammonia would vaporize, with the exhaust being vented overboard through a nozzle in the upper aft fuselage, next to the bottom-right side of the vertical tail.

All three heat rejection systems were used, according to the mission phase. Prior to launch, cooling was provided by the GSE heat exchanger and the thermal inertia of the Freon loops limited the rise in temperature sufficiently to preclude the need for active heat rejection. At solid rocket booster separation, some two minutes into the ascent, the Orbiter was at an altitude at which water evaporation could provide effective cooling. The FES was activated, and served as the primary cooling system through to the post-insertion phase. Due to the high heat load to be dissipated during the ascent because so many Orbiter systems were operating, the Freon flowed first through the high-load evaporator and then the topping evaporator for additional cooling. During the post-insertion procedures the flow was sent to the radiators and the payload bay doors were opened. At that time the radiators became the primary cooling system. If at any point during the mission the Orbiter adopted an attitude in which the radiators were unable to provide the desired cooling, the topping evaporator of the FES would be brought on line to supply the additional cooling required in order to achieve the desired Freon loop temperatures.

During deorbit preparations, cold Freon was stored in the radiators in a process known as "cold soaking". This was accomplished by changing the radiator control temperature from 38°F to 57°F and bringing the FES back on line to cool the loops from 57°F to 38°F. That is, as less cool Freon from the radiator panels was needed to achieve the 57°F radiator output temperature the Freon would remain in the radiator panels for longer, becoming even cooler. After just over an hour, the radiators were bypassed and the FES provided the cooling for deorbit and through to entry interface. Below 175,000 feet the radiators were run through their automatic startup sequence and their flow reinitiated. Below 100,000 feet the atmospheric pressure was too high for the FES to cool effectively, and the radiator cold-soak became the primary source of cooling. If the cold-soak depleted prior to rollout, the ammonia boiler would be activated as the primary means of cooling until the GSE cooling cart was hooked up to take over.

For ascent aborts, the thermal management of the Freon cooling loops was rather different. The FES was still used for cooling after SRB separation but thermal management changed for the re-entry portion of the abort trajectory. In this case the cooling for the lower stages of the abort re-entry would be provided by the ammonia boiler, because at liftoff the Orbiter did not have its radiators cold-soaked and the FES was ineffective below 100,000 feet.

Supply water system

In addition to active thermal control, the Orbiter required water for food preparation and the personal hygiene of the crew. The fuel cells that provided electrical power by reacting liquid hydrogen with oxygen very conveniently issued water as a byproduct. The supply water storage system consisted of four tanks which were pressurized by gaseous nitrogen drawn from the nitrogen tanks of the atmosphere pressure control system.

The water from the fuel cells contained a lot of hydrogen, so to prevent the water tank from filling with hydrogen instead of water a matrix of palladium tubes served as a hydrogen separator. This removed up to 85 per cent of the hydrogen contained in the water. The purified water was then sent to the first tank, known as water tank A, which was sterilized prior to launch. Before entering the tank, the water was further cleansed by passing through a microbial filter. It became potable water for the crew. Initially only tank A was used for storing potable water, but later on a parallel path to tank B was added to provide redundancy. If the primary water path

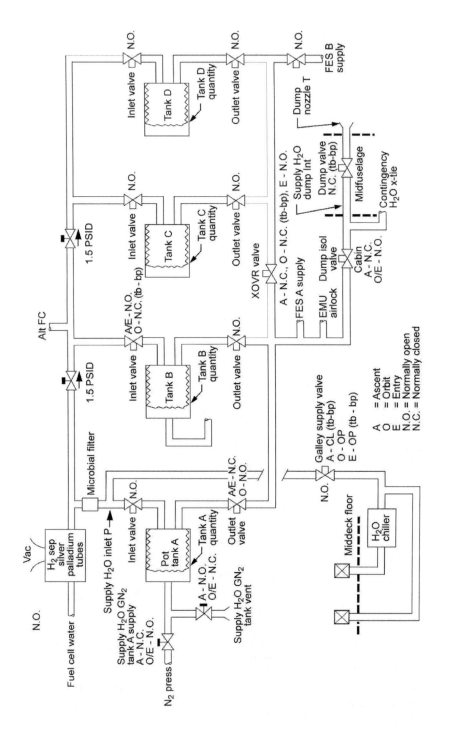

Supply water storage system.

to tank A became blocked, the buildup of pressure would be relieved by the redundant path to potable water tank B. This configuration enabled Shuttle missions to the ISS to use tank B to store potable water for transfer to the station by means of special bags. Once tank A was full, or its inlet valve was closed, the pressure would increase until the 1.5 psid[2] check valve cracked to divert the water into tank B. Once tank B was full, the process would repeat with tank C and tank D. Each tank had an inlet and an outlet valve that could be opened or closed selectively to use water, but the outlet valve of tank A was normally kept closed in order to isolate the treated water from the less clean water in the other tanks. Each fuel cell issued typically 25 pounds of potable water per hour. This was much more than could be stored. For this reason, the crew had periodically to dump the excess overboard. Each tank had a dedicated line and the water dump system used a thermostatically heated nozzle. If the supply water dump failed, or if a mission-specific payload constraint prohibited its use, then ullage in the tanks of the flash evaporator system could be filled. Normally, the plan called for a water dump every twelve hours from tank A and B.[3] Tank C was always kept full for contingency purposes, in particular to guarantee that there would be sufficient water available for the FES in a contingency deorbit.

In addition to the supply water system, the Orbiter had a waste water system with a single tank pressurized by gaseous nitrogen. This tank was identical to those of the supply water system but stored humidity condensate and crew liquid waste (urine) in a sanitary manner until it could be disposed. A dedicated dump line was provided for dumping the contents overboard. The normal procedure was to dump this tank when its contents reached 80 per cent. A crosstie between the dump line of the waste water tank and the dump line of the supply water tanks provided redundancy in the event of one dump line failing. If the waste tank could not be dumped overboard, a collapsible contingency waste water collection bag provided additional ullage. On flights to the ISS, this bag was used to collect humidity condensate on board the Orbiter in order to minimize the number of waste dumps whilst docked to the station, since these would engage its attitude control system and pollute the microgravity environment of the scientific experiments.

WASTE COLLECTOR SYSTEM

"How do you go to the bathroom in space?" is probably one the top three questions asked of astronauts. On the Shuttle this was fairly simple because it carried a toilet similar to those used on Earth. But this waste collector system (WCS) was designed for a weightless environment, which introduced a variety of problems. Henry Pohl, technical director of the Johnson Space Center from 1986 to 1993, explained the issue this way, "To begin to understand the challenges of operating without gravity,

---

[2] The term "psid" means a differential or "delta" pressure across a boundary, in this case of 1.5 psi.
[3] Unless mission-specific payload constraints forced to change this interval.

Heart and lung of the Orbiter

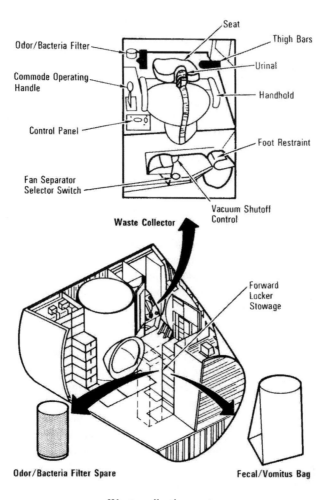

Waste collection system.

imagine removing the commode from your bathroom floor and bolting it to the ceiling, then try to use it. You would then have a measure of the challenges facing NASA." The result was sufficiently small to fit into a 29-inch-wide compartment on the middeck of the cabin immediately aft of the crew ingress/egress side hatch. The system integrated a commode and urinal that served either sex. The commode was 27 by 27 by 29 inches and was used in the style of a conventional toilet. The seat was a contoured, hard plastic material that provided for proper positioning and sealing in order to minimize air leakage. The fecal matter was sucked into a container via a 4-inch-diameter opening in the seat by flowing cabin air through a number of holes beneath the seat at a rate of 30 cubic feet per minute. It was then deposited into a porous bag liner and the air was drawn through the hydrophobic material by a fan separator. After passing through an odor/bacteria filter, this air was fed back into the cabin. The role of the hydrophobic material was to prevent free liquid and bacteria from leaving the storage container. The urinal was simply a funnel with a hose. The fan of the commode was used to create the air flow necessary to transport the urine to the fan separator, where centrifugal force caused the liquid to separate and pass through dual check valves into the waste water tank. The separated air was passed through the odor/bacteria filter and returned to the cabin.

To use the urinal, the astronaut had to adopt a stable position. A bar at the base of the commode had two flexible cylindrical pads on a shaft that could be adjusted to various heights by a pair of locking levers that rotated 90 degrees counterclockwise. The user slipped his or her feet beneath this toe bar to remain in place. After this, the astronaut activated the fan separator and used the urinal. Afterwards the funnel had to be sanitized ready for use by the next person.

Defecation was a lengthier process. First, the astronaut had to adopt the correct position on the commode. For the first four Shuttle flights a seat belt and a fixed foot support were provided, but when these proved inadequate for reliable positioning and restraint other options were tested aboard STS-5. It was decided to secure the feet by an adjustable platform with detachable Velcro straps. The astronaut had to wrap the straps crosswise over each foot and secure them around the back. To accommodate people of different heights, the footrest was adjusted by a lever to various angles and heights. To sit in place, the astronaut placed a thigh bar over their legs. This exerted a preloaded force of about 10 pounds on each thigh. Handholds, some integrated into the top cover of the commode, were provided to assist in positioning. Simple as this might appear, crew members on the early flights reported considerable difficulties in achieving the proper position. As a result, a special training aid was designed. Three-time Shuttle flier Mike Mullane humorously describes this, "Well, the Shuttle toilet, for solid-waste collection, has a very small opening on the top. It has to be small for the air flow to come in from 360 degrees around, in toward the center, and it's being pulled by a fan down inside of the toilet to draw the waste away from your body. To use this thing effectively, aim is critical, because that opening is so small. So NASA built this toilet trainer. Inside it is an upward-pointing television camera and there's a television in front of the toilet. You sit there, looking at unmentionable things on that TV screen, and you wiggle around until you get a bull's eye. And then you memorize where your buttocks and thighs are in relation to the landmarks of the seat so that up in space you can get a bull's eye every time."

Having found the correct position, the astronaut had to raise a handle and wait for 15 seconds while the storage container equalized its pressure with the cabin. Then the handle was pushed forward to open the hole of the commode. Finally, the astronaut could execute the physiological task. Then it was necessary to close the hole in the commode and open a valve to allow the vacuum of space to dry the solid waste in the storage tank. Clearly, going to the toilet was not something that could be done in a hurry!

Along with the toilet trainer, another funny aspect of the WCS was the manner in which the system was developed and tested. As Mullane relates, "Because they knew that they were going to have to address female waste collection, which they'd never addressed before in the history of the program, they had these nurses sit in the back of a KC-135 drinking gallons of iced tea and going up and doing these parabolas and urinating into various designs they were trying – while photographers filmed them so engineers could better understand flow separation and collection and all of this." And as Mullane says, this issue was also addressed in another unusual way, "I heard that at JSC, because the engineers needed data on urine separation from females, they put a camera in one of the toilets in a building there, put a sign on it telling women that if they wanted to volunteer for science, they could use this, and their private parts were going to be filmed while they were urinating." The Shuttle program relied on many unsung heroes. Mullane recalls "a lieutenant that volunteered, male, for solid-waste collection. When he had to go, he picked up the phone and said 'I gotta go', and they scrambled into the airplane and went up there. Here's this guy doing his thing on this toilet while they're doing this zero gravity [parabola], trying to make sure that it was going to work."

Despite the inconvenience, a faulty toilet was not considered sufficient reason for an emergency return to Earth. If a malfunction occurred, a variety of contingency waste collection devices were available on board. For example, fecal collection could be achieved by using Apollo fecal bags which were stowed inside the commode. For urination there was a urine collection device (UCD) for men and an absorbent similar to a diaper for women. Interestingly, these were added only after the maiden flight of *Discovery* as STS-41D in August 1984. Mission specialist Steve Hawley was on his first flight and says, "The ground called us and told us to terminate the supply water dump because they had seen some temperature funnies." An inspection using a TV camera on the robotic arm revealed the presence of an icicle at the exit of the water dump nozzle. Mission Control was worried that the icicle might pose a hazard during re-entry, because if it broke off it could hit either an OMS pod or the vertical tail. It was decided to attempt to detach the icicle using the robotic arm. In case that did not work, two crew members would suit up for a contingency spacewalk that would have been challenging because it had never been envisaged that astronauts would work on that part of the vehicle. Although the arm easily detached the icicle, Mission Control banned further waste water dumps as a precaution against a recurrence. As Hawley explains, "They had done a calculation to figure out how much volume we had left in the waste water tank, and how much of the volume would be used up by the normal condensation from the air that the humidity separator puts into the waste tank, and all that, and they'd concluded that that was about all the room they had left." This meant trouble, "We couldn't use the toilet anymore because there was no room in the tank for the liquid waste. So that ended up being a problem."

Fortunately, human ingenuity is boundless. As Hawley notes, "We did have the old Apollo bags, which are basically just hard plastic. They were really designed for defecation. There is nothing absorbent in them. Actually, the WCS was still usable. It was the liquid waste that we didn't have any place to store." Using an Apollo bag for urination proved to be quite difficult, as Mullane, who was also on board, points out, "The first thing we did was we tried to pee in a plastic bag [but the] urine would hit the bottom of the bag and splash back out, and you would have urine floating around, or you'd be trying to trap it. We figured out that if you put articles of clothing in the bottom of the bags – socks worked great, but any other dirty clothes would work too – and you urinate, the wicking action would still work fine in weightlessness." This required concentration. As Mullane explains, "But you had to regulate your bladder, your flow rate of your urine to not exceed the wicking capacity, because if you did that then it would start splashing again. So you had to be careful, regulate your urine flow rate to make sure it didn't exceed the wicking capacity of whatever it was that you had down in there." This worked quite well, but it required even more attention towards the end of the micturition, since as Mullane says, "As your pressure dropped off, a big ball of urine would stay on you. So then you had to use tissue to mop that off of you. That was a lot fun, let me tell you." The astronauts were able to complete the STS-41D mission successfully, and thereafter every Orbiter carried contingency urine devices.

FIRE DETECTION AND SUPPRESION SYSTEM

A fire is one of the most dreaded situations faced by astronauts in space. Apart from the damage to equipment, a fire would rapidly deplete the breathable oxygen and the smoke would impede recovery actions. The Orbiter had smoke sensors in strategic locations around the cabin and the avionics bays, and there were Halon extinguisher bottles for fire suppression. Fortunately, no Shuttle crew ever had to deal with such an emergency.[4]

The fire detection sensors were active devices which continuously sampled the ambient air for the presence of submicron pyrolitic matter associated with the early (incipient) stage of fire. If either the smoke concentration or smoke increase rate were out of limits, all four master alarm lights would illuminate and a siren would sound off. At this point, two actions were available to the crew. If the system showed fire in one of the three avionics bays, they had to activate the fire extinguisher in that bay to discharge its contents. If fire was in the cabin the crew would use three portable fire extinguishers, two stowed on the middeck and the other on the flight deck, each of which had a nozzle tapered to fit into hole ports in the displays and control panels in order to extinguish a fire in the volume immediately behind a panel.

For a fire during ascent or re-entry, while the crew were suited and in their seats, they would close their visors and activate the suit oxygen before taking steps to deal with the fire. For a fire on-orbit, their first action would be to protect themselves by putting on quick-don masks.

With the fire under control, the WCS charcoal filter, the ATCO, and the LiOH canisters would all be used to cleanse the cabin atmosphere of combustion products. If the crew were unable to purify the atmosphere to a safe level, then an early deorbit would become necessary because the nitrogen supply was insufficient to hold the

---

[4] The worst fire in space occurred aboard the Mir space station in 1997, when NASA astronaut Jerry Linenger was one of six crew members present.

ELECTRICAL POWER SYSTEM

Fuel cells: the Orbiter's power plant

The Gemini capsules introduced the production of electricity by means of small but powerful power plants called fuel cells. This technology was refined by Apollo, then used by the Shuttle.

Fuel cells generate electricity by a chemical reaction between oxygen and a fuel, in this case hydrogen.[5] In each cell, liquid hydrogen and oxygen are mixed together to initiate a chemical reaction with an electrolyte of potassium hydroxide (KOH). On the H_2 side of the cell, the hydrogen reacts with hydroxide ions (OH^-) to produce water and electrons ($2H_2 + 4OH \rightarrow 4H_2O + 4e$). These electrons (e) leave the cell as current to power the various electrical loads of the Orbiter. As the electrons return to the cell from the loads, they react with oxygen and water on the O_2 side of the cell to produce new OH^- ions to replenish those consumed in the hydrogen reaction ($O_2 + 2H_2O + 4e \rightarrow 4OH$). The net result is that two hydrogen molecules and one oxygen molecule are consumed to produce a flow of four electrons, two water molecules and a certain amount of heat. The number of electrons that leave the cell determines the amount of reactants consumed by the fuel cell. This also provides a way of checking whether there are any leaks in the distribution system. If a fuel cell is shut down or no loads are applied to it, a continuing consumption of the reactants from the supplying cryo tanks is clear evidence of a leak.

Each Orbiter had three fuel cells, all of them located in the forward portion of the mid-fuselage beneath the payload bay. As with many other Orbiter subsystems, the fuel cells were designed to be reusable with minimal maintenance between flights. In this case, they had a useful life of around 2,000 hours of on-line service.

Each fuel cell was 14 inches high, 15 inches wide and 40 inches long, which were quite modest dimensions considering that each could supply up to 10 kW maximum continuous power in nominal situations, 12 kW continuous in off-nominal situations (such as if one or more fuel cells failed) and a peak of 16 kW for ten minutes. A fuel cell was split into a power section in which the chemical reactions took place and an accessory section that monitored the reactant flows, removed waste heat and water, and controlled the temperature. The power section consisted of 96 cells grouped into three substacks of 32 cells. Manifolds ran the length of each substack, distributing the reactants and coolant fluid. Within each cell, an oxygen electrode (cathode) and a hydrogen electrode (anode) separated by a porous matrix saturated with potassium

---

[5] Often cryogenic oxygen and hydrogen are referred to as "cryos". As a result, flight controllers, instructors and crews refer to cryo systems, cryo tanks, and so on.

Electrical power system 227

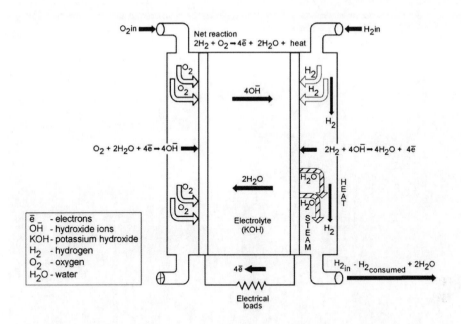

Electrochemical cell. (Courtesy of www.nasaspaceflight.com)

Fuel cell.

228 Heart and lung of the Orbiter

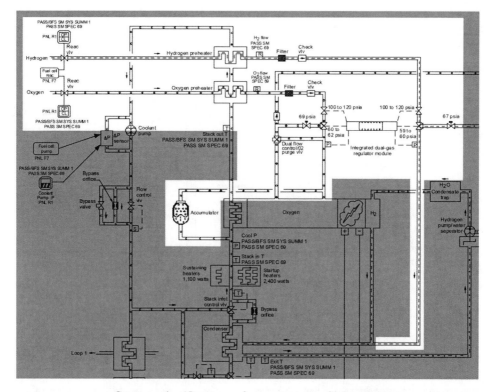

Cryo supply. (Courtesy of www.nasaspaceflight.com)

hydroxide electrolyte facilitated the reaction. Each cell produced electric current at a voltage of 1.15 V with no load applied, or 0.9 V to 1.0 V under normal loads. With the cells connected in series, a substack produced a total voltage ranging from 28 V (the nominal voltage of the on board electric loads) to 32 V; this being the number of cells times the 0.9 to 1.0 V that each cell was able to provide.

The water created as a byproduct of the fuel cell reaction had to be removed from the stack, since otherwise the cell would very quickly become saturated, impairing the efficiency of the reaction. It was directed to tanks for use by the ECLSS, thereby sparing the need to carry large amounts of water into orbit at the expense of mission performance and payload. In order to remove the water, more hydrogen was added so that the portion that was not consumed in the reaction served as a carrier gas, picking up and removing the water vapor. This saturated gas was routed through a condenser, where the temperature of the mixture was reduced sufficiently for the water to form droplets. The mixture of hydrogen and liquid water was then directed to a hydrogen pump/water separator, where a centrifugal separator extracted the liquid water and pressure-fed it into the water storage tanks in the lower deck of the cabin. The first tank to be filled was the tank A, then tank B and so on until all four tanks were full. If all the water tanks were full, or there was a line blockage, a 45-psi relief valve vented the water from the fuel cells overboard. As in the case of many

Electrical power system 229

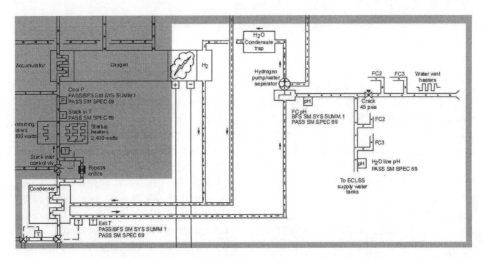

Water loop removal. (Courtesy of www.nasaspaceflight.com)

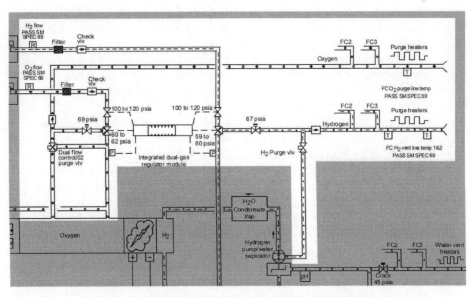

Fuel cell purge system. (Courtesy of www.nasaspaceflight.com)

other vents, this line and its nozzle were equipped with two thermostatically activated heaters to avoid the danger of ice formation. The separated hydrogen was made to flow through a water condensate trap to catch any stray water droplets that may have made it through the pump, and then it was fed back into the stack.

Before reaching the fuel cell, each reactant was first warmed up to a temperature of approximately 200°F and passed through an integrated dual gas regulator module where the oxygen and hydrogen pressures were reduced to values suitable for a fuel cell. At the exit of the regulator module, the oxygen had to have a pressure no higher than 4.5 to 6 psi measured against the hydrogen. The reason for this was that with the hydrogen at higher pressure, the KOH electrolyte in the cells would be forced out of its storage plates, thereby permitting the two reactants to mix (crossover) and create a potentially explosive state. The oxygen feedline was also connected to the coolant circuit of the fuel cell by means of an accumulator, since the oxygen and the coolant had to be at the same pressure to avoid structural damage to the fuel cell. On leaving the pressure regulator, the oxygen entered the stack through two ports. Unlike the hydrogen, the oxygen and was completed consumed.

A closed loop through which fluorinate hydrocarbon flowed, served as coolant to remove the heat generated in the stack. The warm coolant that emerged was used for warming the cryogenic reactants on their way to the fuel cell. It was then sent to two heat exchangers to shed its heat to the Freon loops of the ECLSS. Before returning to the fuel cell, the coolant was used in the condenser of the hydrogen/water vapor separator. Finally, it passed through a heater which was activated only when starting up the fuel cell. This heated the coolant fluid so that it could warm the fuel cell in preparation for an efficient start (it had to be able to warm up the reactants). Once the fuel cell was operating, this heater was switched off and the coolant simply flowed through it. The module had a second heater, known as the sustaining heater, that was used if the fuel cell temperature dropped so low as to reduce the power output.

The fuel cells required periodic purging, otherwise contaminants accumulating in and around the porous electrodes would reduce the efficiency of the reaction and the electrical load that could be supported. A purge could be automatically commanded by the GPCs or Mission Control, or be performed manually by the crew. It involved opening valves to allow the reactants to flow through the fuel cell, carrying away the impurities to be vented overboard through a purging line. The purging lines of each fuel cell were manifolded together in a line leading to a purge outlet, which was sized to allow unrestricted flow from only one fuel cell at a time. If more than one fuel cell were to be purged simultaneously, back pressure could build up in the purge line and cause the pressure in the hydrogen line to exceed that of the oxygen line and create a potential crossover situation. Prior to starting the purging sequence, the heaters on the venting line had to be switched on for at least half an hour to ensure that the reactants would not freeze in the lines. After that, each fuel cell was purged for several minutes one after another. This done, the heaters in the purging line were left on for another half an hour to bake out any remaining water vapor.

During normal fuel cell operation, the hydrogen and oxygen were diffusely mixed to generate electricity. The matrix in each cell, which was a fibrous asbestos blotter device, served to contain the KOH electrolyte to limit mixing between the reactants. But after many hours of exposure to the caustic KOH and the heat generated within the cell, a manufacturing flaw in the matrix or an impurity in the matrix fibers could cause a pin hole to develop and permit direct combination of the reactant molecules at the hole. The heat from that localized reaction would cause the hole to burn and enlarge. As the reaction propagated uncontrollably, it would end in an explosion. At the beginning of the Shuttle program there was no means of monitoring the onset of

Electrical power system 231

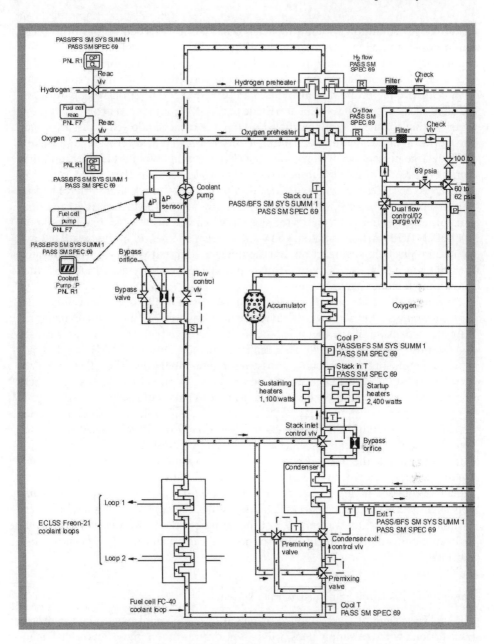

Coolant loop. (Courtesy of www.nasaspaceflight.com)

this crossover condition within a fuel cell, because the degraded output of that cell would be picked up by the remaining healthy cells. This was what happened just two and half hours into the STS-2 mission. The nominal five-day mission was reduced to two days, which was the minimum-time mission permitted by the flight rules. Apart from being disappointing, this also raised an issue that could seriously jeopardize the health of the astronauts. As pilot Joe Engle recalls, "We also had a problem with our water, in that the membrane that failed on the one fuel cell allowed excess hydrogen into our drinking water supply, making it very bubbly. Whenever we'd go to take a drink, a large percentage of the volume was hydrogen bubbles in the water, and they didn't float to the top like bubbles would in a glass here [on Earth] and get rid of themselves, because in zero gravity they don't; they just stay in solution. We had no way to separate those out, so the water had an awful lot of hydrogen in it, and once you got that into your system, it's the same as when you drink Coke real fast and it's still bubbly [with carbon dioxide]; you want to belch and get rid of that gas. That was the natural physiological reaction, but anytime you did that you'd regurgitate water. It wasn't a nice thing, so we didn't drink any water. So we were dehydrated ... when it was time to come back."

Starting with STS-9 in 1983 a third substack was added to each fuel cell in order to produce more electric power because the missions were getting more demanding. With the expansion of the fuel cells, the likelihood of individual cell failures caused by "wear out" or age increased. So a cell performance monitor (CPM) was designed to detect imminent failures such as crossover in individual cells. The CPM compared each half substack voltage and calculated a delta-volts measurement. As all the cells in a substack had to produce the same current, both halves of a substack had to be at approximately identical voltages. A difference of 300 millivolts was defined to be the critical threshold. Early on the STS-83 mission in 1997 the CPM noticed fuel cell #2 go out of limits and restabilize. The available data was insufficient to allow Mission Control to exclude the possibility of a crossover condition, so it was decided to safe the fuel cell and execute a minimum-duration flight. Subsequent testing showed that, instead of one cell having a large degradation, several cells in the same substack had only slightly degraded. If engineers had known this, the mission would not have been terminated. To preclude a recurrence of this situation, the CPM was evolved into the fuel cell monitoring system (FCMS) to monitor performances, cell by cell within a substack.

Power reactants storage and distribution system

The fuel cell reactants were supplied by means of five pairs of oxygen and hydrogen tanks. Along with the supply lines, valves, sensors and so on, these "cryo sets" made up the power reactants storage and distribution (PRSD) system. They were installed beneath the payload bay. Each set was sufficient for about two days of mission time, so the five sets could support a normal mission of nine to ten days. The two types of tank had the same layout of an outer shell with an inner vessel isolated by a vacuum. This enabled the reactants to be stored at supercritical pressures (above 731 psia for oxygen and above 188 psia for hydrogen) and prevented the temperature inside the

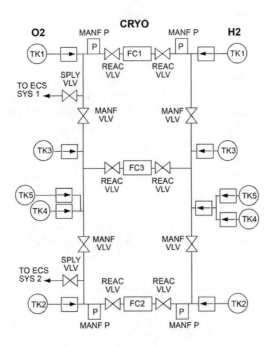

The PRSD system, showing fuel cells (FC), manifolds (MANF), manifold valves (MANF VLV), pressure sensors (P), reactant valves (REAC VLV), supply valves (SPLY VLV) and tanks (TK).

tank from exceeding the boiling point of the reactant (–285°F for oxygen and –420°F for hydrogen). Each tank was fitted with heater assemblies (one for a hydrogen tank and two for an oxygen tank) to warm the liquid and maintain the inner pressure at the same value as the reactants were gradually depleted. This was important, because the reactants were fed into the distribution system solely by the pressures in the tanks. For redundancy, each heater assembly had two identical heater elements, referred to as heater A and heater B.

Each tank had three fluid interfaces: a fill/vent line for loading prior to launch, a relief line in case of overpressure on-orbit, and a supply line to direct the reactant to the fuel cell. The distribution system of the reactants was straightforward. Each set could feed any of the three fuel cells by means of crossover valves, and be isolated from the others to avoid depletion in the event of leaks.

Electrical power distribution and control

When working normally, the fuel cells produced 28 V direct current. This fed all the on board electrical utilities by means of three main buses (MNA, MNB and MNC) with each bus being fed by one fuel cell. For loads needing alternating current there were three other buses (AC1, AC2 and AC3) connected to the main ones. Each main bus was tied to the others by means of tie buses which fed current to a bus whose fuel cell had failed. Loads deemed essential for the flight were powered by three essential buses (ESS 1BC, ESS 2CA and ESS 3AB), which in turn received power from all three main buses. Finally, nine control buses (CNTL AB 1, 2, 3, CNTL BC 1, 2, 3 and CNTL CA 1, 2, 3) supplied power to displays and control panel switches on the flight deck and the middeck. Some ISS missions carried payloads that required direct current at 120 V, so an assembly power converter unit was installed. It was a simple step-up converter that boosted the voltage. Later in the program, Orbiters gained a Station/Shuttle power transfer system that allowed the ISS to transfer current (after reducing the voltage to 28 V, because the ISS operated a direct current at 120 V) to a docked Orbiter to reduce its rate of cryo consumption and permit longer missions.

Operations

During prelaunch operations, the fuel cell reactants were supplied by GSE to ensure a full load in the on board system at liftoff. The vent lines were left open during tank loading to enable vapors to escape, and then closed. The reactants were supplied at a greater pressure than could be achieved by the heaters, some 960 psia for the oxygen and 280 psia for the hydrogen.

By the time that the crew entered in the Orbiter, the fuel cells would be running on GSE-supplied reactants and the electrical load was shared with power provided by the ground. At T-3 minutes 30 seconds, the load was transferred to the on board fuel cells, which started to increase their production of energy. If the cryo tank pressures fell excessively during the ascent, the software would switch on one of the two heater elements (A or B). Only one of the redundant units was used in order to minimize the additional load on the electrical supply during this power hungry phase of the flight. Doing this would prevent the voltages at the SRB interface from dropping too low and prompting the loss of SRB thrust vectoring control. On-orbit, Mission Control would issue daily instructions to the crew specifying valves positions and tank heater activations based on the consumption the previous day. For re-entry, no actions were required for the cryo subsystem.

EXTENDED DURATION ORBITER KIT

When *Columbia* lifted off in November 1983 for STS-9, this marked the start of a new era in space research because the Orbiter carried a Spacelab in its payload bay. This was a European module equipped to enable astronauts to perform a wide variety of experiments. It was accessed via a tunnel linked to the middeck airlock and greatly increased the volume available to the astronauts. Several missions were flown with a number of Spacelab configurations, but it soon became apparent that the potential of the laboratory and other suites of instruments carried in the payload bay could not be fully exploited by an Orbiter whose resources were limited to a maximum duration of ten days.

Extended duration Orbiter kit

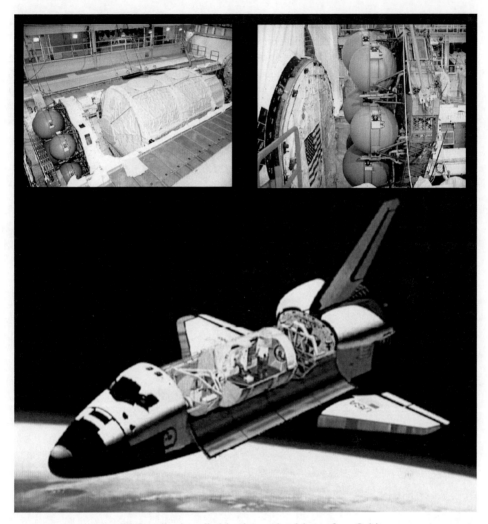

An EDO pallet installed in the payload bay of an Orbiter.

Rockwell developed the extended duration orbiter (EDO) kit to enable an Orbiter to stay in space for up to 16 days, plus a two-day contingency capability. By enabling a single mission to undertake more scientific tasks, this would reduce the number of flights needed to address all of the desired scientific investigations. This would have the beneficial effect of simultaneously cutting risks, costs and vehicle wear and tear. The most visible manifestation of the EDO kit was a set of cryogenic reactant tanks mounted on a pallet that was located at the rear of the payload bay to supplement the fuel cells. The pallet weighed 7,000 pounds and held four sets of liquid hydrogen and oxygen tanks, storing a total of 368 pounds of hydrogen at –418°F and 3,125 pounds of oxygen at –285°F.

Extending a mission to more than two weeks imposed a heavy load on the waste management system, so to avoid unpleasant failures an improved WCS was fitted in the Orbiter. It was more comfortable and more sanitary than its predecessor, and had unlimited capacity. EDO missions also required a different way for scrubbing CO_2. Changing the lithium hydroxide canister twice per day on a mission lasting 16 days was not acceptable due to the large number of canisters required. For EDO missions the Orbiter was fitted with a regenerable carbon dioxide removal system (RCRS) in which the CO_2 was removed by passing cabin air through one of two identical solid amine resin beds. This resin was a polyethylenimine sorbent coating on a porous polymeric substrate. Upon exposure to CO_2-laden air, the resin combined with water vapor in the air to form a hydrated amine that reacted with the CO_2 to form a weak bicarbonate bond. Water was required for the process, because dry amine is unable to react with the carbon directly. While one bed was adsorbing CO_2, the other bed was being regenerated by thermal treatment and vacuum venting. This latter requirement precluded using the RCRS for ascent or re-entry, so LiOH canisters were used during these mission phases.

8

The Orbiter's skin: the thermal protection system

INTRODUCTION

When the first concept studies for a reusable space transportation system began, the Apollo program was still in full swing and therefore it was perfectly natural to look at what this technology had to offer as thermal shielding. The Apollo capsule, like its Mercury and Gemini predecessors, relied on a variable-thickness graphite-epoxy heat shield installed immediately beneath the capsule; which is to say, the part that would have to cope with the thermal stress of re-entering the atmosphere. This shield was designed to be eaten away in a controlled manner without allowing the hot plasma to come into direct contact with the primary structure or letting the temperature of the structure exceed the limit above which it would start to creep, soften and finally melt. Thermal shields that work in this way are called ablatives, and when entering the atmosphere at very high speed they decompose to create an outflow of hot gas that carries away the heat. The ablative leaves behind a porous carbon-char with a low thermal conductivity that prevents the ambient heat from reaching the underlying primary structure.

All Mercury, Gemini and Apollo flights came back to Earth without difficulty, validating the soundness of such thermal protection. This prompted NASA engineers to consider developing a similar shield for the Shuttle Orbiter. However, despite is simplicity and efficiency, an ablative is not reusable. Since each Orbiter was to have a design life of at least 100 missions, it was very soon decided that an ablative shield would not be feasible. Quite apart from the issue of reusability, it would have been difficult to install an ablator on such a large surface and the weight penalty would have been enormous.

Attention therefore switched to a so-called "hot structure" in which the primary components are made using a material capable of withstanding high temperatures without losing its load-carrying capabilities. But even such a structure would not be able to be directly exposed to the heat of re-entry. It would still require a thermal protection system in the form of shingle-like plates installed in the style of the tiles on the roof of a building. The materials chosen for these shingles had to lack heat-sinking capability, in order that the absorbed heat could be re-radiated to give rise to an equilibrium state between the heat absorbed and the heat lost by radiation. Heat would still be transferred to the primary structure, but very slowly thanks both to the low heat-sink characteristics of the material and to an insulator installed beneath the shingles in order to slow the inward flow of heat. In this way, by the time heat started to penetrate the primary structure the vehicle would have passed through the region of peak heating and the structure would have remained well within the limits of its materials. Other requirements were that the shingles be able to expand and contract to accommodate the temperature changes without working loose, and withstand the aerodynamic loads and prevent flutter. Unlike an ablative shield, the shingles would not be eaten up during re-entry, adding to the reusability of the system. Furthermore, maintainability was guaranteed because a damaged shingle could easily be replaced during the turnaround between missions.

Several hot structures had already been developed, such as the well-known X-20 Dyna-Soar, a USAF project intended to develop a reusable spaceplane for strategic missions in low orbit. Research and testing resulted in a primary structure made of René 41, a nickel-chromium alloy that also contained cobalt and molybdenum, with protective shingles of columbium alloy. At that time, René 41 was mainly used for the turbines of jet engines because it could withstand temperatures up to 1,800°F. The columbium alloy shingles could deal with temperatures at least 1,000 degrees higher. The nose cap and the leading edges of the wing of the X-20 would respectively have been made of molybdenum alloy in order to resist temperatures up to 3000° F and Zirconia to survive 4,300°F.

Although the concept of a hot structure was eagerly adopted by the companies that were competing to secure the contract to develop the Shuttle, it soon became evident that shingles posed a major flaw. The alloy used was very susceptible to oxidation at high temperatures, with resultant embrittlement. For this reason, special coatings had to be invented along with appropriate methods of application on the exposed surface of the material. However, this meant that, as Maxim Faget, who played a key role in the early definition of the Shuttle, put it, "The least little scratch in the coating, the shingle would be destroyed during re-entry." Or as Charles Donlan, Shuttle program manager at NASA headquarters said in 1983, "The slightest scratch and you are in trouble."

As the first doubts about using hot structures arose, a new approach to the thermal protection issue became a prospect: a reusable surface insulation (RSI) in the form of tiles. Just like a shingle, a tile would re-radiate the incoming heat to create an equilibrium between incoming and outgoing heat fluxes with such efficiency that the underlying primary structure would be subjected only to a very low thermal load. In fact, it would be possible to build the entire structure of the Orbiter using aluminum, a metal with excellent mechanical properties but a very low heat resistance. Like a shingle, a tile would not be consumed during re-entry and could be replaced when damaged, thereby providing both maintainability and reusability. Further, because the material was already oxidized there was no need to apply a protective coating. In addition, an RSI promised to be lightweight. NASA therefore had little option than to use it as the thermal protection system for the Orbiter.

REUSABLE SURFACE INSULATION TILES

When the final decision was made to use RSI materials, Lockheed was already well ahead in studying the characteristics of this new material, as well as in experimenting with methods for efficient fabrication. In fact, these studies started as early as 1957 and the company investigated a broad range of reusable surface insulation materials, all-silica systems, zirconium compounds, alumina and aluminum silicates (mullites). The results clearly showed that an all-silica system was the best candidate for a fully reusable insulator. This led to the creation of a 15 pounds per cubic foot proprietary material called Li-500 that (in laboratory quantities) achieved the highly desirable theoretical properties of an all-silica system by strict control of material and process. In 1970 Lockheed established a pilot plant to perfect the process and develop the techniques for full production to meet the Shuttle's needs. Its all-silica Li-500 was tested extensively at various NASA centers and Air Force laboratories, and found to meet Shuttle mission and life requirements. Improvements in material production resulted in a new all-silica material called Li-900 that had a density of just 9 pounds per cubic foot and an amorphous fiber silica of an astonishing 99.8 per cent purity. Such high purity was necessary because of the process of devitrification, in which the silica fibers undergo a phase change at high temperatures, transforming them from glassy to crystalline. Subsequent to devitrification, when the material cools down, the fibers undergo a second phase change that produces a sudden shrinkage and imposes large tensile stresses. As a result, the fibers produce internal cracks with consequent degradation of the material properties. A very high purity would keep the fibers in a glassy state, avoiding devitrification and preserving the ability of such a material to resist thermal stress and thermal shock.

Two different kinds of tile were manufactured from the same material and called high temperature reusable surface insulation (HRSI) and low temperature reusable surface insulation (LRSI), with the former being black and the latter white. Wendall D. Emde, a Rockwell supervisor for the TPS development, explains the difference, "The reason we had the LRSI versus HRSI was a thermal consideration. Part of the top side of the vehicle didn't get as hot [during re-entry, so those tiles] were thinner. But when you're on-orbit, sometimes you're looking at minus 250 degrees. When you're looking at space, a black coating didn't offer any heat transfer reflection. They ended up with a white coating on the thinner tiles, which helped prevent heat transfer in the cold condition. That's why you use white tiles from about 700°F to 1,700°F." White tiles were used for those areas of the Orbiter that were mainly exposed to the cold of space and, being thin because they did not have to cope with the intense heat of re-entry, needed a high reflectivity in order to reflect the heat of the Sun. However, as Emde continued, the HRSI tiles "didn't need that thermal reflection because they were thick enough to handle minus 250, but they needed a black coating to reflect the heat during re-entry. Black was much better for reflection of heat."

To better understand this concept it is necessary to remember that silica fiber by itself is white and has a low thermal emissivity, making it a poor radiator of heat. It would be necessary to develop thicker and heavier tiles to accommodate extremely high surface temperatures. But applying a coating that would turn it black in order to achieve a high emissivity would enable it to radiate heat efficiently and remain cool. The selected coating was a borosilicate glass, with silicon carbide to further raise the emissivity. This was named reaction-cured glass. The fact that silica and glass were both silicon dioxide based ensured a match of the coefficients of thermal expansion of the coating and substrate, preventing the coating from cracking under the thermal stress of re-entry. In fact, the glass coating could soften at very high temperatures and heal minor nicks or scratches on the surface of the tile, surviving repeated heating cycles to 2,500°F and thereby offering the essential reusability sought by the Shuttle design.

Tile manufacturing process

Since the white and black tiles were the same material, the fabrication process was basically the same, with just some differences in the finishing. It all started with the supply of "bundles of very micron size fibers of silica," Emde explains. "Since silica is a very high temperature mineral and melts around 3,000°F ... in order to work with it you have to add impurities to it, like oxides of various types of minerals that melt at lower temperatures ... Lockheed would chop up the fibers into a slurry – a mix of pure water and these slight impurities. It's like a blender you have at home, only larger. They'd pour it in a mold roughly 13 by 13 by 13 inches ... After a mold was made it was cured, water dried out. After a couple of tries, they ended up with a very long tunnel oven of about 40 feet. It was heated to around 2,400°F. They'd run this block of fibers through ... During the processing they shrank a little bit. Due to the impurities in the mix, the fibers contacted and melded with each other. So now you had some physical structure to it. Not too strong ... but it had very, very good thermal conductivity."

After being baked in an oven and trimmed of any distortions using a band saw, the block of Li-900 could be cut into smaller cubes using milling machines driven by computers to produce individual tiles. In order not to impair the final coating process these machines used no lubricating oil. For the production of HRSI tiles an additional step was to spray on the top and sides of the tile a mixture of powdered tetrasilicide and borosilicate glass with a liquid carrier. The tile was then heated to 2,300°F in an oven to create a black, waterproof glossy coating that would enable it to withstand adverse weather conditions while the Orbiter was on the launch pad and provide a high radiating surface during re-entry. A waterproof coating was also applied to the white LRSI tiles to protect them. The high temperatures of re-entry would partially melt the coating, filling in the micro cracks which developed on the surface of a tile and preventing erosion. As Emde explains, the coating "doesn't go all the way to the bottom. That design was to allow the tile to vent, because if you had a coating that went to the bottom of the tile it may not have vented in space." This uncoated area was called the "breather area" because it allowed air to vent out during the ascent in order to preclude a loss of coating due to pressure equalization.

Another protection to the integrity of the tiles was provided by waterproofing. As

Terrence R. White, who had a great deal of experience with the TPS, explains, "The tiles are 94 per cent air ... If a tile is not waterproofed, water gets inside and it picks up a lot of weight. Also, on ascent that water expands and actually damages the tile." Due to this, they soon found out that the material used for waterproofing got burned off during every single re-entry. For this reason, after each mission every single tile was individually waterproofed using a micro-hole on the surface into which a needle could be inserted to inject the waterproofing substance.

Initially the machining of the HRSI tiles was accomplished using a straight-sided, diamond-embedded cylindrical cutter one inch in diameter. Although this produced a tile whose pre-coated dimensions were within tolerances, the post-coated or glazed tile assumed a pillow shape as a result of excessive shrinkage at the top corners. This was because the upper surfaces received more radiant heat than the lower surface, which sat on a silica plate, resulting in more shrinkage at the upper surfaces than at the lower or inner mold line surface. This meant that tiles larger than or equal to two inches thick would exceed the specification allowance just due to the coating glazing cycle. As Emde recalls, "We said, 'Why don't we just make a cutting tool that has a taper to it so that it cuts the bottom a little shorter than the top? Then when we run it through the furnace it'll have virtually a straight side.'" The tile shrinkage was then eliminated using a conical tool with a 0.25 degree cone half angle to cut the tile prior to the coating application process. In this way, the tile would be cut with the upper surface larger than the lower one. Shrinkage of the upper surface during the coating restored the tile to a shape with all the vertical sides straight.

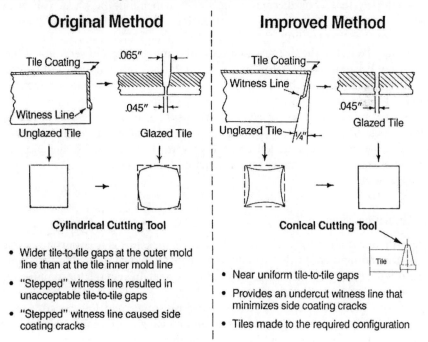

Cutting methodology for HRSI tiles.

The Orbiter's skin: the thermal protection system

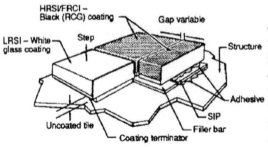

Tile installation configuration.

This process was the methodology used by Lockheed in manufacturing all the thousands of tiles installed on each Orbiter for delivery to NASA. After each flight it was always necessary to replace tiles that had suffered damage. The manufacture of a specific tile was accomplished in a facility at the Kennedy Space Center. The same fabrication process was used, but on a much smaller production scale.

Originally NASA intended that all tiles should have the same shape, and differ in thickness depending upon the position on the Orbiter. In this way, the manufacturing process would be simple and fast and it would be possible to hold a stock of spare tiles. But it was soon realized that this would produce a faceted surface on the curved face of the Orbiter that would promote the onset of turbulence in the airflow with a higher rate of heating that would have necessitated thicker and hence heavier tiles. A better option was to machine every single tile to closely match the Orbiter's contour. This meant each tile had to have a unique size and shape. As Emde explains, "The problem was that the structure had the outline of an airplane, but the requirements to keep the airplane cool would say that your thermal protection – depending on where you were on vehicle – had to be some thickness and some shape. The only thing that was common with the tile, was that the bottom was supposed to match the shape of the outside of the vehicle. The rest of the tile might be very thick in the nose and very thin at the top of the wing. It had its own contour based upon the aerothermal and thermal heating data. The key thing was that we had to design to keep the aluminum below 350 degrees. If you went over that temperature, it would weaken the structure for the design of 100 missions." Having each tile only as thick as was necessary at that point on the vehicle minimized the weight of the thermal protection system, but the fact that every tile was unique added considerably to the workload and hence to the turnaround costs. It was simpler to manufacture replacement tiles as necessary than to keep a stock of spares.

Tile installation configuration

Along with a manufacturing methodology for the tiles, it was necessary to develop a means of bonding them to the structure of the Orbiter. The difficulties encountered in researching this problem resulted in an important change in the tile manufacturing process.

Owing to their extremely high temperature resistance and low density, tiles are most useful for a spacecraft where protection and weight are critical factors. But as Emde explains, "The aluminum structure stretched a lot during re-entry ... If you just bonded the tile to the structure, the tile virtually has no thermal expansion at all.

You can heat it all over the place, it doesn't go anywhere ... It would just shear itself off." Hence an intermediate layer called a strain isolation pad (SIP) was installed beneath each tile to prevent stress failure by isolating it from the deflections, expansions and acoustic excitation of the Orbiter's structure.

The SIP material chosen was Nomex, a flame resistant material, that "looks like tennis ball fuzz" as Emde puts it. "They would comb it out into a wide rug and make a very thin layer. These layers would be overlapped several times and then the result run through a reverse machine. Imagine a whole bunch of torture needles with barbs going ... downward. It would push the fibers down and give them some strength in the through-the-thickness direction. The top side had to be good for about 550°F to keep the structure at 350°F on the side bonded to the aluminum." A room temperature vulcanizing (RTV) methylphenyl silicone adhesive was used for bonding the SIP pad to the structure, as well as bonding the SIP to the tile. Selection of this material was based upon the fact that it retains resiliency to temperatures as low as –170°F, so that it does not become brittle when in shadow on-orbit. The silicon adhesive was applied to the structure in a layer approximately 0.008 inch thick. Such a very thin bond line reduced weight and minimized thermal expansion at a temperature of 500°F during re-entry and thermal contraction at temperatures below –170°F in shadow on-orbit. Coating the lightweight tiles with the very viscous adhesive created many problems, particularly because the thickness requirement for this adhesive layer was 0.0075 +/– 0.001 inch. A solution to this problem was to hold down the lightweight, fragile tile and flexible SIP with a metal wire grid or a nylon net of specific thickness, and then coat them with the RTV adhesive using a short nap roller. This provided a uniform coating without edge smearing in a relatively brief time, and cleanup and handling problems encountered early in the adhesive application to the tile-SIP assembly were also reduced. Tiles coated with RTV using this technique could be rapidly applied within the 0.001-inch thickness

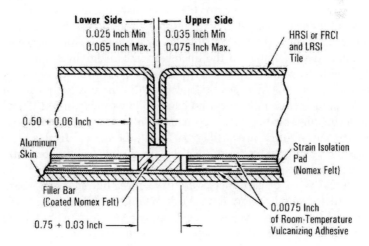

SIP and filler bar arrangement.

tolerance. The tile/SIP bond was applied at room temperature under pressure applied by vacuum bags.

Each SIP pad was sized 0.5 inch smaller than the tile planform on each side, so that the edges of the pad and the RTV adhesive would not degrade from the higher temperatures encountered in the tile-to-tile gap. These gaps were necessary in order to prevent tile-to-tile contact due to thermal expansion of the underlying structure. But the presence of these gaps posed other thermal problems that were related to the orientation in which the tiles had to be installed on the vehicle. As Emde says, "We found out we didn't want to band the tiles in line with the air flow. We had to put them catawampus to the flow so that you didn't have in-line cracks that would cause gap heating. It's like a river, it runs slow in a big wide basin. With the same amount of flow, you narrow the gap and it goes fast. And fast is hotter. So we had to change the design to turn the tiles wherever we could to reduce in-line gap heating." Without this attention to detail, the tile-to-tile gaps were channels by which the plasma could penetrate to the underlying structure. One means of reducing the amount of plasma that could slip into the gaps was to diverge them as far as possible from the direction of the plasma flow. However, this in itself was insufficient. "We found out that this gap heating ... still got hot in the crevice," Emde points out. To completely protect the structure, it was decided to use stripes of SIP material called "filler bars". In simple terms, a filler bar was a strip of waterproofed Nomex that was placed at the bottom of a tile-to-tile gap, between the SIP pads, having a thickness about 10 per cent greater than the SIP pad so that it could serve as a seal. They were normally bonded by mean of RTV adhesive to the structure, but in areas subjected to high aerodynamic shocks they were also bonded to the tiles in order to avoid their being ripped out by the high aerodynamic loads.

Much to the engineers' dismay, the problem persisted despite the presence of the filler bars. Robert C. Ried worked on the design of the TPS. "On an exterior surface, the tiles can take extremely high heating rates and the way they get rid of the energy is to radiate it away. In between the two tiles, they just radiate back and forth to each other, so there's no heat lost and they're very poor conductors. So with a little bit of heating between the gaps the temperature goes real high ... and the gas stays real hot." The solution was similar to the filler bar, since it involved filling the spaces between tiles with "gap fillers". As Ried explains, "These were pieces of astroquarz or quartz fabric ... maybe two pieces depending [on the position]. They had to be bonded down in the between the tiles and had to allow for expansion. They were put in the gaps in the high heating areas up near the nose, and on the underside except where the aero guys figured just a filler bar would be okay." In this way, the gap fillers created an additional seal to protect the aluminum structure beneath the gaps by preventing the influx of plasma.

Many different types of gap fillers were developed, but generally speaking they formed two categories: pillow and layer, both of which bonded to the surface of the filler bar beneath. The pillow-type were sufficiently resilient to maintain a seal as the gap between the tiles opened and closed in reaction to structural and thermal stresses. They consisted of an envelope of ceramic fabric stuffed with a resilient ceramic fiber batt, and sometimes with a metal foil added for reinforcement. The

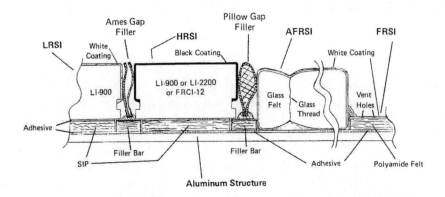

SIP, filler bar and gap filler arrangement.

fabric, which at times was coated with an emittance layer, was sewn together at the ends and bottom with quartz thread. Pillow gap fillers were usually installed in tile-to-tile gaps having a width greater than 0.1 inch and could accommodate a total excursion of 0.04 inch. On the other hand the layer gap fillers (also known as Ames gap filler because it was a NASA-developed material) was used for tile-to-tile gaps having a width of less than 0.1 inch, and usually in the range between 0.003 and 0.06 inch. Various types of Ames gap filler were developed, but basically they were single or multiple plies of ceramic cloth impregnated with one or two RTV silicon-type coatings depending on the temperature to be endured. Unlike the pillow-type of filler the layer-type did not follow the gap movement, and because the RTV in this filler tended to char upon re-entry their useful life was only a given number of flights. Pillow and layer-type gap fillers could also be combined as a "composite gap filler" to increase the thickness of the pillow filler type. It should be noted that the width of the gaps was important not only for the localized overheating and plasma ingestion, but also to avoid tripping the aerodynamic boundary layer too early in the re-entry and imposing higher heat loads due to a premature transition from laminar to turbulent flow.

Tile densification

It was determined from material characterization tests that the insulating tiles had a minimum tensile strength of 13 psi, the SIP had a minimum tensile strength of 26 psi, and the RTV bond had a tensile strength of 400 psi. In the beginning it was believed that the pull-out force acting on the tiles due to aerodynamic loads would not exceed 2 psi. With this 2-psi force in mind, and knowing that each single element of the tile configuration had a good minimum tensile strength, NASA focused its attention on the thermal properties of the insulating shield. But failure to fully investigate the real loads acting on the tiles proved to be a mistake that potentially jeopardized the entire project.

When the installation of tiles began on the first spaceworthy Orbiter, *Columbia*,

246 The Orbiter's skin: the thermal protection system

detailed analysis revealed that only a very small percentage of the thousands of tiles would be subjected to a pull-out force of 2 psi; the majority of tiles would have to withstand much higher forces. And worse, further experimentation showed that the bond between the tile and the SIP was much weaker than expected. In particular, it was determined that an interfacial failure could begin at a pull-out force of just 6 psi. As Thomas L. Moser, one of the engineers, remembers, the reason for this was soon understood, "The strain isolation pad was just a loosely woven felt material ... This loosely woven felt did [the job] beautifully. To keep the felt together, [i.e.] give it some integrity, every [few inches] a stitch was put through [the felt. This was so] it would not come unraveled. Well, when you attach a tile to the SIP, which is attached to the structure, and something tends to pull the tile away from the structure – like aerodynamic loads or vibration – all these little individual stitches acted as a stiff spot ... These little stiff spots would cause a stress concentration. Or it would cause the bond to fail right there. When that happened, being a very [brittle and] fragile material that the silica fiber is, [the failure] would propagate. So once you started a failure, it's like a crack in the windshield of your car. Once those suckers start, they

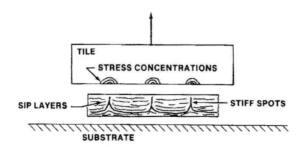

Stress at SIP/tile bonding surface.

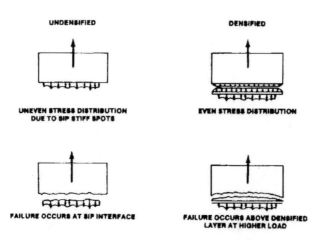

Breaking modes for undensified (left) and densified (right) tiles.

just go. That's the way these little cracks would [propagate], and all of a sudden you'd lose the bond integrity."

Having already installed thousands of tiles on *Columbia*, and with the program over budget and overdue, an effective, cheap, non-invasive solution was essential. "An engineer by the name of Glenn [M.] Ecord in the materials group one weekend came up with the solution. This falls under the title of 'necessity is the mother of invention'. We had to find a solution. Glenn took real, real fine silica powder, like talcum powder fineness, and put it in distilled water, nothing more than that, and then [brushed] this colloidal solution of silica powder on the bottom of a tile. Little grains of silica would just embed themselves amongst all the fibers in the tile, then when they dried it was like putting a little barrier of real densified silica right there. [The powdered silica] adhered to the silica fibers, and that provided the strength that we needed on the bottom of the tile. It literally doubled or tripled the effective strength of the tile by dissipating that concentrated load into the rest of the tile."

To summarize, the design of the SIP was such that it caused an uneven stress distribution at the surface interface with the tile, causing spots of stress well above the maximum tensile strength of the tile even for very low pull-out loads. The result was failure of the tile/SIP interface. To overcome premature failure, it was found that increasing the tile density of the surface in contact with the SIP served to provide a stronger layer acting as a plate that distributed the concentrated SIP loads into more evenly distributed loads where they reached the unmodified silica fibers of the tile, forcing the failure surface above the densified layer and to higher pull-out loads.

A densifying slurry was therefore developed using an aqueous colloidal sol and a fused silica slip. While the fused silica acted as reinforcement of the solution (as sand behaves in concrete) the colloidal sol served as a cement or bonding agent for the fused silica. Application of this slurry was accomplished by simply brushing the solution on the tile face, permitting the silica slip particles to fill the interstices in the surface of the tile. The densified tile was air dried for 24 hours and further dried in an oven at approximately 150°F for two hours. After waterproofing, the tile received a final heating in an oven at approximately 400°F for a minimum of two hours. It was then ready for installation. Generally speaking, the depth of the densification layer ranged from 0.06 to about 0.11 inch. Analysis showed that tiles having a density of 9 pounds per cubic foot after densification exhibited no significant interfacial failure. Rather, the tile material failed internally at a minimum tensile stress of 13 psi; that is to say, the nominal tensile strength of the material itself, which was more than double the 6 psi pull-out force that caused failure in non-densified tiles.

Tile installation

A casual glance at the tile arrangement on the surface of the Orbiter might suggest that they were installed in a random manner. In fact, they were arranged in a highly organized fashion, as Terrence R. White explains, "In Palmdale they installed tiles, the majority of them, in a method called array bonding. There is a fixture that's about four feet by four feet and holds in the neighborhood of forty to fifty tiles, and

248 The Orbiter's skin: the thermal protection system

Preparing to install a tile.

they put all of those tiles up at once, so you're actually bonding a lot of tiles at one time."

Array bonding was a very effective method for installing tiles for the first time on a brand new Orbiter, but for replacing individual tiles the only effective way was to manually install them. The process began by applying a green epoxy to the skin as a corrosion inhibitor. Next was a fit-check of the tile to be installed. If necessary, the tile could be sanded in order to bring it within the required tolerances. Once the fit-check test was confirmed, the filler bars were glued to the skin. The tile was then sent to a separate room for SIP installation. Glue was applied, then the tile/SIP assembly was weighed to make sure there was the proper amount of adhesive present. Then the tile/SIP was left to cure at room temperature for several hours. Meanwhile, in the cavity in the TPS where the tile was to be installed, adhesive was spread on the skin with a thickness of 0.005 to 0.007 inch. The tile was installed, and held in position by means of mechanical and vacuum devices to cure for several days and complete the bonding process.

Almost all of the tiles were installed well before liftoff, but several on the ingress hatch and on the fuselage around the hatch cutout were put in place only after crew ingress and hatch closure. These were located over the hatch locking mechanism, and could be installed only after having securely locked the hatch. To do so, cover plates that had tiles bonded to them were screwed onto the hatch and surrounding fuselage structure. To permit the screw to be inserted each tile had a hole, and this was then filled with a protective compound. The process was reversed once the Orbiter was back on Earth. The compound was removed, the cover plates disassembled, and the locking mechanism exposed to allow ground personnel to open the hatch. In looking at a picture showing the ingress/egress area, it is easy to recognize these tiles by the white compound that fills the holes.

FLEXIBLE REUSABLE INSULATION

Rigid tiles were not the only material used to thermally protect the Orbiter. It was soon realized that the use of tiles with the required minimum thickness of 0.5 inch would overprotect surfaces that were not expected to exceed 700°F. This constituted unnecessary weight. For example, the upper surfaces were effectively shielded from maximum heating during re-entry due to the 40-degree angle of attack of the vehicle, but they were exposed to maximum solar radiation while on-orbit.

Even before *Columbia*'s maiden flight, NASA engineers had started to develop a so-called flexible insulation material. It was called felt reusable surface insulation (FRSI) and consisted of a waterproofed Nomex fibrous pad similar to the SIP. Its thickness varied from 0.16 to 0.40 inch and it was used to protect those parts of the Orbiter that faced temperatures in the range 350°F to 700°F. The felt consisted of sheets of 3 to 4 feet square that were cut to fit in proximity to closeout areas. Coated with a white silicone rubber paint for waterproofing and the required thermal and optical properties, each blanket was bonded directly to the Orbiter by the same RTV adhesive as used for the SIP.

When *Discovery*, *Atlantis* and *Endeavour* rolled out of the Rockwell assembly plant in Palmdale, most of the FRSI initially used on *Columbia* and *Challenger* had been superseded by another type of flexible insulation known as advanced flexible reusable surface insulation (AFRSI). In this case it was a low-density fibrous silica batt made up of high-purity silica and 99.8 per cent amorphous silica fibers. The batt was sandwiched between an outer woven silica high temperature fabric and an inner woven glass lower temperature fabric. "To put the composite together, they sewed it on a multiple needle sewing machine one way, then turned it 90 degrees and sewed it the other way," Emde explains. In this way, once sewn the blanket assumed a quilt-like appearance. Ceramic colloidal silica and high-purity silica fibers (known as C-9) served as a coating to provide endurance. Its thicknesses varied from 0.45 to 0.95 inch, and the thickness of each blanket was determined by the heat load that it would

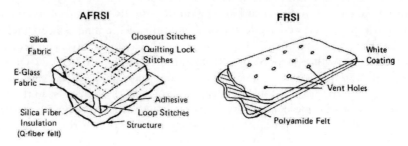

Flexible insulation.

encounter during re-entry. Bonding was performed using the usual RTV adhesive, gluing the blanket directly to the Orbiter. *Columbia* and *Challenger* were both later retrofitted with AFRSI blankets.

REINFORCED CARBON-CARBON

During re-entry some parts of the Orbiter, most notably its nose cap and the leading edges of its wings, had to withstand temperatures as high as 2,300°F, far in excess of the thermal strength of the black HRSI tiles. These parts were therefore protected by another type of reusable surface insulation called reinforced carbon-carbon (RCC). Carbon was not a new material for the space industry. During World War II the Nazi V2 had graphite vanes in its hot rocket exhaust to provide a means of vectoring the thrust for steering. Graphite was also the preferred material selected for the nose cap of the X-20 Dyna-Soar spaceplane. Unfortunately, although carbon had very good refractory thermal properties and was lighter than other such materials, it was brittle and rather susceptible to damage, which made it inappropriate for use in thin-wall structures. Nevertheless, the Dyna-Soar and later Apollo programs funded research to overcome its poor mechanical properties. The result was a composite material in which both the matrix and the fibers were made of carbon, prompting the name of reinforced carbon-carbon. It preserved the light weight and refractory properties of graphite and, with an appropriate coating, was resistant to oxidation. Its strength and low coefficient of thermal expansion provided excellent resistance to thermal shock and to the physical stresses of temperature changes. In addition, it had much better damage tolerance than graphite, and was easily shaped.

The fabrication of RCC began with a rayon cloth graphitized and impregnated with a phenolic resin. This was laminated and cured in an autoclave. The laminate was then pyrolyzed to convert the resin to carbon. It was impregnated with furfural alcohol in a vacuum chamber, and then cured and pyrolyzed again to convert the furfural alcohol to carbon. This process was repeated until the desired carbon-carbon properties were achieved.

Carbon-carbon needs its own coatings, as by itself it will rapidly oxidize above 1,500°F. If protected, it can withstand temperatures exceeding 3,000°F. To provide resistance to oxidation in order to achieve reusability, the outer layers of the RCC had to be converted to silicon carbide. To do this, the RCC was packed into a retort with a dry pack material made up of a mixture of alumina, silicon and silicon carbide. The retort was placed in a furnace, and the coating conversion process took place in argon with a stepped-time-temperature cycle up to 3,200°F. During this cycle, a diffusion reaction occurred in which the outer layers of the carbon-carbon were converted to silicon carbide (whitish-grey color) with no increase in thickness. It was this silicon-carbide coating that protected the carbon-carbon from oxidation. Since mismatching differential thermal expansion would give rise to cracks in the surface of the silicon-carbide coating, additional resistance to oxidation was provided by impregnating a coated RCC part with tetraethyl orthosilicate (TEOS).

Reinforced carbon-carbon 251

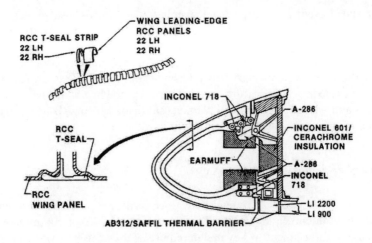

RCC wing leading edge installation.

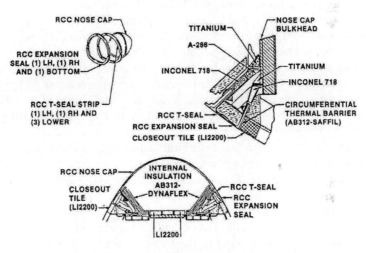

Nose cap system components.

When cured, this left a silicon dioxide residue sealant (sodium silicate/silicon carbide mixture) to fill any remaining surface porosity or micro cracks.

From a structural point of view, the RCC laminate was superior to a sandwich arrangement because it weighed less and was rugged. It also promoted internal cross-radiation from the hot stagnation region to cooler areas, thereby reducing stagnation temperatures and thermal gradients around the leading edges.

While the nose cap consisted of a single large piece, the protection for the leading edge of a wing consisted of 22 panels that were mechanically attached with a series of 22 "T-seal" floating joints made of RCC in order to minimize loading on the panels due to wing deflections. This segmentation was necessary not only to facilitate the high-temperature fabrication process, but also to allow for thermal expansion during re-entry and prevent large gaps from opening. The T-seals also prevented the hot boundary-layer gases from penetrating into the leading edge cavity of the wing during re-entry.

Because carbon was a good thermal conductor, the metallic attachments and the adjacent aluminum were protected by internal insulation blankets in order to prevent them from exceeding temperature limits. This was cerachrome insulation, contained in formed and welded Inconel foil, blankets fabricated from AB-312 ceramic cloth, saffil and cerachrome insulation, and ceramic tiles.

ORBITER TPS CONFIGURATIONS

The TPS of the Orbiter was intended to operate over a spectrum of environments typical of both aircraft and spacecraft, with an intended life of 100 missions and the minimum of refurbishment. A key design requirement was to behave as a radiator and reflector in order to dissipate heat and as an insulator to block the remaining heat from reaching the structure. As a radiator the TPS had to emit the ascent and re-entry heat, while on the upper surfaces it had to behave as a reflector to reject most of the incident solar radiation on-orbit. For example, over 90 per cent of the radiation heat generated at maximum temperatures during re-entry was dissipated at the surface due to the emissivity of the black coating of the HRSI. The remaining heat was blocked by the very low conductivity LRSI or AFRSI. Furthermore, the transfer of the heat pulse through the coating was delayed by about 33 minutes, with the result that the structure of the Orbiter was exposed to its maximum temperature only after landing.

The TPS had to protect the structure of the Orbiter on a variety of missions, with critical mission parameters including orbital inclination, altitude, size of payload to be returned, and the down-range and cross-range requirements during re-entry. It was also required to resist localized heating from the SSMEs and SRBs, and the RCS and OMS engines. And it had to withstand the acoustics of launch, the structural deflection generated by aerodynamic loads, the low temperatures of space (cold soak) on-orbit, and of course the natural environments such as salt, fog, wind and rain. Although the size and shape of the Orbiters were the same, there were some differences in the TPS configurations of *Columbia* and her siblings. Generally speaking, HRSI tiles were used in areas of the upper forward fuselage, including around the front windows, the entire underside of the vehicle, portions of the OMS and RCS pods, the leading edge of the vertical stabilizer, the wing glove areas, the elevon trailing edges and the body flap. LRSI tiles were used on the upper wings, OMS and RCS pods, and some parts of the forward, mid and aft fuselage sections. Especially for the three youngest Orbiters, AFRSI was used almost everywhere with the exclusion of the areas protected by the black tiles.

Special cases

Some areas of the Orbiter, such as the landing gear doors, interfaces with the nose cap, ET/Orbiter umbilical doors, vent doors, and the leading edge of the vertical

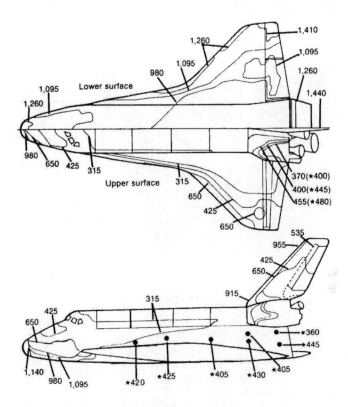

Maximum surface temperatures experienced during normal ascent (*) and re-entry.

stabilizer were subjected to high concentrated heat that could induce shrinkage and density increases that imposed unacceptable dimensional changes on the tiles. Lockheed created a new grade of silica that was given the commercial name of Li-2200. It had a density of 22 pounds per cubic foot and the tiles made from it provided the strength and insulation properties required for these areas of high thermal stress, albeit with an undesirable weight penalty. This prompted NASA to develop fibrous refractory composite insulation tiles, known by the commercial name of FRCI-12. This type of tile was made by adding an amorphous alumina-boro-silicate fiber called Nextel to the pure silica slurry. In essence, the Nextel "welded" the micron-size fibers of pure silica into a rigid structure during sintering in a high-temperature furnace. The result was a composite fiber refractory material composed of 20 per cent Nextel and 80 per cent silica fiber with entirely different physical properties from the original 99.8 per cent pure silica tiles. Having an expansion coefficient 10 times that of the original silica, Nextel acted like a preshrunk concrete reinforcing bar in the fiber matrix. The reaction-cured glass (black) coating of the FRCI-12 tiles was compressed as it was cured to reduce the coating's sensitivity to cracking during handling and operations. In addition to the improved coating, the FRCI-12 tiles were about 10 per cent lighter than the HRSI

254 The Orbiter's skin: the thermal protection system

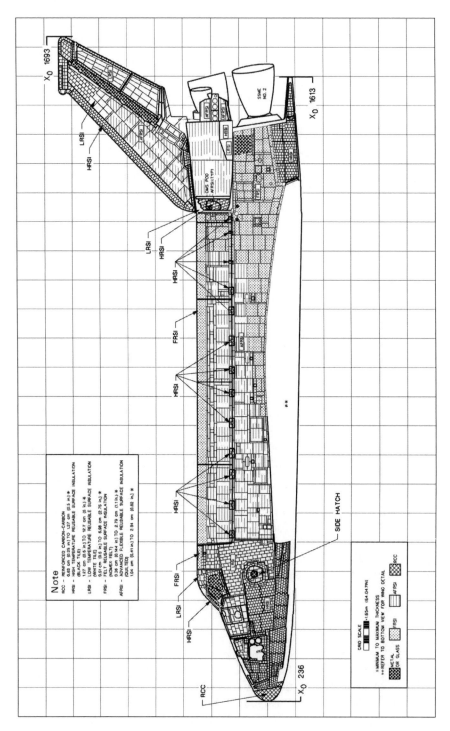

Columbia TPS arrangement (left-hand side). Notice the use of LRSI tile on the full surface of the forward fuselage section.

Orbiter TPS configurations 255

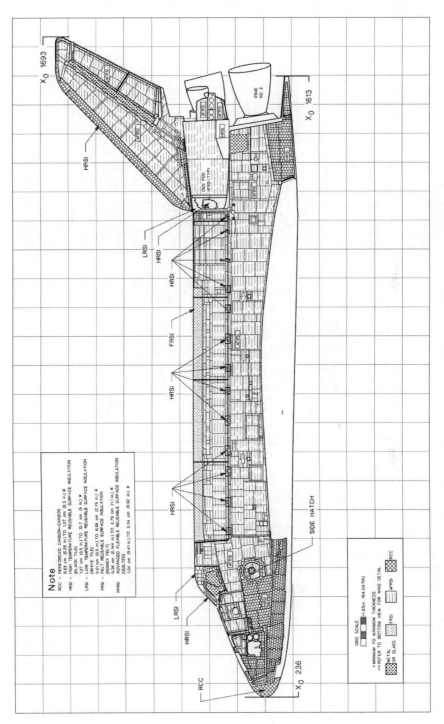

TPS arrangement for *Discovery*, *Atlantis* and *Endeavour* (left-hand side).

256 The Orbiter's skin: the thermal protection system

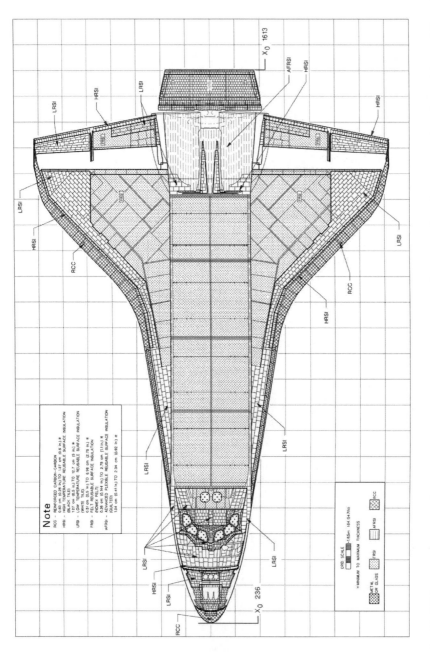

Columbia TPS arrangement on the top surface. Notice the use of LRSI on the upper wing glove area and on the periphery of the wings between the leading edge RCC panels and the area covered with flexible insulation material.

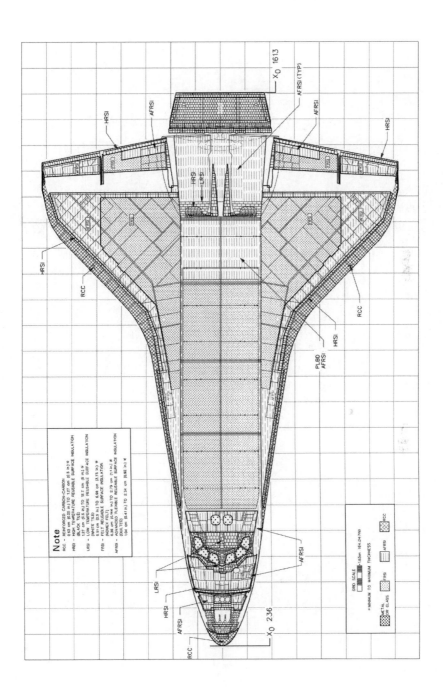

TPS arrangement on the top surface of *Discovery*, *Atlantis* and *Endeavour*.

258 The Orbiter's skin: the thermal protection system

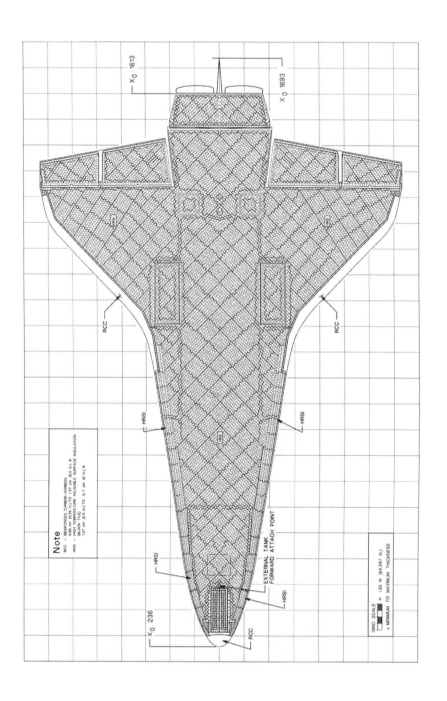

TPS arrangement of the underside of all Orbiters.

It was easy to tell *Columbia* apart from her sisters because of the black tiles that covered the upper surface of the forward wing box area, the black tiles covering a good portion of the sides of the tail fin, and the tiles installed on the surface of the Orbiter forward section.

tiles, with a demonstrated tensile strength at least thrice that of the HRSI tile and a temperature limit approximately 100°F higher.

During the certification and test program, it was realized that the tiles bordering the front windows of the Orbiter were subjected to high stagnation pressures which tended to lift the tiles and drive the SIP bond line stresses above the tolerances of the silica material. This reduction in strength was attributed to the orientation of the silica fibers inside the tile. As Emde says, "Most of the fibers ... were layered horizontally 'in-plane'. Not very many went vertically, or 'through the thickness'. There wasn't a lot of conductive thermal transfer through the thickness because most of the fibers were layered horizontally."

This horizontal fiber orientation (that is to say, parallel to the skin of the Orbiter) was a thermal requirement to minimize the conduction of heat from the outer mold line to the inner mold line. As a result, whilst horizontal fibers created the incredible heat resistance of the tiles, vertically fibers (running perpendicular to the skin of the Orbiter) provided structural strength to the tile and transferred loads to the structure. With most of the fibers being orientated horizontally, it is easy to appreciate why the tiles were so fragile. To give more strength to the windshield tiles,

260 The Orbiter's skin: the thermal protection system

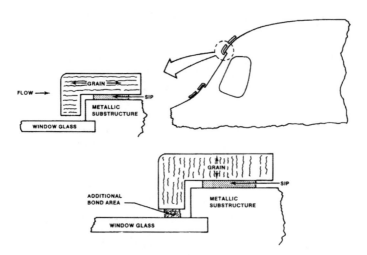

Windshield tile grain orientation.

it was decided to machine them with most of the fibers running perpendicular. This would double the nominal strength but at the risk of overheating the structure. A thermal analysis was performed on such tiles and indeed more heat reached the underlying aluminum. On the other hand, the heavy framing around the windows acted as a large heat sink and prevented unacceptable temperatures. It was therefore decided that the windshield tiles would be produced with the majority of the fibers running perpendicular to the skin of the Orbiter. But whilst this significantly increased the strength of the tiles, the margin of safety was inadequate. A further improvement was obtained by bonding the "overhanging" portions of these tiles to the outer glass panes. This extra area, in combination with the fiber orientation, provided an acceptable margin of safety for flight.

During calculation of flight stresses for the body flap tiles, it was determined that the trailing edge corner tiles did not have an adequate safety margin for the expected vibration and acoustic loads during liftoff. This was later confirmed during acoustic testing of the body flap when both corner tiles (weighing approximately 6 pounds) failed within seconds of startup. These Li-2200 corner tiles were overhanging in two directions, thereby creating a large overturning moment on a small SIP area, with correspondingly high tension stress. But even more significant was the inability of the SIP to withstand this high cyclic loading. The challenge was therefore two-fold: (1) to prevent failure caused by high tensile stress, and (2) to prevent the SIP from failing due to the high cyclic loading experienced on the body flap. The solution was to twist a mechanical attachment system called auger into a tile and then attach this to the substructure using bolts. The important aspect of this design was to use Belville washers to preload the auger with a tension load, whilst at the same time inducing a compression stress in the SIP. The preloaded washers (serving as a soft spring) and the compressed SIP (serving as a stiff spring) then functioned together as a parallel spring system. The stiff SIP, being greatly compressed, took most of the

Orbiter TPS configurations

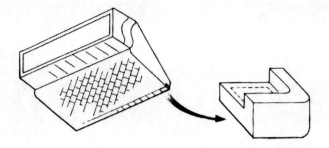

Body flap corner tile.

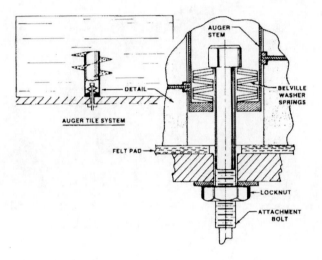

Auger system.

external cyclic loading, and the Belville washers helped to prevent significant cyclic loads from being induced by the auger into the intolerant silica material. The auger system was preloaded sufficiently for the SIP to remain in compression throughout the flight environment, and the tension-sensitive bondline never experienced a high tensile loading.

After a few flights a particular modification was made to the tiles that covered the OMS pods. As White explains, "The OMS pod is a composite structure ... We have to keep that structure less than 250 degrees. So in their analysis – and *Columbia* had a lot of instrumentation – they realized that we were getting higher than that in the concentrated area where it was coming up over the wings, down the sidewalls, and coming right up against that OMS." Interestingly, the thermal analysts had already discovered that this area was critical temperature-wise. But as White says, "It took a while to convince the program that it really was a problem." The solution was simply to replace some of the white tiles with black ones. Since this area of black tiles had a

262 The Orbiter's skin: the thermal protection system

The "bull's eye" on the left-hand OMS pod.

sort of circular shape when viewed from a distance, the modification was dubbed the "bull's-eye mod".

Shortly prior to STS-1 it was discovered that the revised combined loads on the OMS pod structure would produce considerably higher deflection than previously anticipated. Since these tiles were quite thin, this increase in deflection loads might cause them to fracture and possibly even detach from the pod. The solution involved a "dicing" procedure. As White explains, "Rather that have an entire eight-by-eight [inch] tile on the upper surface, we actually cut that eight-by-eight into sixteen pieces and they were still held together by that SIP... So basically this thing is able to flex a lot without cracking the tile." Of the five Orbiters built during the Shuttle program, only *Columbia* and *Challenger* were provided with "diced tiles". In fact, as White points out, "Diced tiles were all over the vertical stabilizer, the tail of the vehicle, down the mid-fuselage sidewalls, payload bay doors, and crew module." Following the loss of *Challenger* in 1986, all these diced tiles were replaced by blankets.

After STS-5, extensive damage caused by gap heating, tile slumping, gap filler degradation, subsurface flow, and localized aluminum melting was found at the RCC nose cap/tile interface. For this reason, this area was subsequently redesigned

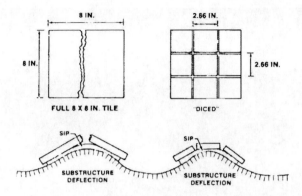

Diced tile concept.

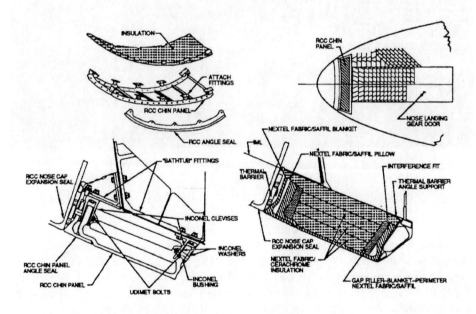

RCC chin panel system components.

with a stiffer support structure and all the tiles were replaced with a section of RCC dubbed the "chin panel".

THE FLIGHT EXPERIENCE

Despite the conceptual simplicity of an Orbiter having a TPS of lightweight tiles, it became evident on the very first flight that there would be problems.

Robert Crippen made his first trip in space aboard *Columbia* during her maiden

264 The Orbiter's skin: the thermal protection system

Missing tiles on *Columbia*'s OMS pods during STS-1.

voyage, and probably could never have imagined that he would see what he saw as the payload bay doors opened. As he recalls, "I went ahead and did the procedure on the doors. I opened up the first door, and at that time I saw, back on the OMS pod, that there were some squares where obviously the tiles were gone. They were dark instead of being white. We told the ground, 'Hey, there's some tiles missing back there,' and we gave them some TV views of the tiles that were missing. Personally, that didn't cause me any great concern, because I knew that all the critical tiles, the ones primarily on the bottom [of *Columbia*], we'd gone through and done a pull test with a little device to make sure that they were snugly adhered to the vehicle. Some of them we hadn't done, and that included the ones back on the OMS pods, and we didn't do them because those were primarily there for reusability, and the worst that would probably happen was we'd get a little heat damage back there from it. But, of course, the big question on the ground was, 'Well, if some of those are missing, are there some on the bottom missing?' So I knew there was a lot of consternation on the ground about, 'Hey, are the tiles really there?' There wasn't much that we could do about it if they were gone, so I didn't worry about it. And I don't think John [Young] worried about it very much. But it was one of the items in the flight that got a lot of publicity here on the ground, I know." *Columbia* landed two days later, inaugurating the new era of the National Space Transportation System.

Atlantis is landing at the end of STS-27. Visible are the missing tiles on the bottom surface (from the nose along the full length of the wing glove).

But *Columbia*'s missing tiles were not an isolated case. Virtually every mission reported some kind of damage to the protective shield. The worst non-fatal instance afflicted *Atlantis* on STS-27, a Department of Defense mission. Despite the veil of secrecy that still shrouds that mission, it is not a secret that the Orbiter TPS sustained an incredible degree of damage. As mission specialist Mike Mullane says, "After we got in orbit, [Mission Control] asked if we saw anything go by the window during launch ... Those are the type of calls you don't want to get. And we said, 'No. Why do you ask?' And they said on one of the engineering cameras that they looked at after launch, they saw what appeared to be something breaking off the tip of the solid booster and flying down and they were wondering if we'd seen it. And then they wondered if it had hit the Orbiter's belly. And so they sent up instructions for me to use the robotic arm to bend around in order to look under the belly, and we saw a lot of damage to the heat tiles ... That gave us a little bit of concern." As the commander of that flight, Robert "Hoot" Gibson, recalls, "I will never forget, we hung the arm over the right wing, we panned it to the damaged location and took a look. I said to myself, 'We're going to die.' There was so much damage. I looked at that stuff and I said, 'Oh, holy smokes, this looks horrible, this looks awful.'"

And it was awful indeed, as Mullane continues, "One place looked like a tile was completely missing, but it looked to us like there was a lot of damage on the belly of this thing." Oddly enough, Mission Control actually downgraded the severity of the damage. "We told [them] and they just kind of seemed blasé about it, like they were looking at the video and just didn't have the sense of urgency that we expected them to have. It kind of baffled us. We said, 'Why aren't they more concerned about this?' It was obvious to us that there were probably hundreds of damaged tiles." Despite the vivid concern expressed by the astronauts, Mission Control reported more than once that the damage was no more severe than damage seen on previous missions. Based on their analysis, the vehicle would not have any problem during re-entry. There was nothing that the astronauts could do except hope for the best. During the earliest part of re-entry on 6 December 1989, Gibson kept his eyes fixed on a gauge that showed the deflection of the elevons. "I knew that if we started to burn through ... we would change the drag on the right wing and we would start seeing right elevon trim, which means putting down the left elevon ... I knew we'd start developing a split [between right and left wing elevon positions] if we had excessive drag over on the right side. The automatic system would try to trim it out with the elevons. That's one of the things we always watched on re-entry anyhow, because ... if you had half a degree of trim, something was wrong ... normally, you wouldn't see even a quarter of a degree of difference on the thing. So I knew that that's what I was going to see if it started to go. And that told me that I'd have at least 60 seconds to tell Mission Control what I thought of their analysis."

Luckily for Gibson, his crew, *Atlantis* and the program, re-entry and landing were performed flawlessly at Edwards Air Force Base in California. When the astronauts walked around the vehicle they saw the full extent of the damage, and finally NASA engineers realized how wrong their judgment had been. But the apparent lightness with which NASA addressed this issue had a reason. "It turned out that the video was such poor quality with the Sun shining on those black tiles that they really couldn't see what we were seeing. They saw a few scrapes and scratches and stuff and didn't think it was all that big of a deal". And due to the military nature of this mission, the TV video sent to Mission Control had to be encrypted, further degrading its quality. With the Orbiter on the runway, engineers could see the real extent of the damage, and as Mullane continues, "I think everybody was shocked when the vehicle landed, and I think they ended up changing out like 700 heat tiles or something. It was a lot of heat tiles they had to change out that were damaged on that thing." One black tile on the belly was completely missing, with the underlying aluminum showing severe signs of burning. But the loss of some tiles, even on the belly, did not constitute a real hazard during re-entry because the high conductivity of the aluminum would have dissipated the localized heat rapidly through the structure, precluding a concentration of thermal loads. Obviously, the loss of several tiles in the same region or in a region that was in any case subjected to high thermal loads would have compromised safety. It was later established that the damage was caused by thermal protection material detaching from the right-hand SRB during ascent.

Although STS-27 was a close call, indicating to NASA how vulnerable the tiles were, no countermeasure were undertaken. On 1 February 2003 *Columbia* started its return to Earth after a highly successful mission. As per procedure, Rick Husband, the mission commander, would have monitored the elevon gauge. Unlike Gibson, he was not expecting to see any significant divergence. When this occurred, he would have known that he was in trouble, and why, but there was no way to overcome the problem and *Columbia* was lost.

On-orbit TPS repairs

Although it was evident that tiles could be easily damaged, for the first two decades of the Shuttle program there were no on-orbit TPS repair techniques available. Many missions suffered minor damage that never raised much concern, but then *Columbia* failed to return home from STS-107. After that, NASA became so concerned about the integrity of the TPS that while each mission was on-orbit the implications of even the smallest damage were thoroughly analyzed. In some cases, prudence and a desire to keep the safety margins well above their minimum, prompted the agency to devise contingency EVA procedures and tasks for declaring the TPS safe for re-entry. One of the most thrilling and spectacular of these contingency EVA tasks occurred during STS-114, the first mission after the loss of *Columbia*. In this case *Discovery* lifted off on 26 July 2005 on a mission to resupply the International Space Station and test some possible TPS repair techniques.

Prior to docking with the ISS, the flight plan called for a new maneuver called the rendezvous pitch maneuver that required the Orbiter to hold position some 600 feet directly below the station and execute a nose-forward, three-quarters-of-a-degree-per-second back flip. As the vehicle reached the 145-degree point, pilot Mark Kelly informed station commander Sergei Krikalev and flight engineer John Philips, who visually inspected for damage to the TPS. In addition, pictures were taken using a 400-mm lens with a resolution of 2 inches to detect damage to the upper surface of the Shuttle and frames were taken using a camera with an 800-mm lens with a 1 inch

Orbiter TPS belly as viewed from the ISS.

268 The Orbiter's skin: the thermal protection system

The protruding gap fillers on *Discovery* during STS-114.

analytical resolution of the nose landing gear door seals, the main landing gear door seals, and the elevon cove.[1] This new procedure proved its usefulness immediately, since two gap fillers were detected protruding from the belly of *Discovery*. This was not a new issue, because gap fillers had been found to extend above the tile surface after several earlier missions.

Although a large piece of foam was seen to detach from the external tank during launch, in this case the protruding gap fillers were caused by the vibrations of launch. This time, several factors prompted action. One gap filler protruded 1 inch and the other for 0.6 inch, whereas in the past the largest gap filler protrusion in that area was only 0.25 inch. And position-wise, these gap fillers were different from in the past. In particular, one was near the centerline of the vehicle, far forward by the nose landing gear door, and the other was a little bit farther back and off to the side of the vehicle. Previously, this problem had always occurred well aft on the belly. And, of course, after the loss of *Columbia* no one wanted to take any risks with the spacecraft and its crew. The really disturbing thing was that one, if not both gap fillers, could "trip the boundary layer" during re-entry. Tripping the boundary layer is a technical term for prematurely transitioning the thin layer of insulating plasma (the boundary layer) which encases the vehicle during re-entry from a laminar state to a turbulent state. In addition to inducing more drag, a turbulent boundary layer will increase the heat load on the tiles downstream of the transition point. If this were to occur when

---

[1] The cove is that part of the elevon which is partially inside the wing.

the Orbiter was still at high altitude and hence still in the peak-heating period, the thermal load (as a combination of heat and exposure period) would exceed the design limits, causing damage not only to the tiles but also to the underlying structure. The boundary layer transition usually occurred when the Orbiter had slowed to a speed in the range Mach 12 to Mach 8. Anything above Mach 13 was regarded as an early boundary transition and caused a 15 to 25 per cent increase in the heating loads. Another risk associated with premature boundary layer transition was that if its source was off the centerline, this would yield an asymmetric flow around the Orbiter that could result in control issues.

As aerodynamicists could not guarantee a problem-free re-entry, and despite an initial concern from the crew, on the ninth day of the mission spacewalker Stephen Robinson mounted the station's robotic arm to attempt the very first on-orbit TPS repair. The plan was to gently remove both fillers from their inserts. If this did not work, plan B was to cut away as much material as possible in order to minimize the protrusion. Though conceptually straightforward, the task involved many unknowns. As Robinson said in an interview the day prior to the spacewalk, "Like all kinds of repairs, it's conceptually very simple but it has to be done very, very carefully. The tiles, as we all know, are fragile and an EVA crew member out there is a pretty large mass. I'll have to be very, very carefully. But the task is extremely simple and we predict it won't be too complex. We're ready to do this in a very careful manner. And besides, it's not just me. We'll also have a camera... looking at me, and trying to look at the clearance between me and the Orbiter's belly. So we'll have lots of ways to be very, very conservative. It's going to be like watching grass grow. Nothing's going to happen fast. It will be a gentle pull with my hands. And if that doesn't work, I have some forceps and I'll give a slightly more than gentle pull. And if that doesn't work, I'll saw it off with a hacksaw."

Astronaut Robinson working on *Discovery*'s belly TPS during STS-114.

270 **The Orbiter's skin: the thermal protection system**

Technique for removing the protruding gap fillers on *Discovery*.

Astronaut Andrew Thomas radioed from the station to Robinson, "We don't want any inadvertent contact with the tiles on the belly of the Orbiter. You've got a lot of things still hanging onto you even though we have cleaned you up, so try to keep an eye on where they are ... And under the Orbiter, we'll probably have comm drop outs. We may lose wireless video. So we'll need continuous communications protocol while you're doing the job, so we can be assured it's going properly." If the hacksaw had to be used to cut the gap fillers off, Thomas advised, "The serrated edge is also going to be sharp, so you'll need to watch that. If by any chance you need to contact the tile with your hands, we'd require only gentle hand reaction alone. We want you to distribute the load over several fingers, or the backs of the fingers." Robinson's reply did not leave any doubt about how he would conduct this delicate task, "Copy all, particularly the hand touch. My goal of course, is not to touch the tiles at all, but I have touched the tiles at KSC with my work gloves on, so I know what to expect. I'll use a very gently touch." As the arm swung him in towards his first target, Robinson took a moment to describe what he was seeing in front of him, "I'm about eight feet, maybe seven feet, looking straight down on it. It looks to be about close to three inches on one side and about an inch and a half on the other side. The corner looks like it is bent over, presumably by air loads." He put his gloved hand on the gap filler and carefully pulled it off. "I'm grasping it. I'm pulling. It's coming out very easily. Okay, the offending gap filler has been removed." Ten minutes later the second gap filler was similarly removed.

Another issue, perhaps less known than Robinson's spacewalk, was a damaged insulation blanket just below commander Eileen Collins's left cockpit window. But unlike the protruding gap fillers which could affect the distribution of the heat on the belly during re-entry, this blanket posed a different risk. During a press conference, Wayne Hale, chairman of the mission management team, explained the problem in this way, "Even if that blanket completely came off during entry we're perfectly safe from the local area thermal effects; that is not a problem at all for us in that local area ... This is just a question of could [the blanket] fly back and hit something on the after part of the vehicle? This would not be a concern if it came off at a high Mach number, above Mach 6 ... At lower Mach numbers, where the air is thicker, then there is some transport mechanism that folks are going to go off and look at ... They are concerned about a physical impact. The heating concerns rapidly go away

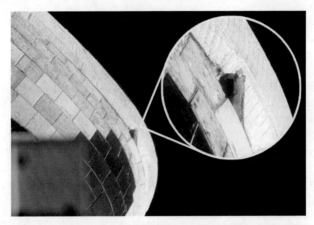

Protruding blanket on *Atlantis*'s left-hand OMS pod during STS-117.

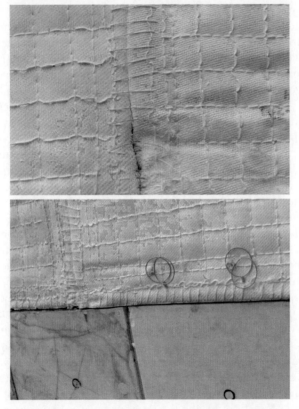

Close-ups of the repaired blanket on *Atlantis* during STS-117.

later in the trajectory. So it wouldn't be a high heating, high Mach number kind of concern. This would be a low Mach number impact concern." After several days of exhaustive analysis, Canadian astronaut Julie Payette, serving as Capcom, radioed the results to *Discovery*'s crew, "We have new analysis that shows debris transport would be no issue, and we came to the same conclusion with the Ames [wind] tunnel test." In space, Japanese astronaut Soichi Noguchi replied, "That's, I would say, good news." Indeed it was, and *Discovery* returned safely to Earth.

Four missions later, the STS-117 crew was called upon to make another manual intervention. During the routine post-launch inspection the astronauts flying *Atlantis* noticed a protruding blanket on the left-hand OMS pod. It seemed likely that this had been pulled loose by the force of the airstream in the early part of the ascent. Again, this was nothing new. As John Shannon, chairman of the mission management team said, "We have flight history of damage to these OMS pods. On STS-1 we lost some tile material and on STS-6 we lost at least one blanket in that area." But in the wake of *Columbia*, NASA was unwilling to take a chance with something as apparently minor as this.

Shannon added, "When we talked to the engineering guys and the structural team, they were a little bit uncomfortable ... because if you get down to that honeycomb, that structural area, you're losing some margin, you're not exactly in the flight parameters you expect to go fly ... You have the heating of the vehicle in this area that goes up and is significant for a 15 to 20 minute period, but you really don't have any aerodynamic loads on it at that time, it's very, very low dynamic pressure on that blanket area." If for STS-114 the concern was the possibility of the damaged cockpit blanket detaching and striking the aft fuselage, for *Atlantis* the motivation was not to allow the heat of re-entry to damage the underlying OMS pod honeycomb structure, with consequent repairs once back home. "Because we do not want to damage flight hardware, we sent a [team of engineers] to work with the EVA team to assess some options." One option stood out. "We have limitations in our ability to analyze this," Shannon said. "You're going to have a temperature that exceeds that top face coat [outer skin honeycomb panel of the OMS pod structure] capability. How long it would take to completely erode is very questionable, no one could give me that answer. So the right answer here ... is to go and put it down and secure it. We think the astronauts will be able to go out there and just push it right back down. They're working right now on different ways to attach it to the blanket that's beside it, or maybe to attach it to the tile face that's in front of it." And so Danny Olives, like Stephen Robinson two years previously, found himself making a TPS repair with the precision of a surgical procedure. Standing on the end of *Atlantis*'s arm, Olives first gently laid the peeled-back blanket down flat and then secured it into position using a stapler that drove steel pins through the blanket into nearby tiles. The first example of sewing in space was completely successfully in about two hours. Seven days later, *Atlantis* safely landed at Edwards Air Force Base.

ON-ORBIT TPS REPAIR TECHNIQUES

Even before the first Shuttle flight, the possibility of TPS damage during ascent had prompted NASA engineers to devise ways for a crew to make repairs on-orbit. Several studies were initiated with the aim of developing techniques for repairing a wide variety of damage.

But as the launch of STS-1 approached, the effort to develop an array of tools and materials to enable an astronaut to restore the thermal protection capabilities of the TPS was canceled due to various technical problems and a renewed confidence in the tiles themselves. At the time, the RCC was considered particularly resilient and little thought was given to a repair capability. As Robert Crippen recalls, "Well, the tiles were a big concern. Initially, when they put the tiles on, they weren't adhering to the vehicle like they should. In fact, *Columbia*, when they brought it from Edwards to the Kennedy Space Center the first time, it didn't arrive with as many tiles as it left California with. People started working very diligently to try to correct that problem, but at the same time people said, 'Well, if we've got a tile missing off the bottom of the spacecraft that's critical to being able to come back in, we ought to have a way to repair it.' So we started looking at various techniques, and I remember that we took advantage of a simulator that Martin [Marietta Corporation] had out in Denver [Colorado], where you could actually get some of the effect of crawling around on the bottom of the vehicle and what it would be like in zero-G. I rapidly came to the conclusion I was going to tear up more tiles than I could repair, and that the only realistic answer was for us to make sure the tiles stayed on."

Moser's words help to better understand NASA's thinking at that time, "After the first few flights we did some stuff on repair techniques, but we dropped it just because it didn't look like anything was panning out. [There were] places on the vehicle you couldn't get to. We even looked at a little fly-around vehicle to inspect it, and even have some capability to do repairs. But we were getting more flights under our belt, and we thought we're not going to ever go there [with a capability to repair the TPS]."

Unfortunately, on a Saturday morning some 22 years later, NASA received its second wake-up call and realized that it would have been wiser to have continued the development of those studies for TPS repair on-orbit. It was the day *Columbia* was lost. A few minutes before 8:00 am CST on 1 February, the residents of East Texas were startled by a long, low-pitched rumble that signaled the passage of fragments of *Columbia* through the upper atmosphere at nearly 12,000 miles per hour.

An independent review board was formed to determine the cause of the accident and to suggest operational improvements for NASA's space program. The extensive tests in the ensuing months identified the cause of the accident without any doubt. As clearly and succinctly stated in the Columbia Accident Investigation Board (CAIB) report, "The physical cause of the loss of *Columbia* and its crew was a breach in the Thermal Protection System on the leading edge of the left wing, caused by a piece of insulating foam which separated from the left bipod ramp section of the External Tank at 81.7 seconds after launch, and struck the wing in the vicinity of the lower half of Reinforced Carbon-Carbon panel number 8. During re-entry this breach in the Thermal Protection System allowed superheated air to penetrate through the leading edge insulation and progressively melt the aluminum structures of the left wing, resulting in a weakening of the structure until increasing aerodynamic forces caused loss of control, failure of the wing, and breakup of the Orbiter ... in a flight regime in which, given the current design of the Orbiter, there was no possibility for the crew to survive."

After examining more than 30,000 documents, conducting 200 formal interviews, hearing testimony from dozens of expert witnesses and reviewing more than 3,000 inputs from the general public, the 13-member Board issued recommendations. Nine of the recommendations addressed issues with the thermal protection system. Two of these read:

"Develop and implement a comprehensive inspection plan to determine the structural integrity of all Reinforced Carbon-Carbon system components. This inspection plan should take advantage of advanced non-destructive inspection technology." [R3.3-1]

"For mission to the International Space Station, develop a practicable capability to inspect and effect emergency repairs to the widest possible range of damage to the Thermal Protection System, including both tile and Reinforced Carbon-Carbon ... For non-Station missions, develop a comprehensive autonomous (independent of Station) inspection and repair capability to cover the widest possible range of damage scenarios. Accomplish an on-orbit Thermal Protection System inspection, using appropriated assets and capabilities, early in all missions. The ultimate objective should be a fully autonomous capability for all missions to address the possibility that an International Space Station mission fails to achieve the correct orbit, fails to dock successfully or is damaged during or after undocking." [R6.4-1]

NASA lost little time in beginning work on viable repair strategies. Despite the huge effort that teams put into the development of inspection and repair techniques, a comprehensive repair capability for all possible damage was not developed. Tile and RCC techniques proved to be far more challenging than the CAIB had thought. Moser is emphatic that the Orbiter "was never ever designed for any debris". But the problem-solving set in motion by CAIB recommendation R6.4-1 led to the development of two basic repair categories: mechanical and chemical. The former relied on prefabricated materials and the fasteners to connect them to the existing protection systems. The latter required materials which would be applied in a raw form and spread upon the existing TPS in order to form a chemical adhesive bond after a curing period.

From the collaboration of six NASA centers, eleven contractors and the US Air Force, two RCC repair concepts emerged under the names of NOAX and RCC plug repair. Both concepts have limitations in terms of damage characteristics and damage location. Generally speaking, it is worth noting that the main challenges in repairing an RCC panel were maintaining a bond to the RCC coating during entry heating, and meeting stringent aerodynamic requirements for repair patches and fills.

On-orbit TPS repair techniques

In parallel, several techniques were developed for repairing damage on the tiles of the thermal protection system.

RCC repair techniques: NOAX

In a December 2004 interview Paul Hill, the lead flight director for STS-114, said, "Until about six or seven months ago we thought entry-critical damage required a penetration of the RCC, not just [glass] coating damage or even small damage to the substrate on the outside." But arc jet testing proved differently. As Hill continued, "More recent arc jet testing has us worried that coating damage alone, if it's large enough and combined with internal damage, delamination between the layers, could be entry-critical." All previous tests had suggested that a leading edge RCC panel could withstand penetrations one quarter of an inch across. But the flaw in these tests was that the penetration was thought to happen at high speed, punching a clean hole in the panel. In fact, impact velocities during launch proved to be much lower,

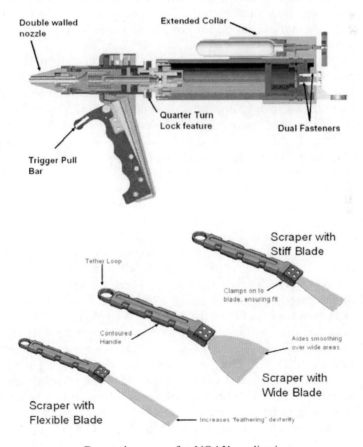

Gun and scrapers for NOAX application.

with damage scratching the surface of the panel and raising the prospect of delamination. What does happen in this case? Hill explains, "If the coating is gone and underneath that coating you're delaminated, then picture the RCC itself from a side view like a cross-section. Now you've got this void between layers. What you've done is, you've significantly reduced the density of this RCC that's exposed to the heat load, and it burns faster."

To validate the findings, engineers deliberately damaged an RCC panel and after confirming that it had developed delamination they tested it. "They put that bad boy under the arc jet, and it burned like there was no tomorrow. The whole area that covered the delamination burned off like a fuse," explained Hill. "An impact that has enough energy to cause a delamination and takes off the coating, doesn't have to be big [in order] to be catastrophic. From an RCC damage perspective, that looks like a penetration."

This astonishing finding prompted the development of NOAX, an acronym for non-oxide adhesive experimental sealant. It was a pre-ceramic polymer designed specifically to repair small cracks and coating losses on RCC panels, fixing the most likely type of damage from small pieces of foam detaching from the ET. It could be used at any location that did not require any physical modification of the RCC prior to making a repair.

On 30 July 2005 STS-114 spacewalker Stephen Robinson performed the first heat shield repair test assigned to the mission. At a pallet in the payload bay loaded with a variety of damaged tiles and RCC samples he used a space-hardened caulk gun to apply NOAX to the samples in much the same manner as if he were plastering his house's walls. Then he used scrapers similar to putty knives to work the material into deliberate cracks and gouges. "It seems to be well behaved," Robinson commented while he was working. "I see just a very little bit of bubbling ... It's about like a pizza dough. Licorice flavored pizza dough." He also had a suggestion, "I'd recommend if we were to do this for real, to use lots of spatulas. You can't clean it." The repaired samples were then tested on Earth to determine how well they would perform during re-entry.

On-orbit experiments for this repair technique continued on the second return-to-flight mission, STS-121 in July 2006. Piers Sellers and Mike Fossum were able to further investigate the behavior of this material. Sellers explained the objectives of this second round of NOAX trials as follows, "There are straight cracks, which is like mechanical damage, which we're trying to fill in, and there's coating loss, where the thin layer of glass over the top of the RCC has been eroded off. So we are going to try applying this material in different ways to repair these different kinds of damage. Now the 'different conditions' part, is that this material doesn't seem to work very well when it's very, very cold. Likewise, when it's very hot it behaves badly. So we're going to try and catch it under the optimal conditions between extreme heat and extreme cold, when the surface is cooling, and see how well we do. Mike and I are going to be very careful ... on how to get the material to the right temperature and apply it to a set of samples. It really is 'lab work' and we are going to do the best, most careful job we can." There was another aspect of NOAX behavior that is worth mentioning. As Fossum explains, "We have a special space

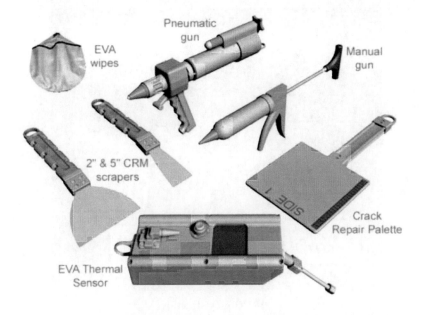

RCC repair tools tested on STS-121.

caulking gun that we'll use to squirt out a small amount of this repair material." The funny thing was, "In the vacuum of space, the stuff boils. It's just amazing to see it in a vacuum. The black goo literally boils and sputters. We work it with a putty knife until it settles down and becomes more workable. We'll then use that to make repairs, filling small cracks and holes in the sample tiles." This bubbling posed a distinct inconvenience. In tests in a terrestrial vacuum chamber the NOAX outgassing manifested itself as bubbles which exploded on the surface, but in the weightlessness of space things were different. In weightlessness the bubbles could merge together inside the material to form a void during the hardening phase. The mission's lead spacewalk officer, Tomas Gonzalez-Torres, explained the consequences as follows, "You don't want a direct path down thorough into the damage area. The repair itself, you're trying to have a protective layer, a protective coating. [During re-entry] you have some slight erosion. If you have that erosion and you get into that void, and that void actually goes all the way down to the damage, you haven't enacted a good repair. So we want to minimize those voids and make sure whatever voids we do have, if we've done a good repair, they shouldn't be one big void [because] that wouldn't be able to protect us." Arc jet tests of the STS-114 and STS-121 repair samples showed that the material lived up to its role of repair agent, so all subsequent missions carried a NOAX kit just in case it was necessary to repair RCC damage.

278 The Orbiter's skin: the thermal protection system

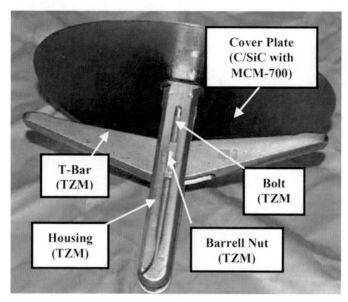

RCC plug structure.

RCC repair techniques: RCC plug

Although NOAX was effective at restoring the surface capability of an RCC panel to withstand the heat of re-entry by preventing delamination from further decreasing the heat resistance capability of the panel, it was limited because it could not be used to repair a hole. The so-called repair plug technique was developed for this task.

The RCC plug system was developed to repair the panels on the leading edges of the wings in the event of damage caused either during the ascent or by microscopic debris on-orbit. This was a round, flexible carbon-silicon carbide cover plate with a silicon carbide coating. It was 7 inches in diameter, 0.03 inches thick, and was able to flex up to 0.25 inch in order to conform to the shape of the RCC leading edge panels of the wings. A coating of MCM-700 was brushed over the silicon carbide coating for increased oxidation protection. A mechanical titanium zirconium molybdenum (TZM) system was used to conform and attach the flexible plug to the RCC curved surface.

Based on the dimensions of the damage, astronauts would select the appropriate cover plate at the work site, connect it to a TZM bolt, and then insert the folded bolt through the hole. By tightening a fastener that extended through the cover plate to the TZM bolt, the astronaut would unfold the toggle inside the RCC panel and tighten it until the cover plate conformed to the exterior shape of the panel. After ensuring that any gap between the plate and the panel was within tolerances, the astronaut would apply a thin trail of uncured NOAX sealant around the edge of the repair as further protection against plasma infiltration.

This plug system had the potential to repair a hole up to 6 inches in diameter. For a hole that was less than 1 inch across, the astronaut would use a pistol grip tool and

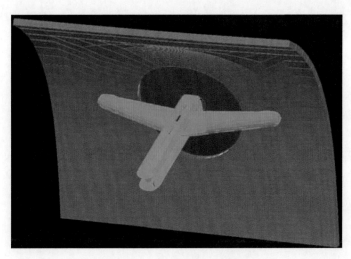

RCC plug installation configuration.

a special bit to drill out the hole. It was estimated that 20 to 30 plug sizes and shapes would be needed to cater for all possible RCC panel damage locations. Fortunately, this repair was never required by any of the post-*Columbia* missions. The only time it was attempted was during STS-114. Because the repair was straightforward, mission planners did not require it to be performed during a spacewalk. It was validated by having the astronauts demonstrate installation techniques for a variety of plugs inside *Discovery*'s cabin.

Tile repair: emittance wash

One important quality concerning the performance of the TPS is a physical parameter called emissivity, which defines the ability of a material to reject heat. The tiles on the Orbiter's belly bore the main duty of protecting the structure from reentry heat. Their outer surface was coated with a thin black protective layer of reaction-cured glass, a substance with a high emissivity value. Any breach of this thin coating would expose the white silicon, which had a much lower emissivity value, especially at high temperatures, thereby significantly reducing the protection of the underlying structure.

NASA therefore developed a tile repair technique known as "emittance wash". As Lora Bailey, an EVA planner at the Johnson Space Center explained, "The idea of emittance wash is to apply a coat of a thick kind of dark gray paint to areas where the black tile coating has been cracked and removed." In a nutshell, the science behind the emittance wash repair called for replacing a damaged tile's coating to restore its ability to reject the high temperatures of re-entry. In this manner the emissivity of a damaged tile could at least be partially restored, increasing the heat rejection through radiation and preventing small gouges from becoming deeper holes.

280 The Orbiter's skin: the thermal protection system

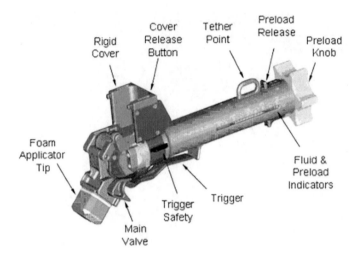

Emittance wash applicator.

The substance utilized for this task was a sticky, caulk-like mixture of fine-grit silicon carbide granules and a room temperature vulcanizing material. A special tool was developed. This emittance wash applicator resembled a dauber-like applicator. It was complemented by brushes to cover larger areas and wipes to remove excess material from spacesuits, gloves and tools. On the first spacewalk conducted during STS-114, Japanese astronaut Soichi Noguchi demonstrated the practicality of this concept.

Tile repair: tile ablator

Given the broad range of damage dimensions that could impair the TPS, another tile repair technique was developed. Shuttle tile ablator 54 (STA-54) revived efforts in the 1970s to develop an ablative material to fill gaps caused by tiles that were lost or damaged during launch. Ablatives such as those used for the heat shields of the Mercury, Gemini and Apollo capsules were designed to partially burn away during re-entry. Engineers improved the earlier formula for a silicone-based, cure-in-place ablative material to fill cavities in tiles or to substitute for missing tiles.

This material consisted of a two-part room temperature material – a base material and a catalyst – mixed together during application with a caulk-like gun. Astronauts would have to apply the material into the damage to fill it in. Since the material was sticky with a consistency similar to cake frosting when dispensed, and was to adhere to tile, tools such as foam brushes and tampers would be used to smooth the repair material without sticking. For very large or very deep gouges, the emittance wash could be used as a primer for STA-54, since its ability to wick into the tile substrate encouraged a stronger bond between the tile and the STA-54 repair material, as well as providing protection along the edges of the repaired area when they were under-filled in order to allow for ablative swelling. To accommodate the curing time of 24

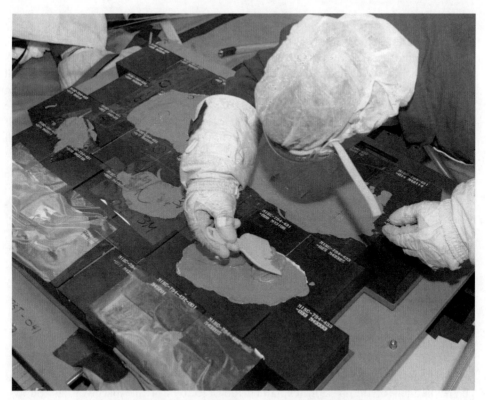

An astronaut tests techniques for STA-54 application.

to 48 hours, a second spacewalk would be necessary to inspect the repair and test its hardness. This information, along with photographs of a dissected "test bed" created at the same time as the repair, were required in order to certify the repair for re-entry.

Born from the CAIB's recommendations to develop reliable means of repairing the TPS, this STA-54 was not an easy material to tame. Initially expected to be tested during the very first return-to-flight mission post-*Columbia*, NASA had to cancel the test due to difficulties in working with the material in a terrestrial vacuum chamber. The real problem was vacuum bubbling. Hill, the lead flight director for STS-114, described the issue as follows, "The biggest concern we have right now is ... we have started seeing bubbling in the material in tests in vacuum. This material has little micro balloons of silica, little glass balls, in it. The little glass balls have air in them. One of the concerns was that as we put this material in the flight gun, the pressures we're operating at as we push it through the static mixer ... causes those little micro balloons to break and the very small quantity of air in those micro balloons, we've shown on paper, would be enough to show the bubbles we're seeing."

One solution was to reduce the pressure in the gun, but this meant a decreased capacity for the material to bond with the damaged tile. In itself, the bubbling was not a serious problem on Earth, since bubbles tend to rise to the surface, pop and then disappear. Things are different in the weightlessness of space. "In orbit, there is some concern amongst the materials guys that as these bubbles form inside the material, they're going to coalesce together and they may form a big pocket," Hill explained. "Then during deorbit, instead of getting a nice, glassy external surface, instead we'll burn down a little bit and hit that pocket and then the plasma will be channeled down into that damaged cavity and will dig that patch in the damaged tile right out." He continued, "There are two main technical questions about this repair technique. One, if we can't make this bubbling go away like we thought we had, can we tolerate it? Or will it cause large voids that will end up burning through? Secondly, when we extrude this into tile in zero-G and in vacuum and in the orbital thermal environment where we could be swinging significantly in temperature, are we confident we are going to get a good enough bond to the powdery tile surface so that the patch isn't going to break loose?" Due to the many uncertainties in this repair technique as the launch date of STS-114 approached, astronaut Andy Thomas said, "The problem that the office has, is not actually ... the idea of squirting the material out, it's the material itself. It's not performing the way we'd like because of the bubbling, it's outgassing and so on. So I think a lot of work needs to be done on the material itself." This was backed up by chief astronaut Kent Rominger, "It is a really challenging problem. But we're at the point ... where we think we're better off without it on STS-114."

The on-orbit test happened on STS-123, a mission whose main objectives were to deliver to the ISS the first part of the Japanese laboratory and the last element of the station's robotic arm. Spacewalkers Bob Behnken and Mike Foreman were assigned experiments in *Endeavour*'s payload bay to try out STA-54. Zeb Scoville, the lead EVA planner for the mission, explained the objectives of the spacewalk as follows, "The are a number of different sizes and shapes of samples, and really this correlates to the different objectives we're trying to get out of this test. Some of our tests are going to be involving a study of the material itself, how it adheres to tile substrates, how it expands, if it bubbles, what sort of density it's going to have. Other objectives of this test are to focus on how well the crew can operate and perform. It's one thing to be able to repair a very evenly machined sample. It's another thing to have a divot or pock mark that's been cut by an ice impact of foam damage." The bubbling issue was fundamental. "One of the big questions we have in zero-G is, are those bubbles going to rise to the surface or are they going to act more like a bread loaf as it bakes with the gas expanding in the material and being evenly distributed bubbles that then cause the surface to rise up? Surface smoothness is a big key in understanding how this will react during a re-entry scenario. If you have a lot of bubbles and expanded ridges and what not, this can ... cause a turbulent flow transition, which can cause downstream heating and damage to the Orbiter. So understanding how this material is going to react and expand, and what we can do to control that, is really one of our primary objectives of the this test."

Mike Foreman was designated to carry out the battery of tests using a special tool called the tile repair ablator dispenser (T-RAD) to apply the STA-54 to a number of samples mounted on the outside of the *Destiny* laboratory module. The T-RAD was a caulk-like gun in which a single carbon dioxide pressurized vessel was separated into two main sections. The mixing and delivery system involved a static mixer, a 3-foot-

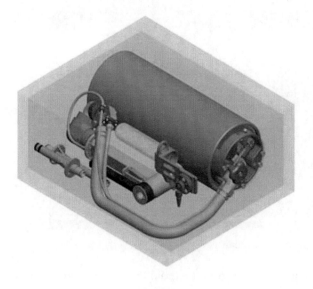

T-RAD.

length of hose, and an applicator gun which controlled the flow of the extrusion. The assembly was about the same size as a hand-held vacuum and weighed 55 pounds. It replaced the cure-in-place ablative applicator (CIPAA) which was carried on STS-114 without being tested. The cumbersome CIPAA consisted of an EVA backpack with tanks that held the base and catalyst components under pressure, with paired hoses to transport the components to the damaged area. But the applicator gun was incorporated directly into the T-RAD design.

The experiment was a success, and Foreman had time for some fun. "Looking for a success here," he said as the work began. "You're Captain T-RAD, Mister Goo," Richard Linnehan told him from inside *Endeavour*. "You are in control today." The STA-54 started to bubble as expected, but Foreman used pads to tamp the material down. "You're going to be our tile and grout specialist," Linnehan said as Foreman worked the material into a cavity. "I hope we don't need one," Foreman replied. He later said, "I can see the bubbles under the surface, those little nodule-type bubbles, they're still forming in the material but it's not building as much as it was ... It's still bread loafing, but it seems like when I hit it with the tamping it doesn't bounce back quite as quickly, or as much." The test ended successfully, leading to the certification of this repair technique. This closed out the experiments begun by STS-114, providing a variety of repair options for the few missions that remained in the program.

Tile repair techniques: overlay repair concept

The tile overlay concept developed by the Johnson Space Center was for on-orbit repair of extensive damage to the thermal protection system in order to facilitate safe re-entry. It was a flexible carbon/silicon carbide plate 25 inches long, 15 inches wide and 0.04 inch thick, backed with a layer of fibrous insulation to cover the expanse of damaged tiles. Around the edges, between the overlay and the existing tiles, a fabric gasket would prevent plasma from penetrating beneath the overlay. The gasket and the plate would be secured by auger-like fasteners pushed through holes at discrete locations around the perimeter of the plate.

STS-114 astronaut Andy Thomas explained the concept as follows, "It's actually

284 The Orbiter's skin: the thermal protection system

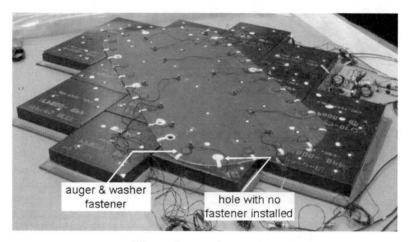

Tile overlay repair concept.

pretty interesting what they've come up with. It's a material that you would basically screw on, and they've come up with methods of making fasteners that can take all the heat ... The material is impregnated with a glass, so it kind of liquefies a bit and fills the voids. It potentially has applications for RCC." From a given point of view, this technique could be judged superior to the chemical method of applying the STA-54 because it was a mechanical fix and therefore immune to the space environment (for example, it would not bubble). Furthermore, unlike the STA-54, the overly plate was able to be stored indefinitely because it did not have a shelf life.

This mechanical repair concept was not new. It was something that the aviation industry had learned to do a long time ago. If a dent or a gouge on the skin of an aircraft fuselage exceeds the limits set by the manufacturer, the damaged section is reinforced by means of an external (or internal, if applicable) patch or "doubler" that restores the static strength of the skin panel. Obviously there were pros and cons. "It would have some pluses and minus," Thomas said in an interview. "You don't have the problem of the material flowing – the adhering problem, which is really a very tricky problem. But you've got a material edge that you wouldn't want air flow to get under, so you've got to somehow button that down. The fasteners that I've seen are actually pretty slick, though. They're low profile, very smooth and flush mounted. They're made of RCC-type materials. And by all accounts they seem to work very well." As part of the development, a number of vibration and flutter analyses of the overlay panel were made in order to determine the onset of flutter and identify the uncertainties in the analyses.[2] These analyses indicated that at Mach numbers above 1.0 there were adequate panel flutter margins within the Shuttle's flight envelope.

During the development of the flight plan for the first return-to-flight mission

---

[2] Panel flutter is a self-excited dynamic-aeroelastic instability of thin plate or shell-like components of a vehicle occurring frequently, though not exclusively, in supersonic flow.

post -*Columbia*, it was initially intended to include an in-cabin demonstration of this tile repair concept, but it was rejected owing to safety issues concerning the fine particles that could pollute the air or choke the environmental control system. Shuttle program manager William Parsons said of this decision, "We've decided to carry the overlay and tools up, but not pull them out and do a demonstration. The repair technique is so straightforward that we don't believe the demo is really necessary. We may pull them out and show them, who knows, but we won't demo." As it turned out, STS-114's astronauts did not need to repair any extensive damage to their TPS and this concept was never attempted.

In-flight Orbiter inspection

One of the CAIB recommendations was to provide "an on-orbit Thermal Protection System inspection, using appropriate assets and capabilities, early in all missions." The response to this recommendation (R6.4-1) was the development of the Orbiter boom sensor system, which was a 50-foot extension of the robotic arm along with specific sensors.

Although the robotic arm was potentially capable of inspecting part of the TPS on its own, the OBSS extended that reach to all critical areas of the leading edges of the wings and the underside. The OBSS was assembled by MD Robotics in Brampton, Ontario, Canada, which also manufactured the remote manipulator systems for both the Orbiter and the International Space Station. The OBSS combined two 20-foot-long graphite cylinders (originally manufactured as Orbiter arm replacement parts) joined by a rigid fixture. It had a modified flight releasable grapple fixture. The boom was installed on the right-hand payload bay sill, using the same type of positioning mechanism as the robotic arm on the other sill. The arm would grasp the boom using a grapple fixture on the forward side of the boom, and maneuver it so that the battery of imaging sensors on the other end of the boom could inspect the areas of interest.

The imaging system, powered by cables running along the boom, consisted of a laser dynamic range imager (LDRI), an intensified television camera (ITVC), and a laser camera system (LCS). The first two were on a pan-tilt unit for pointing, but the last one was hard-mounted to the side of the boom, just behind the other instruments.

Manufactured by Sandia National Laboratories in Albuquerque, New Mexico, the dynamic range imager comprised an infrared laser illuminator and an infrared camera receiver. It generated two-dimensional or three-dimensional video imagery that was downlinked to Mission Control via the Orbiter's high bandwidth Ku antenna system. The television camera was less sophisticated, being the same low-light, monochrome camera used in the Orbiter's payload bay. These two imaging systems could not be used simultaneously. The laser camera manufactured by Neptec in Ottawa, Ontario, Canada, was a scanning laser range finder which generated computer models of the scanned objects with a resolution of several tenths of an inch. The camera used a scanning modulating laser beam, produced images on a CCD, and processed them to provide depth information. In this case, the output

OBSS.

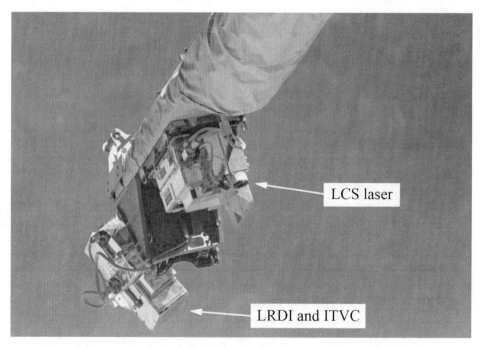

Camera sensors on the OBSS.

was not video but data files stored on a dedicated laptop. For STS-121 another digital camera was added to the boom's sensor package. This integrated sensor inspection system digital camera (IDC) was packaged with the laser camera system and had sufficient resolution to see minuscule damage on the wing leading edge panels. It was to help to distinguish "brown spots" and gap fillers as observed by inspections on STS-114, the first OBSS mission. The data was downlinked to the ground for processing. Since the LDRI and LCS had to remain within 10 feet of their target for acceptable image quality, and the arm and boom had not to contact any of the Orbiter's surfaces in the process, a combination of automated and manual arm operation modes were used. For each specific area, such as the nose cap, wing leading edge, and so on, an auto sequence was loaded and executed by the arm with the astronauts monitoring the survey and ensuring that the boom never strayed too close to the surface of the vehicle. The only times that the astronauts could manually operate it was at the end of a given sequence, to place the boom in position to start the next survey sequence.

Although the OBSS was mainly regarded of as a means of inspecting the most vulnerable parts of the TPS, the flight plan for STS-121 called for testing the boom as a platform for a contingency EVA in the event of the OBSS sensors failing or if it proved necessary to perform an on-orbit TPS repair, especially for flights that were not to the ISS. Ground simulator models had been developed to train crews, but there was no engineering data from an actual EVA in space to validate those models and certify repair hardware and procedures.

An artistic concept of Orbiter TPS in-flight inspection.

Steve Lindsey, the commander of STS-121, explained the issue as follows, "So, the scenario is you have a problem, you want to go repair a tile or a leading edge panel or something like that. Can you put one or two spacewalkers out on the end of this boom and maneuver them underneath the vehicle? Is the platform stable enough to allow them to do repairs? If an EVA crewmember puts a load or pushes at the end of the boom, how much flex do you get in that boom? How much flex do you get in the arm? Do the joints on the arm back drive? How much do they back drive? And is it stable enough to be able to do precise work? The EVA crewmember will simulate doing a repair and say 'Can I do a repair with the arm in this configuration?' So, the goal is to be able to figure out at what points underneath the Orbiter do we need an additional means of stabilizing the crew, like putting something to stabilize them on the bottom of the vehicle itself? Or, can we get away without doing that at all?"

To answer these intriguing questions, the EVA plan included several different combinations of crew member configurations, simulated tasks, and arm positions to gather information on as many variables as possible. The EVA would start from a stable configuration (stiffer arm position with a single astronaut) with fairly benign operations and progress to less stable configurations (less stiff arm positions with two astronauts) and more aggressive operations. To collect the loading and acceleration data, special test hardware was installed on the foot restraint attached to the OBSS. The spectacular first EVA of the mission, performed by British-born astronaut Piers Sellers and astronaut Mike Fossum, successfully proved the OBSS as a stable platform should the need arise for carrying out a contingency TPS repair in orbit.

TPS GROUND MAINTENANCE

The spacewalks of STS-114 and STS-117 represented the exception rather than the rule in TPS maintenance. Much more typical and routine was the maintenance of the thousands of tiles on the ground during the turnaround between missions. Several well-established techniques for inspection and repair were available. Before starting a repair, it was necessary to assess the integrity of the tiles by specific inspections. As White explains, "Our first one is a quick look at the runway. Of course, we can't see all the vehicle at the runway because we just can't get up next to it, but we basically walk around underneath. There are some specific areas we measure while they're still hot, where the gap's expanded ... looking for large damage ... We actually photograph everything before we leave." Once the vehicle was in the OPF, "We do what's called a micro inspection, up close, very personal, shining light down inside the gap, 100 per cent of the vehicle." During these inspections tiles could be repaired.

For minor damage a repair commonly used

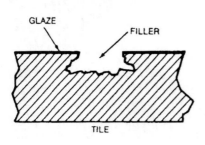

Small damage to a tile.

 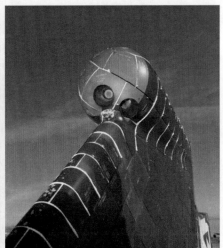

The SILTS pod on *Columbia*'s vertical fin.

The tile modified for the BLT transition experiment.

was the so-called putty repair, which involved drilling the damage out of the tile, filling the void with a paste of the same material and curing it on the ship for about three and a half hours. Once the curing was completed, the repair was sanded down to reinstate the exact contour and finally waterproofed. As straightforward and efficient as this seems, there was a downside. "The majority of the tile material is 9 pounds per cubic foot ... but the repair material is 68 pounds per cubic foot. So you're adding a lot of weight," White notes. For this reason, this repair technique was employed only for minor damage. For larger cases, "We drill a hole into it on the ship, and we have another piece of tile material that is a cylinder and we glue it with ceramic cement ... So basically we just put a new plug of tile material in the damaged area."

In the aftermath of *Columbia,* hundreds of tests were carried out to understand the effects of one of these repairs coming off the tile. The results were not satisfactory at all. As White recalls, "We learned that a large repair ... could cause severe damage to anything aft of it. If we lost a repair forward, it could potentially take out one of the wing leading edge RCCs. It could cause severe damage around the main landing gear. It's like a chunk of rock coming out at a very high speed." As a result, all of the repaired tiles on the remaining Orbiters were replaced with brand new tiles.

TPS FLIGHT TESTING

After STS-9 in 1983, *Columbia* was returned to the manufacturer for a series of upgrades. One modification was the installation of a battery of experiments called the Orbiter experiment (OEX) whose aim was to obtain data for physics, aerodynamics, thermodynamics, structures, and materials in flight and on-orbit.

One of these experiments was the Shuttle infrared leeside temperature sensor (SILTS) to obtain high resolution temperatures of the upper (leeside) surfaces during aerodynamic flight and re-entry. It made use of an infrared camera in a pod placed on top of the vertical stabilizer, from where it could produce a thermal map of the upper surface. Data from this experiment helped to improve the operational capability of the vehicle by reduction of the upper thermal protection system, saving weight and refurbishment costs. The SILTS camera was removed after the OEX program was completed but the pod was left atop the fin.

Another interesting experiment was developed in the aftermath of the *Columbia* accident in order to better understand the phenomena of boundary layer transition. Any step or protuberance on the underside of the Orbiter could initiate the transition of the boundary layer from laminar to turbulent flow, creating an even hotter plasma downstream that could damage the thermal protection system. This boundary layer transition had been studied using many complex mathematical models, but it was decided that data was required from real flight. A manufactured protuberance tile was fitted on the port wing of *Discovery* and flown for STS-119, STS-128, STS-131 and STS-133, and on *Endeavour* for STS-134. Additional instrumentation was installed in order to obtain more spatially resolved measurements downstream of the protuberance. The data from these flights helped to validate the mathematical models and to refine them for future studies of re-entry vehicles.

9

Auxiliary power unit and hydraulic system

INTRODUCTION

When Stan M. Barauskas was given the assignment of developing an auxiliary power unit (APU) system for the Shuttle, his initial thought was, "'Oh, auxiliary power unit doesn't sound very important to me.' Auxiliary means you may need it, you may not, it's an occasional use type device." He was correct with regard to the meaning of the word "auxiliary", but the simplicity stopped there. In fact, the difficulties in devising, designing, testing, and operating the APU would keep him busy from the onset of the approach and landing tests using *Enterprise* up to the very end of the program in July 2011; fully 37 years devoted to this seemingly simple component of the Orbiter.

When the Shuttle was approved for development, it came with the requirement of a hydraulic system. That was because the Orbiter had aerosurfaces, something that no previous spacecraft included. To minimize development costs, it was decided to use the proven technology of a conventional aircraft auxiliary power unit. Unfortunately, the power supplies employed by aircraft to drive their hydraulics are not designed to operate in the space environment. The challenges imposed upon Barauskas were, for example, to develop a hydraulic power unit that could operate at temperatures in the range $-54°C$ to $107°C$, with power levels in the range 8 to 148 horsepower, enduring accelerations ranging from 3.3 g (boost) to 0 g (on-orbit) and 1.5 g (landing shock). They were to be capable of at least two restarts during each mission. And they were to be usable for the 100-mission life of the Orbiter. With these requirements in mind, Barauskas changed his mind about the scale of the task, "The more I learned about it, the more I realized that [auxiliary] was a misnomer. It ought to have been called the *primary* power unit because it does provide all the power for the Orbiter to operate like an aircraft on landing, and it's got a critical phase during ascent where it controls the actuators that move the main engines and operate the engine valves. Its area of responsibility is far greater than what an APU does on an aircraft." As he explains, "On an aircraft, typically an APU starts the engines and thereafter provides auxiliary power for operating the air conditioning system, lighting, and that type of thing. But the function of the APU on the Orbiter is to provide the hydraulic power needed to control the flight surfaces and the flight controls during its ascent stage as well as descent. During ascent it supplies power to turn the main engine valves on and off and to throttle them. Sometimes the engines are throttled down to 67 per cent, and sometimes all the way up to 104 per cent. That is achieved hydraulically with power from the APUs. Once the ascent phase is over, the APUs remains dormant. Then they are required to operate in the descent, where they power all the elevons, the vertical tail, and the speedbrake. And the landing gear, the nose wheel steering, and the brake system are hydraulically powered through the operation of the APUs."

APU SYSTEM DESCRIPTION

The configuration initially chosen involved installing four APUs on the aft section of the fuselage of the Orbiter, since as Barauskas says, "It was deemed that three APUs were more than adequate to satisfy all the requirements for the Orbiter and that we'd need another APU as a backup. In case one the three went down ... the standby unit was ready to be put on line within a period of several milliseconds. It would be tied into the fuel system of the other APUs, and would take over the function of the one that failed." But this proposal was rejected when NASA eliminated the standby unit for the sake of simplicity and to save weight.

"Of course," Barauskas says, "when you speak of the APU you're really speaking of the APU *system*. This is composed of the APU unit itself, which is supplied by the Sundstrand Corporation, and the rest of the system that supports the APU is the fuel system and the water system. There are multiple other components, such as the tank, control valve, filters, and servicing quick disconnects. All of those come from other suppliers to make up the entire subsystem."

The final configuration consisted of a hydrazine fueled turbine driven power unit that generated the mechanical shaft power needed to drive the hydraulic pump which would supply pressure to the hydraulics. For each APU system, about 325 pounds of hydrazine were provided, more than sufficient for either a nominal operating time of 90 minutes or 110 minutes of continuous running during an ascent abort. There was a diaphragm in the fuel tank that was inflated by gaseous nitrogen to provide a positive force to expel the hydrazine into the fuel distribution line to maintain a positive fuel supply throughout the period of operation, and a fixed-displacement gear-type pump raised the pressure to that necessary to accomplish the reaction in the gas generator, what Barauskas calls "the heart" of the APU. "That is where the fuel is directed, and that's where the catalyst bed is. The fuel goes in and is sprayed onto the catalyst bed, and immediately on contact of the catalyst the pressure shoots up as high as 1,400 psi and the temperature to about 1,200 degrees or so. That happens in seconds."

This rapidly expanding hot gas makes two passes through a single-stage turbine wheel, passes over the outside gas generator chamber for cooling, and is then vented overboard through exhaust ducts near the base of the vertical stabilizer. After night landings, the red hot flames from these exhaust ducts led naive observers to suspect a

APU system description

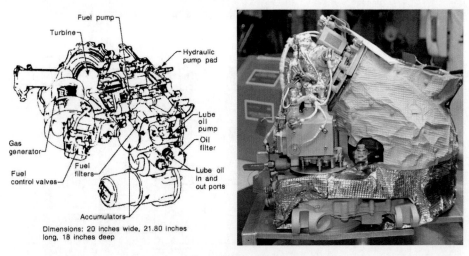

Orbiter Auxiliary Power Unit.

The APU turbine exhausts issued visible flames during night landings.

fire in the aft structure. After being reduced by a speed reduction gearbox, the shaft power generated by the spinning turbine was used to drive the fuel pump, hydraulic pump, and lube oil pump.

The operating speed was controlled by an electronic controller, speed sensors, and the gas generator valve module (GGVM); an assembly of a primary and a secondary

Auxiliary power unit and hydraulic system

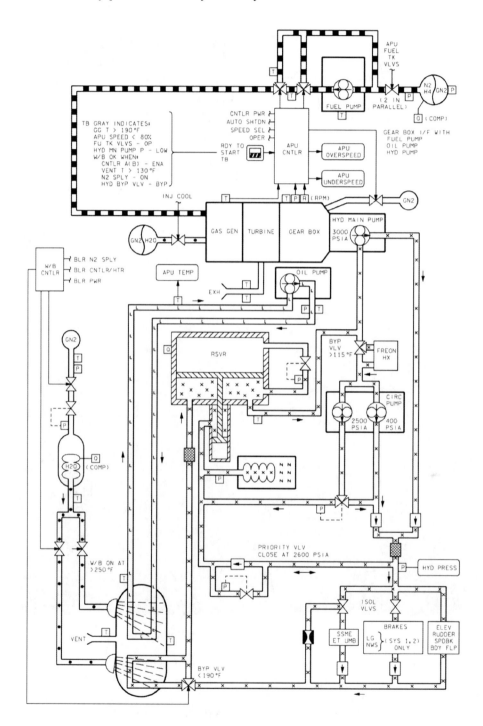

APU/hydraulic systems overview.

fuel control valve installed in series downstream of the fuel pump. Each APU was started just 5 minutes prior to the scheduled time of liftoff. As the fuel pump was not initially driven, fuel expelled from the tank was passed through a bypass line directly to the GGVM and then into the gas generator. There the fuel decomposed through a catalytic reaction to create hot gas that was directed to the turbine to set in motion the gearbox, which in turn started the fuel pump, lube oil pump, and hydraulic pump. As the APU built up to its operating speed, the underspeed logic check was inhibited for 10.5 seconds after the start command was issued. Then the shutdown logic watched for a speed of less than 80 per cent. The overspeed logic was activated at startup so that in the event of an overspeed condition the controller would automatically close the tank isolation valves and the valve module. When the turbine speed exceeded the control target during the startup, the bypass line was closed and the fuel sent directly to the fuel pump.

The APUs continued to run for 5 minutes after orbit insertion in order to allow the main engine fuel lines to be purged and to shut down the engine valves. For the rest of the orbital flight, the APUs were inactive. Twenty-four hours prior to re-entry, one APU, selected by Mission Control, was activated for a hydraulic system checkout to make sure that all of the flight control systems were functional. Five minutes before the deorbit maneuver, an APU was switched on to guarantee that at least this one unit was fully functional. Thirteen minutes prior to the entry interface, the other two were started, and all three would continue to operate well after wheel stop. "That's fairly unusual," Barauskas explains. "Most of the subsystems in the Orbiter, they're pretty quickly shut down after wheel stop upon landing. When you are in Mission Control in Houston, you'll see the various engineers at their consoles get up and leave shortly after wheel stop. But the APU engineers intently watch their data for another 15 to 18 minutes because the APUs continue running to allow the systems to be reconfigured in what is called a "rain drain" configuration. The body flap is allowed to go down to allow the engines to go down. The reason they call it "rain drain" is in case rain does happen on the runway, the rain does not enter in the injectors of the main engines."

APU water spray boiler and lube oil system

The APU system had a water spray boiler (WSB), an evaporative heat exchanger in which APU lubricating oil and Orbiter hydraulic fluid could be cooled down. There were three identical independent WSBs, one for each APU/hydraulic system, located in the aft fuselage. In each boiler, hydraulic fluid and APU lubricating oil lines were sprayed with water, stored in a bellows-type tank that was pressurized with gaseous nitrogen. On coming into contact with the hot lines containing the fluid, water evaporation cooled the lines and the fluid inside.

Prior to launch, the WSB was preloaded with water mixed with propylene glycol monomethyl ether (PGME), an additive for what was called the "pool mode". In the pool mode, the boiler held enough water mixture to immerse the hydraulic fluid and lubricating oil cooling tubes. This mode was used during ascent and re-entry. During ascent, the APU lubricating oil heated up. About 10 minutes into the ascent the pool water boiled off and the boiler went into the spray mode. The hydraulic fluid usually did not heat up sufficiently during the ascent to require water spray cooling. At an altitude of about 17,000 feet during re-entry, the boiler core temperature reached the boiling point of water for the given altitude or internal pressure, and the boiler was returned to the pool mode. The sprays were turned off when the water reached the liquid level sensors in order to prevent the boiler overfilling. The steam was expelled overboard through a vent situated to the right of the vertical stabilizer. The water boiler, water tanks, and steam vent were equipped with heaters to prevent them from freezing on-orbit.

APU development challenges

Numerous obstacles had to be overcome to provide the APU system that ultimately went on to demonstrate high efficiency throughout the Shuttle program.

The first technological challenge was "heat soakback", a problem that would arise if a restart was initiated to perform an emergency re-entry shortly after the APUs had been shut down on-orbit. As Barauskas explains, "The gas generator operates at well over 1,000°F. Once it shuts down it immediately begins to cool, but what happens is that the temperatures soak back into the rest of the APU hardware, specifically into the valve and the fuel pump – and those components reach fairly high temperatures. We found out through development testing that if you attempt to start the APU at the elevated temperatures, the fuel contains such high energy that explosions occur." The solution to this problem was to add a water system to cool the components so that the APU could be safely restarted in an emergency. Water cooling was required for the fuel pump, gas generator valve module, and gas generator. But Sundstrand devised a passive cooling system consisting of additional fins, standoffs, and isolators which reduced the peak temperatures on the fuel pump and valve module so efficiently that the water cooling system for those components was able to be deleted, with resultant weight reduction. "The thing that they retained was the injector cooling," Barauskas says. "That was still considered critical. The injector cooling tank was a fairly small tank. It only contained about 9 pounds of water, as opposed to 20 pounds for each of the other tanks. That was retained only for an emergency, in case the APU had to be restarted within seconds or minutes of its shutdown."

"Another problem was the seal design," Barauskas continues. "The fuel pump seal design was a common seal between the fuel and oil [lubricant] systems. We found out, depending upon the pressure balance, often the fuel side of that seal was higher pressure than the oil side, allowing some small seepage of the fuel into the oil. The fuel would go into the oil and form a wax-like substance. Depending on the ratio of the fuel and the oil mixing, this waxy substance could be as fluid as molasses or as hard as candle wax. It would even fracture. If you were to hit it with a hammer, it would fall apart in granules." It is not difficult to appreciate the issue. Should a fuel leak develop in the oil system, this wax-like substance could very easily clog the oil system and prevent proper cooling of the high speed bearing within the gearbox. This problem was solved in the improved APU simply by having two separate seals, one on the fuel side and the other on the oil side.

The worst APU problem during the 135 missions of the Shuttle program occurred on board *Columbia* at the conclusion of STS-9. When the APU 1 & 2 data started to misbehave during the "rain drain" procedure it was unclear whether this was merely an instrumentation problem, so it was decided to send someone to take a look. This fell to Barauskas. When he climbed inside the aft fuselage he could not believe his eyes. As he says, "The fuel pump on APU 1 was blown to pieces. These were pretty heavy-duty walls, three-quarters of an inch thick. They were totally broken and their parts were just all around. The valve covers had blown off, there was instrumentation cabling that had burned off completely, there was evidence of fire all around the area around APU 1. Then I climbed up a little bit higher to take a look at the condition of APU 2. It was the same thing all over again. There was quite a bit of damage. There was fire all around, and I knew a major catastrophe had occurred." An investigation revealed that the failure for both APUs was due to cracking of the injector tube. But the root cause was considerably more interesting. It was discovered that there was a direct correlation with the downtime between flights, a parameter that was to become a critical factor for the APUs. "The exhaust gases actually created a very, very harsh environment within the injector system. Chemicals like carbazic acid played a role in that. Any slight leakage from the GGVM into the gas generator created these gases as well. They were in very minor amounts, but over long periods of time, like months in between, they came to be so corrosive that they would actually attack some of the materials within the APU itself." The combination of the installation procedure that was used and chemical attack by these gases had created a highly stressful environment that resulted in the explosion of these two APUs.

As a result of the investigation, several improvements were applied to the injector tube. "A chromizing was done, to introduce a protective layer in the internal part of the tube. A procedure was initiate that very carefully controlled the installation of the injector tube and measured how much strain was being introduced into the tube as it was being assembled into the valve. They had to develop special mechanical devices that very carefully inserted that tube into the valve to prevent these high stresses from being created, so that it was in a stress-free environment from that point on."

As a result of these changes, the APU system never showed any further problems during the remainder of the program.

HYDRAULIC SYSTEM

The hydraulic system involved a hydraulic main pump, a reservoir and accumulator, a circulation pump, a Freon heat exchanger to warm the hydraulic fluid, and heaters. The main pump was the *raison d'être* for the APU, since it was the means by which pressure was provided to the hydraulic fluid in order to be able to throttle and gimbal the main engines, retract the ET umbilical plates, move the aerodynamic control surfaces, lower the landing gear, apply the brakes, and provide nose wheel steering.

298 Auxiliary power unit and hydraulic system

The hydraulic line for each APU was provided with a reservoir to ensure positive head pressure at the main and circulation pump inlets. This not only provided for thermal expansion and contraction of the hydraulic fluid, it would also counteract a hydraulic leak if one occurred. An accumulator pressurized the reservoir through a 40:1 differential area piston. It also dampened out pressure surges in the system that arose from the varying demands. When the hydraulic pump was not running on-orbit, because the APU was shut down, the accumulator maintained pressure on the main pump inlet so that the system could be restarted in weightlessness. If this pressure were absent, the pump would have suffered cavitation, resulting in a loss of hydraulic power.

In addition to the main pump, each hydraulic system had a circulation pump that was used on-orbit when the main pumps were off, to maintain accumulator pressure and for hydraulic thermal conditioning. These pumps were activated when either the hydraulic lines were cold and therefore in need of thermal conditioning, or when the accumulator had to be repressurized. The circulation pump actually consisted of two fixed-displacement pumps in parallel, using a single driving motor. One was a high pressure (2,500 psia) low volume pump to repressurize the accumulator. The other was a low pressure (200 psia) high volume pump to pass hydraulic fluid through a heat exchanger on an ECLSS Freon coolant loop to prevent the hydraulic fluid from getting too cold. The hydraulic lines in the various aerosurfaces were also warmed by heaters that were automatically controlled by thermostats to maintain the hydraulic line temperatures within a specified range.

10

Fundamentals of the Shuttle GNC

GUIDANCE, NAVIGATION & CONTROL

If we were asked to specify as briefly as possible what spaceflight is, we could just use three simple words: guidance, navigation and control. But if we want to be a little more prolix, we could say that it involves answering two questions, *"Where am I now?"* and *"How can I get from here to there?"* It is remarkably difficult to answer these deceptively simple questions. During a dynamic phase of a mission, such as launch where enormous velocities and forces are at play, even the smallest mistake in determining the position of the vehicle can have disastrous consequences. The same holds true in re-entering the atmosphere, where, for example, a small miscalculation in measuring the position of a vehicle along the re-entry path can cause it to miss its intended landing point by thousands of kilometers and finally crash. Orbital flight, on the other hand, is much less dynamic, and for this reason can be likened to sailing a ship on the ocean. A sailor must know where he is, and in which direction he should travel to reach his destination. Similarly, a spacecraft in orbit must know its position precisely in order to avoid becoming lost in the vastness of the orbital ocean.

Before delving deeper into the subject of spaceflight for the Shuttle, a few general definitions are appropriate.

Navigation is determining the so-called state vector of the vehicle; that is to say, its position and velocity relative to a system of coordinates suitable for measuring its motion. This answers the question, "Where am I now?"

Guidance is the first step in answering the question, "How can I get from here to there?" Knowing the current state vector as a result of navigation, and the state vector that is desired at a given future point in time, guidance performs all the computations for the actions needed to fly the spacecraft to there.

If navigation and guidance together represent the brain of the spacecraft, then it is **control** that performs the sequence of maneuvers required to implement the changes computed by guidance. Flight control software converts guidance computations into effector commands to point and translate the vehicle, often using navigation data as a means of determining whether the desired state vector was achieved.

300 Fundamentals of the Shuttle GNC

Each of these disciplines is sufficiently tricky to keep dozens of engineers busy for a considerable period when planning a space mission, irrespective of whether it is a sounding rocket to fly a suborbital arc or a probe for deep space. And despite the generality of the definitions, the practical manner by which spaceflight is achieved differs from one spacecraft to another, and in some cases for different missions by a single type of vehicle.

This chapter outlines some basic foundations of guidance, navigation and control (GNC) for the Shuttle. Subsequent chapters will provide a more extensive discussion of each of these disciplines during launch, orbital flight, and re-entry.

COORDINATE SYSTEMS

The basic task of the navigation system of the Orbiter was to maintain an accurate estimate of the state vector with respect to time. In addition, during a rendezvous, it had also to update an estimate of the target's state vector to enable on board guidance to compute the commands to fly purposefully. The data also enabled Mission Control and the flight crew to monitor the status of the navigation system.

The navigation system employed the standard equations of motion, along with the information from a plethora of on board sensors and software models, and predicted ahead by a process called state vector propagation. In defining the state vector above, it was noted that this would be with reference to a mission phase-dependent frame of reference or set of coordinates. Since the equations of motion used by the navigation software (NAV) are based on Newton's laws of motion, any

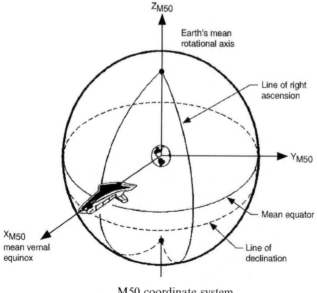

M50 coordinate system.

reference frame must be inertial, meaning that it is neither accelerating or rotating. Although astronomically speaking everything in the universe is accelerating, meaning that there is no *absolute* inertial frame, it is possible to devise a specific inertial reference for the purpose of spaceflight without any detrimental effect.

The main reference used by both the PASS and BFS to establish and maintain the inertial position and velocity of the Shuttle for all flight phases was the Aries Mean of the 1950 Cartesian system, also known as the M50 system. In this reference frame the X axis points toward the mean vernal equinox (the apparent point on the celestial sphere where the Sun made its northbound crossing of the Earth's equator) in the year 1950.

The vernal equinox is sometimes referred to as the "first point of Aries" because in ancient times its position was within the constellation of Aries. However, owing to the precession of the Earth's axis of rotation, the vernal equinox is now in Pisces. An additional effect known as nutation adds a wobble to the rotational axis, creating a shift in the point of the vernal equinox equal to 9 minutes of arc over a period of 18.6 years. In defining M50, the term "mean" averages out this nutational motion. For this reason, the Z axis points along Earth's mean rotational axis of 1950 with the positive direction toward the north pole. The Y axis completes the right-handed system, with the corresponding X-Y plane matching the plane of the equator. It is important to be aware that although the origin of this inertial frame lies at the Earth's center, the M50 system is completely independent of the Earth's rotation.

Using this system guaranteed that the normal equations of motion would be valid. But some computations are much simpler if performed in other coordinate systems. For example, when landing, the position of the Orbiter relative to the runway is more meaningful than its position in a coordinate system fixed in space. Consequently, the on board navigation system used a number of different coordinate systems in order to simplify the various inputs, outputs and computations. When necessary, the software would use coordinate transformations to convert between frames.

One of the most popular reference frames was the *Body Axis Coordinate System,* based upon the Orbiter's center of mass. The X axis was parallel to the X_0 axis of the structural body (i.e. the Orbiter plus the external tank and the solid rocket boosters) in the Orbiter's plane of symmetry and positive towards the nose. The Z axis was in the Orbiter's plane of symmetry, perpendicular to the X axis and

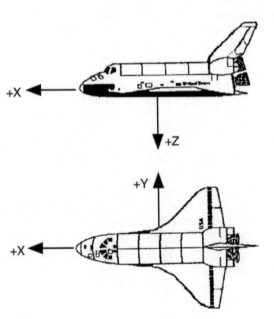

Body axis coordinate system.

302 Fundamentals of the Shuttle GNC

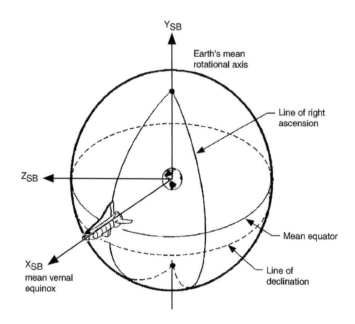

Starball coordinate system.

positive out through the belly. The Y axis pointed out along the right wing. Accordingly, pitch was defined as the motion around the Y body axis, roll as rotation about the X body axis, and yaw as rotation about the Z body axis.

Early in the space program there was concern about the ability of astronauts in space to locate and identify stars as a way of determining the attitude and position of their spacecraft in certain failure situations. A simple system was needed that would enable them to look at a star chart and then point their vehicle at a selected star. The solution was to define an inertial reference frame rotated 90 degrees about the X axis of the M50 coordinate system. This was called the *Starball Coordinate System*. The X axis of this system matched the X axis of M50, but the Y axis was parallel to the north pole of the Earth's axis in the direction of M50's positive Z axis. This meant the pitch angle was equivalent to right ascension and the yaw angle was equivalent to declination. Although it would simplify pointing towards a given star, this system was seldom (if ever) used.

A more frequently used reference frame was the *Local Vertical/Local Horizontal Coordinate System (LVLH)*. The Z axis pointed towards the center of the Earth along the geocentric radial vector of the vehicle (local vertical). The X axis (local horizon) was aligned with the velocity vector for a circular orbit. This right-handed coordinate system was completed by pointing the Y axis along the so-called negative angular momentum vector of the vehicle in orbit.

Since Shuttle orbits were invariably almost circular, the +X axis was very nearly in the direction of the velocity vector for all Shuttle orbits. Because the plane defined by the X and Y axes was perpendicular to the Z axis or local vertical, this plane was

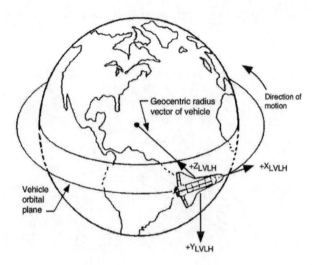

LVLH coordinate system.

parallel to the plane which was tangent to the Earth's surface point. When the Orbiter was in LVLH attitude with its pitch, roll and yaw all nulled relative to the reference frame, it was said to be flying in an "airplane" orientation with its belly pointing at the ground, its nose pointing in the local horizontal direction, and with its right wing pointing towards the south side of the orbit. Since this frame rotated along the orbit in order to maintain the X-Y plane parallel to the surface, it was not inertial; it rotated at approximately 4 degrees per minute with respect to the inertial frame.

For launch, landing and on-orbit Earth observations, it was advantageous to know the position and velocity of the vehicle in a coordinate system that was fixed to the Earth. A number of non-inertial systems were related to M50 by transformations that depended on time, the rotation rate of the Earth, and the latitude and longitude of the point of interest. One good example of such a frame is the runway coordinate system whose origin was at the approach threshold of the runway. Its Z axis was normal to the ellipsoid model of the Earth through the runway centerline and positive towards the center of the Earth. The X axis was perpendicular to the Z axis and in a plane that contained both the Z axis and the runway centerline, being positive in the direction of the landing. The Y axis completed the right-hand rule. This system was used by the navigation software and the crew during the ascent, re-entry and landing phases of a mission.

SHUTTLE NAVIGATION HARDWARE

The Orbiter had sophisticated hardware to perform the functions of both navigation and flight control. Although each piece of hardware was independent of the others, each was hard wired to at least one of the eight flight-critical MDMs, connected to all five GPCs. The hardware outputs were therefore combined at software level,

304 Fundamentals of the Shuttle GNC

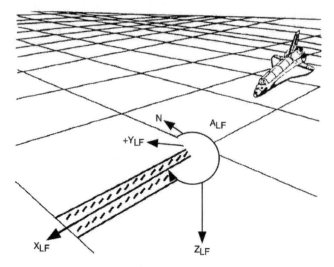

Runway coordinate system.

where data processing took place to determine the state vector and execute any maneuvers necessary for navigation purposes.

The navigation hardware included:

- Inertial measurement units (IMU) to sense the attitude of the vehicle in inertial space and accelerations along the inertial axes.
- Star trackers to automatically sense line-of-sight vectors from the vehicle to stars or orbiting targets.
- Crew optical alignment sight (COAS) to manually set up line-of-sight vectors from the vehicle to stars or orbiting targets.
- Tactical air navigation (TACAN) to sense the position of the vehicle relative to a ground-based TACAN station.
- Air data system to provide air temperature and pressure in free stream and air disturbed by the vehicle.
- Microwave landing system (MLS) to determine the position of the vehicle relative to ground stations alongside the landing runway.
- Radar altimeters to determine the altitude of the vehicle above the ground. If this data was not used for navigation or guidance, it could be displayed for monitoring purposes.
- Global positioning system (GPS) to measure the position and velocity of the vehicle by means of signals provided by a constellation of satellites.

Depending on the flight phase, one or more of these hardware elements could be used to supply data to determine the state vector.[1]

---

[1] Since some of these pieces of hardware were exclusively for the re-entry and landing phases, they will be discussed in Chapter 13.

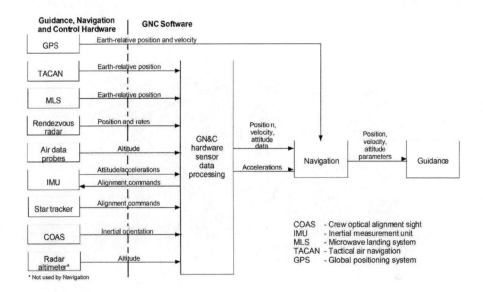

Navigation interfaces.

The inertial measurement unit

A mission could be achieved using only one IMU but for redundancy there were three units in what was called a navigation base, a structurally stable platform located inside the crew compartment, forward of the flight deck control and display panels. As the orientation of this base relative to the vehicle was known with great accuracy, it permitted the definition of coordinate transformations to relate measurements in the IMU frame of reference to any other frame used by the NAV software. The function of the IMUs was to supply inertial attitude and velocity data to the GNC software functions. In particular, the navigation software used the processed IMU velocity and attitude data to propagate the state vector. In parallel, the guidance software used the state vector from the navigation software along with the IMU attitude data to develop steering commands for the control software, which used the attitude data to calculate aerosurface, engine gimbal (thrust vectoring) and RCS jet firing commands.

Each IMU consisted of a platform that was isolated from any vehicle rotations by four gimbals that enabled the platform to maintain a fixed (inertial) orientation. The platform was directly connected to the innermost gimbal, called the azimuth gimbal. The inner roll gimbal was for redundancy to provide an all-attitude IMU that would not suffer gimbal-lock, a condition that can occur with a three-gimbal system and can cause the inertial platform to lose its reference. Disturbances to the inner roll gimbal were used to drive the outer roll gimbal, so that the inner roll gimbal could remain at its null position, orthogonal to the pitch gimbal.

306 Fundamentals of the Shuttle GNC

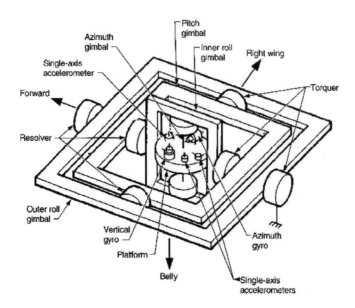

IMU platform assembly.

Each platform was equipped with three accelerometers, one on each of the X, Y and Z axes. An accelerometer is basically a force rebalance-type of instrument which causes a pendulum mass to displace when subjected to an acceleration along its input axis. This displacement is measured by a pick-off device that generates an electrical signal that is proportional to the sensed acceleration. The signal is then amplified and returned to a torque within the accelerometer, which attempts to restore the mass to its null position. Using the sensed accelerations to resolve the equations of motion, the software could determine the three position and three velocity components of the state vector. The GPC software read the IMU acceleration data at a rate of 6.25 hertz and transformed it from IMU platform coordinates into M50 coordinates.

Along with the accelerometers, each platform also had two gyroscopes possessing two degrees of freedom to determine the attitude of the vehicle. Exploiting the well-understood property of a gyro that angular motion about any axis perpendicular to its spin axis will induce a rotation about a third orthogonal axis, any time that such an angular disturbance was sensed the appropriate gimbals were torqued in order to null out the gyro rotation, with a two-fold result: the platform remained stabilized in the inertial frame, and the measured angles between the gimbals could be used to define the attitude of the vehicle. Inverting this gyroscopic principle, by applying inputs to torque the gimbals it was possible to put the platform into a new inertial orientation.

It is noteworthy that the IMUs were arranged in a so-called skewed orientation, meaning that they were not aligned with themselves or along any of the Orbiter's axes. This was because the IMUs were designed to provide a fail operational/fail safe

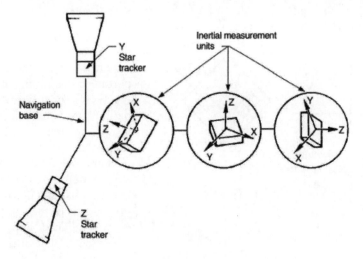

IMU skew.

level of redundancy. The first failure would be detected and isolated in the standard manner by comparing the outputs of the three units. The skewed arrangement was to enable a second failure to be identified. With the skewing, each accelerometer and gyro on the inertial platform sensed inputs along or around a unique axis. An input sensed by a single gyro or accelerometer in one platform could be compared with a composite value from components along the same axis sensed by a combination of instruments in the other platform. By making a series of comparisons with different combinations of sensors, it was possible to identify a malfunctioning instrument. As the system matured and analyses and test data were obtained, it became apparent that certain obscure failures at the dual-redundancy level were unable to be isolated. For this reason, an attempt was made to install a fourth IMU but the presence of the star trackers did not allow room for this. So it was decided to keep the triple redundant IMUs and exploit the built-in-test-electronics to the maximum in order to be able to detect up to 90 per cent of all possible failures.

To achieve the IMU performance requirements, very precise thermal control was maintained by means of internal heaters and a forced air cooling system. The internal heaters were fully automatic. They were activated during the IMU startup sequence, and continued to operate until the IMUs were powered down. The forced air cooling system had three fans that serviced all three units. The IMUs were kept in operative mode from the time the flight crew entered the vehicle through to landing, unless it was necessary to switch them off to minimize power consumption. As the accuracy of an IMU would deteriorate with time, the operative system compensated for most of the inaccuracy using a software calibration based on known errors. Nevertheless, it was sometimes necessary to realign the IMUs by means of procedures that involved using the star trackers, COAS, or head-up display (HUD).

308 **Fundamentals of the Shuttle GNC**

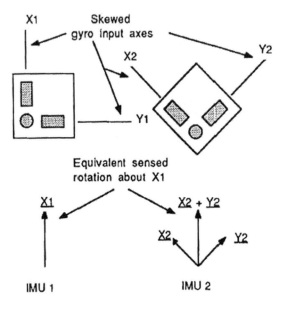

Skewed IMU approach.

Star trackers

As an extension of the navigation base, two star trackers were mounted outside the crew compartment as a means of aligning the IMUs for attitude determination based on the line-of-sight vectors relative to 100 stars in the sky chosen for their brightness and overall coverage of the sky, as well tracking targets for rendezvous calculations. The optical axis of the $-Z$ star tracker was aligned approximately along the $-Z$ axis of the Orbiter, and the axis of the $-Y$ star tracker approximately along the $-Y$ axis of the Orbiter.

Two different types of star tracker were used during the program, namely image dissector tubes (IDT) and solid-state star trackers (SSST). The IDT was the first type of star tracker used, and was a photomultiplier device in which the sensitive area was sampled using an electronic/magnetic scanning technique. It was superseded by the SSST, which had a CCD sensor identical to those used in digital cameras. Despite the differences in how the output was produced, both types of star tracker consisted of an electronic assembly that was mounted on the underside of the navigation base and a light shade assembly. The star tracker itself did not move while the sky was scanned in search for a particular target or star, and each operated autonomously without any redundant management system. Each tracker could be commanded to scan the entire field of view or a smaller field of view (1 degree square) that was centered on a point defined by horizontal and vertical offsets. In either case, when an object exceeding a predetermined brightness threshold (chosen either by the crew or by the GPCs) was located, the vertical and horizontal scan was switched to a cruciform pattern centered on that object. The intensity of the object was monitored and the position of the cruciform pattern altered to keep the pattern centered on the point of highest intensity until the object passed out of the field of view.

The purpose of the light shade assembly was to define the tracker's field of view (10 degrees square). A shutter mechanism could be operated manually by the crew or automatically by a bright object sensor or by target suppression software. This option was included to prevent the sensor being damaged by a very bright object such as the Sun or the Moon.

The star trackers were mainly used to align the inertial platforms of the IMUs. To do this, both star trackers had to acquire and track a given star in order to establish

Shuttle navigation hardware 309

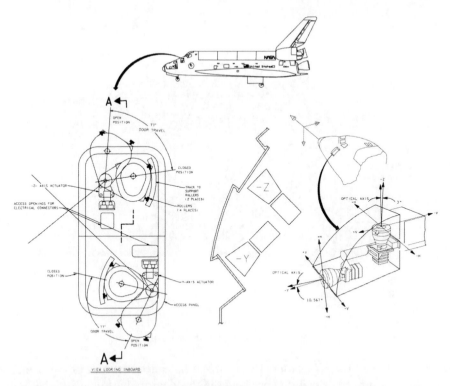

Star trackers.

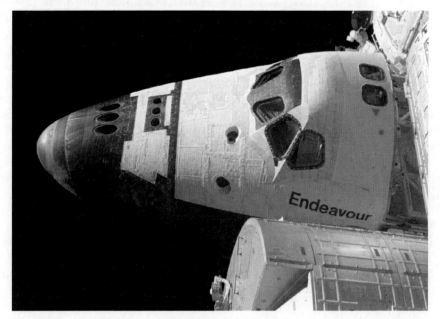

Star tracker doors open.

310 Fundamentals of the Shuttle GNC

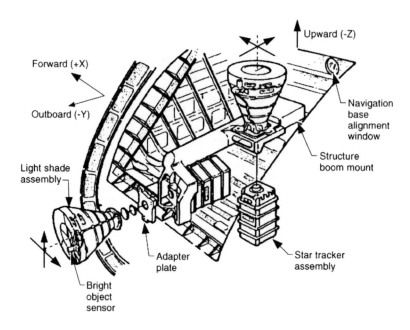

View into the star tracker cavity.

two different line-of-sight vectors. This enabled the inertial attitude of the Orbiter to be readily determined. The difference between the measured attitude and that sensed by a given IMU provided a measure of the inaccuracy (or drift) that had accumulated since the previous alignment. Torqueing commands could then be calculated and the platform driven back to the correct inertial attitude. It is worth noting that the IMU alignment using star trackers was practicable only if the difference between the two sensed attitudes did not exceed 1.4 degrees. For an angle larger than this, the attitude measured with the star tracker was considered greatly in error, since the angles given to the tracker to acquire a star were based on the current knowledge of the attitude of the Orbiter, which in turn was based on the data provided by inaccurate IMUs. This meant that if the angular difference exceeded 1.4 degrees the stars that were being tracked by the star trackers could not be the correct ones. In this case, the crew would use either the HUD or COAS to realign the vehicle within the 1.4-degree threshold so that the star trackers could be used to precisely align the IMUs. In addition, procedures were written to use the star trackers to obtain angular data of a target relative to the Orbiter during rendezvous and proximity operations.

Each star tracker unit was protected during launch and re-entry by a door driven by two motors and rotated underneath the outer tile and metal surface of the Orbiter. Studies performed by Rockwell had shown that if one of these doors remained open during re-entry, the plasma stagnation in the open cavity would result in only minor damage to the structure of the Orbiter. Studies of what would have happened if both doors remained open during re-entry were not pursued, but it was presumed that this would result in catastrophic failure of the structure due to ingress and circulation of plasma through the two open cavities. Flight rules therefore dictated that if one door jammed open during orbital flight, the good one must be closed immediately in order to prevent a possible second failure leading to a condition with both doors open. The same held true in the event of a failure of the primary system commanding the door motor. One door had to be promptly closed to prevent a situation in which the backup system also failed. Flight rules also stated that during orbital flight both doors had to be kept open and the star trackers powered up in order to exploit all possible star-of-opportunity alignments. In other words, the star trackers had to continuously scan the sky, taking line-of-sight measurements of any of the 100 stars known to the GPCs as soon as they entered the field of view of a tracker. The advantage of operating in this manner was that it would not be necessary to maneuver the Orbiter to view specific stars for an IMU alignment since the GPCs always had in memory line-of-sight data for at least two known stars to perform a quick IMU realignment.

The crew optical alignment sight

The crew optical alignment sight (COAS) was a small optical device with a reticle focused at infinity projected on a combining glass. The reticle was inscribed by a 10-degree-wide circle with 1-degree marks, illuminated by a variable-brightness lamp. Two mounting points were available, one at the commander's station to view along the +X axis and the other at the overhead starboard window of the aft flight deck to view along the –Z axis. The COAS was developed in order to assist in aligning the Orbiter with a target in the final and most critical part of a rendezvous or proximity operations. In addition, it could be used as a coarse means of aligning an inaccurate IMU inertial platform.

In either case, the Orbiter had to be maneuvered so that a reference star or target would cross the center of the reticle. At the precise instant of that crossing, the crew member viewing the COAS would press a button to tell the software that a mark had been obtained. On making a mark, the software stored in memory the gimbal angles of the three IMU inertial platforms. This process could be repeated as many times as

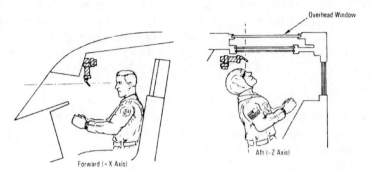

COAS.

Overhead window COAS installation.

necessary up to when the crew member felt that an accurate mark had been obtained, at which point the software was told to accept the mark along with the gimbal angles. With good marks on at least two stars, and knowing the exact position of the COAS relative to the structure of the Orbiter, the software could compute two line-of-sight vectors and determine the attitude of the vehicle in an inertial coordinate system. Comparing this measured attitude with that given by the IMUs enabled any error to be computed for a coarse IMU alignment. This would then be refined using the star trackers.

It was essential that the software know the precise position of the COAS at the time of a mark. The positions of the two mounting points relative to the IMUs were stored in memory prior to launch, but once on-orbit the crew had to perform a recalibration to take account of the slight changes to the structural configuration of the flight deck as a result of the dynamics of the ascent.

SHUTTLE FLIGHT CONTROL SYSTEM HARDWARE

Along with the hardware to enable the navigation software to calculate and propagate the state vector, there was hardware to enable the flight control system to execute the commands issued by the guidance software. This included sensors for flight control data, systems to respond to software commands, and also systems to provide manual guidance commands. The sensors for flight control data included four accelerometer assemblies (AA), four rate gyro assemblies (RGA) on the Orbiter, and two rate gyro assemblies on each of the solid rocket boosters.

The four identical AAs each contained two separate single-axis accelerometers, with one to sense vehicle acceleration along the lateral (left and right) vehicle Y axis and the other to sense acceleration along the normal (vertical) Z axis. The role of the AAs was to provide acceleration feedback to the flight control software to augment stability during first-stage ascent, aborts and re-entry. In addition, during first-stage ascent, the AAs provided data to compute the steering commands to relieve loads on the vehicle during the period of maximum dynamic pressure. The lateral acceleration readings enabled the software to null side forces during both ascent and re-entry. The normal accelerations provided feedback to control pitch and relieve normal loads. It is important to note that although these accelerometers worked in the same way as those on the IMUs, they were used for different purposes. While the AAs measured accelerations for feedback to the flight control software, the accelerometers on the IMUs were used by the navigation software to compute the state vector.

The four Orbiter RGAs contained three identical single-degree-of-freedom rate gyros arranged so that each sensed rotation about only one of the vehicle axes. They were used to provide feedback to the flight control software during ascent, insertion, deorbit and re-entry in order to maintain control in all three axes. Because a fuselage can be fairly flexible, the gyros had to be placed as close as possible to the center of gravity of the structure to prevent them from picking up spurious motions due to a deformation of the structure. In fact, it was not possible to put them precisely at the center of gravity because this moved during the flight as propellant was consumed and cargo was deployed or retrieved. They were placed on the aft bulkhead of the payload bay, the nearest viable structural component that reasonably approximated the desired conditions. The initial location of the RGAs was a mount on each of the four corners of this bulkhead, as far apart as possible for redundancy isolation. But ground vibration tests revealed local resonances that made this unacceptable. For this reason, the desire for physical separation of the redundant sensors was abandoned in favor of a dynamically identical signal and the mounting location was changed to the central base of the bulkhead.

The RGAs on the solid rocket boosters were identical to those on the Orbiter and were used exclusively to provide feedback to identify rate errors prior to SRB separation. As the SRBs were more rigid than the Orbiter body, these rates were less susceptible to errors created by structural bending and hence were very useful in thrust vectoring control. Both assemblies were mounted on the forward ring, within the forward skirt and near the attachment point for the external tank.

Flight control was achieved not only by software but also via manual inputs by one of the following items of hardware:

- The rotational hand controller (RHC) provided commands for rotational rates and accelerations about a body axis.
- The translational hand controller (THC) provided commands for vehicle translations along a body axis.
- The rudder pedal transducer assembly (RPTA) generated an electrical output signal proportional to the rudder pedal displacement. During gliding flight this controlled the position of the rudder and during rollout it controlled the nose wheel steering.
- The speedbrake and thrust controller (SBTC) was used during ascent to command the SSME throttle settings between 67 per cent and 109 per cent, and during re-entry to command the position of the speedbrake installed on the rudder.

Finally, two items of hardware were used to respond to software commands:

- The ascent thrust vectoring control (ATVC) took position commands from the GPCs during ascent and converted them into position error commands to drive the SSME and SRB nozzles.
- The aerosurface servoamplifier (ASA) took position commands generated by the GPCs during gliding flight and converted them into position error commands to drive the aerosurfaces.

As all of these items of hardware were used in specific flight phases, they will be described as appropriate in the following chapters.

STATE VECTOR PROPAGATION

As defined above, the state vector is a representation of the state of motion of the vehicle in terms of three components of position (X, Y and Z) and velocity (V_x, V_y and V_z) in the M50 coordinate system. In fact, it is also necessary to know the time at which the state vector was applicable, expressed as Greenwich mean time. At regular intervals, the navigation software computed the state vector using the equations of motion and data inputs from various on board sensors in a process called state vector propagation. In an ideal world this process would be fairly easy to accomplish, since the equations of motion are very well determined. But a spacecraft's orbit is affected by a variety of perturbations, and the manner in which these can alter the trajectory must be accounted for in determining the state vector. Furthermore, intrinsic errors in the navigation hardware impair an accurate determination of a spacecraft's trajectory.

Super-G navigation model

Any navigation software written for an orbiting spacecraft has as its basis the well-known equations of motion derived by Newton that are functions of acceleration and time. For the Shuttle the accelerometers placed on the inertial platforms of the IMUs fed values into a mathematical model known as Super-G which assumed that in any time interval the Orbiter would undergo uniform, linear accelerations. The algorithm worked as follows: given the simplifying assumption of uniform, linear acceleration, an estimate of the new position was calculated based on the position, velocity and acceleration of gravity of the current state vector. The new velocity was similarly computed from the velocity of the current state vector, the sensed accelerations, and the average of the acceleration of gravity over the considered time interval. Next, the value of the acceleration of gravity at the new position was calculated. Finally, a new more accurate position was determined. The GPCs used this technique to update the state vector every 3.84 seconds in what was called the NAV cycle.

During orbital flight a spacecraft is subjected to very minor accelerations, and most of the time the sensed accelerations are the result of small vibrations or biases. It is evident that basing the calculations of orbital motion on spurious accelerations would lead to erroneous state vector propagation. But this is impossible during highly dynamic phases of flight, such as ascent and re-entry, because the sensed accelerations far exceed any spurious measurements. Hence the navigation software could rely on the values sensed by the IMUs during ascent and re-entry. To prevent the erroneous accelerations from polluting the state vector on-orbit, the GPCs performed a simple check: if the sensed acceleration was below a given threshold (which could be set as appropriate by Mission Control or by the crew) then a sophisticated atmospheric drag model was used to determine the true accelerations on the vehicle, otherwise the accelerations sensed by the IMUs were accepted. This is well represented by OMS burns or RCS firings.

Drag acceleration model

The amount of breathable air and the partial pressure of the oxygen at the altitudes at which the Shuttle flew are at such low levels as to be incapable of sustaining human life. In terms of aerodynamics, the density of the air is so low that it is impossible for a wing to generate the lift necessary to counteract an aircraft's weight. However, one must bear in mind that velocity also plays a very important role in enabling a body to generate lift. This is the key to understanding why space is not a void from the point of view of a spacecraft.

Despite being extremely rarefied, impacts with air molecules at a typical orbital velocity of 16,770 statute miles per hour produced a tiny but not negligible force that acted constantly on the Orbiter in a retrograde direction to produce a steady decrease in the orbit's height day after day. These negative accelerations could be sensed by the IMUs, but being of the same order of magnitude as the spurious accelerations mentioned above they were not used by the navigation software in propagating an accurate state vector.

Hence in non-dynamic phases of orbital flight (i.e. coasting), if the accelerations sensed by the IMUs were lower than a threshold of 1,000 µg, the navigation software used a model of aerodynamic drag to derive the accelerations to propagate the state vector. The sophistication of the model was such that it took into consideration the variation of density with altitude, time of the day and time of the year, along with information concerning the Orbiter angle of attack, sideslip angle, and reference area. In addition, since the vehicle's reference area could significantly change when a large payload was moved around, Mission Control could uplink a corrective factor (called KFACTOR) to account for the increase or decrease in the calculated drag for that configuration. This factor also adjusted for atmospheric conditions that differed from those incorporated in the model. Fifteen seconds prior to an OMS burn on-orbit, the navigation software was again fed with data from the IMUs. When the maneuver was complete, the navigation software resumed using the drag model.

Gravity model

Although in the past people risked their lives by asserting that the Earth was a sphere, today we can acknowledge that there was some truth to the medieval belief that the Earth is not round. We know that the diameter of our planet is some 13 miles larger at the equator than it is at the poles. And of course topographical features spoil the perfection of a sphere: the summit of the tallest mountain is about 5 miles above sea level and the floor of the deepest oceanic trench is 6.8 miles below sea level. Furthermore, the composition and distribution of matter in the interior is not homogeneous. All of this means that the acceleration of gravity, and by extension the force of gravity acting on an orbiting vehicle, is not constant. As a result, the orbit is altered day by day in a quite significant manner. The Orbiter's IMU accelerometers were not able to measure these tiny but not negligible alterations. This was for a very simple reason. Being physically connected, the vehicle and the IMUs were subjected to the same force of gravity. The acceleration on the IMUs was the same as that on the vehicle, so gravity alone could not displace the pendulums of the accelerometers relative to the vehicle.

To circumvent this problem, in orbital flight the navigation software made use of a precise model of the variation of the acceleration of gravity all across the surface of the planet. The model was a very complex analytical formulation based on harmonic expansion. It began by supposing the Earth to be a perfect homogenous sphere, then added a series of very complicated terms called zonal coefficients, sectoral terms and tesseral terms that yielded gravitational variations in terms of latitude and longitude. The number of terms used varied with the precision desired. There were two models for the Orbiter. The one used on-orbit was more complex and precise but required a lot of computational power. Owing to the significant forces that act on the vehicle during ascent and re-entry, a simpler model was used in order to save on processing time.

Single state vector scheme

With all the information provided above, it is now possible to describe the process by which the state vector was propagated through a mission. Given that three different IMUs were each providing a state vector, two schemes were designed to enable the navigation software to derive a single state vector.

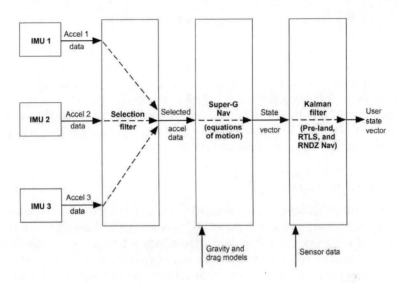

Single state vector scheme. (Courtesy of www.nasaspaceflight.com)

The single state vector scheme was used for all mission phases except re-entry. The total velocities calculated by the IMUs were selected to provide a unique set of accelerations, component by component. The selection process was performed by a filter that worked as follows. If three IMUs were available, the filter would select the median of the three X-direction, three Y-direction and three Z-direction components of the total velocities. If only two IMUs were available, the filter would average the total velocity components. If only one IMU was available, its velocity components would be used directly.

The selected accelerations were fed into the Super-G navigation model, which was integrated with the gravity and drag models to compute the new state vector. For the Return To Launch Site abort, rendezvous, and pre-landing phases of flight, the computed state vector was elaborated by a Kalman filter, with data from external sensors being used to refine the state vector. The requirement for a Kalman filter stemmed from the fact that neither the Super-G algorithm nor the IMU data were perfect, causing the calculated state vector to diverge from the real one (that is to say, the true position of the Orbiter). For critical phases of flight, this divergence was unacceptable. The role of the Kalman filter was to correct the Super-G state vector in order to meet mission requirements. In a Return To Launch Site abort data from the TACAN, air data system and microwave landing system would be added; during rendezvous data from the star trackers, COAS and rendezvous radar would be added; and in the pre-landing phase data from the microwave landing system would be used.

The final state vector output by the Kalman filter was referred to as the user state vector. Since external sensors were not incorporated (meaning the Kalman filter was not incorporated) during the ascent and non-rendezvous orbital phases, at these times the user state vector was simply the Super-G state vector. The user state vector was primarily used by the GNC software but the 3.84-second computational cycle of the navigation software was too long to enable the guidance software to maintain smooth control during dynamic flight. To overcome this, a guidance software module called user parameter processing (UPP) integrated the equations of motion over an interval of 0.16 seconds. At the start of each navigation cycle, UPP reinitialized to the new user state vector and propagated it using a simplified algorithm that was satisfactory over the short term. The user state vector was also supplied to various DPS and flight instrument displays that assisted the crew to fly the Shuttle. It was also downlinked for comparison with Mission Control's estimate of the position and velocity of the vehicle.

Three state vector scheme

The three state vector scheme resembled the single state vector scheme but with the difference that a state vector for each IMU was calculated separately (always using Super-G and the gravity and drag models) and then subjected to Kalman filtering as described above. This scheme was used only during re-entry, in order to provide a more accurate state vector than could have been achieved by the single state scheme in the event of multiple IMU failures. This precaution reflected the small margin of error for a successful landing – after all, the Orbiter had just one chance to make the runway!

To appreciate the three state vector scheme, consider what would happen in the event of multiple failures. If all IMUs were giving good results, the selection process in the final step of the scheme would give the best optimized state vector available. If one IMU failed, the process would average the two good state vectors. If a second IMU failed, the process would average a good state vector with a bad one to produce a degraded user state vector. If we consider this scenario with the single state vector scheme, at the second IMU failure the selection process would average two good and

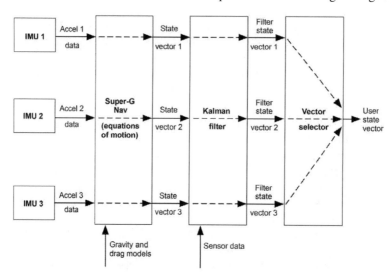

Three state vector scheme. (Courtesy of www.nasaspaceflight.com)

two bad acceleration values for each component, creating a degraded state vector. On recognizing the second failure, the filter would select the good measurements. But because the Super-G model is iterative, in this case the good data would be used in generating a new state vector based on the degraded state vector calculated during the previous cycle, thus degrading the user state vector. In other words, Super-G would continue to update a degraded user state vector with good IMU data, resulting in a still degraded user state vector. In the three vector scheme with two IMU failures, the user state vector was never degraded. In fact, since each state vector was computed separately, the first time that a bad state vector was calculated it was averaged with the good one and degraded the state vector. In the next NAV cycle though, the GPCs would recognize that another IMU had failed and would prohibit it from providing a state vector. Now only the remaining valid IMU was used, and the user state vector was calculated exclusively from data provided by that. That is to say, in the single vector scheme all measured accelerations from all IMUs were combined to provide the user state vector, which was always calculated in an iterative process where each new state vector was based on the previous one. In the three vector scheme however, three different state vectors were calculated and the user state vector derived using the appropriate averaging technique. Each state vector was calculated based on the measurements obtained in the given moment, rather than from the previous one. In other words, whilst the single state vector looked at the past to compute the user state vector, the three vector scheme looked only at the present. The disadvantage of the three state vector scheme was that it required much more software processing than the single state vector scheme to maintain the three independent state vectors. Hence it was used only for re-entry, where accuracy was mission critical.

State vector updates

Despite the sophistication of the IMUs, inexorably the inertial platforms tended to drift from their preset inertial orientation with a typical rate of approximately 0.006 degrees per hour. And the external navigation sensors were not immune to errors and biases. Neither were the sophisticated mathematical models used by Super-G NAV. All of this caused the state vector provided by an IMU to diverge from the true state vector of the vehicle. For this reason, Mission Control monitored the true position of the Orbiter throughout the mission by means of TDRS ranging data, radar tracking data, and state vector prediction computations. Its state vector was then compared to that downloaded from the Orbiter and if the divergence exceeded a preset threshold, which was a function of altitude and mission phase, an updated state vector could be uplinked and either automatically or manually inserted into the flight software. But the preferred solution later in the program was for the crew to supersede the degraded state vector with one independently calculated by GPS.

Updates could be communicated in two ways. In the first case, known as "whole state vector update", Mission Control calculated a propagated state vector and it was transmitted to the Orbiter's navigation software along with a time tag. A navigation algorithm called the "precise predictor" was then used to propagate the uplinked state vector forward or backwards to the current time. The data supplied by the IMUs were not used. Whole state vector updates were periodically executed on-orbit to maintain the navigation software within the limits set by the flight rules and thereby guarantee the continuous presence of a good state vector for use in the event of a contingency deorbit due to a loss of ground link. In addition, whole state vector updates could be planned to satisfy mission or payload requirements.

During ascent or re-entry, whole state vector updates were not used because this involved taking the uplinked state vector and "predicting" it forward or backwards to the current time. In such a dynamic environment where the vehicle was considerably accelerating (or decelerating), this propagation could introduce errors in the new state vector. If updates were required, they were made by the "delta state vector update" method. In this case, Mission Control computed the differences (deltas) between the ground computed and the on board computed state vectors, and uplinked the deltas to be algebraically added to the on board state vector at a specified time. This was faster and safer than uplinking a whole state vector. For example, a plus or minus delta error in one of the components would have less severe impact on the state vector than an entirely incorrect component.

If the capability to uplink was lost but voice uplink and downlink was retained, and GPS was unavailable, the deltas could be read up to the crew to manually update the navigation software. Mission Control typically specified a time at which the crew were to execute the "load". The crew would make the update as close to this time as possible in order to increase the accuracy of the new state. In practice, voice updates were hardly ever performed, being deemed too specific and too susceptible to errors.

THE ORBITER FLIGHT DECK

The most fascinating place in a flying machine has to be the flight deck. The myriad of lights, switches and instruments fill us with awe that the machine can be harnessed and steered through the skies with such seeming simplicity. Among the many flying machines built since the first flight of the Wright brothers, the Shuttle's flight deck was one of the most advanced because it was for a vehicle that operated both as an aircraft and a spacecraft. Development of the display and controls (D&C) system, as it was called, posed a challenge because it had to provide within its pilots' reach and vision such a wide variety of controls.

As is customary in the development of a new system, several requirements were listed as guidelines for the design of the D&C system. Normal operations (with the exception of payload management) had to be able to be executed during all mission phases by a crew of two only astronauts. It was essential that a single pilot be able to perform a safe return from either position on the forward flight deck. Of course, the pilots required accessibility to all the controls for vehicle or subsystem management during ascent and re-entry whilst seated. And it was essential that the crew be able to override automatic critical command functions and select between the automatic and manual fight guidance and control modes.

The design of the D&C began as soon as Rockwell International was awarded the contract to develop the Orbiter. One agreement that proved to be very valuable was the creation of the D&C work breakdown structure by Rockwell and the Johnson Space Center. In total, thirteen major reviews were held in which representatives of the flight crew, flight operations, engineering, reliability, safety, program office, payloads, software, Kennedy Space Center, and the US Air Force were present for an average of 46 people per session, including at least four astronauts. Each review took two or three weeks, with the first week being given over to Rockwell engineers to describe the vehicle subsystems and the D&C engineers to give the D&C concept for each subsystem. In the second week, Rockwell engineers would work with flight crew to determine the D&C for each subsystem and the most appropriate nomenclature. A full-scale foam mockup of the flight deck was constructed and drawings were used to arrange the panels. In addition, cutouts were used to determine the proper location within the reach and visibility requirements derived from other studies. After each review, Rockwell produced a D&C configuration drawing and updated the mockup ready for the next review.

Given the many activities that would take place on the flight deck, it was decided to split it into forward and aft sections. The forward section would be similar to an airline flight deck, with all the controls for piloting the vehicle. The aft section would serve as the command center for all other activities, such as rendezvous and docking, and operating the robotic arm and payloads. Both sections required displays and controls in a variety of forms, with toggle, push button, thumbwheel, rotary switches, circular meters, rectangular dials, and rectangular tapes. To prevent inadvertent activation of any of the controls, several different protection techniques were used. For example, toggle switches were protected using wicket guards, lever lock switches were used where inadvertent action would be detrimental to flight operations or could damage equipment, cover guards were used on switches where inadvertent actuation would be irreversible, and switches were recessed into panels that the astronauts were most likely to step on.

Early in the D&C panel configuration design, it was decided to group subsystem controls by function. But associated circuit breakers were separated owing to concern that if the high power circuits behind the breakers suffered either an electrical fault or mechanical damage, the entire subsystem might be affected. To reduce the training impact on the crew the breakers for each system were positioned at the same relative location on different panels or section of a panel, and an identification scheme was devised to enable the astronauts to readily find controls. In particular, each panel was numbered from left to right or from forward to aft and letters added, such as 'R' for right side, 'L' for left side, 'O' for overhead, 'C' for center console, 'A' for aft, and 'M' for middeck. The full-scale foam mockup was an invaluable tool for establishing the best location for every control on the flight deck. After taking into account human factors in producing the D&C system design, components were located by priority, with precedence assigned to the systems that needed to be reached and viewed during maximum ascent acceleration. Then the other controls that had to be accessible while the crew were seated were positioned. The lower order subsystem D&C components were placed on the aft flight deck. A three-dimensional computer model was used to depict the pilots' access to different D&C components during the most demanding moments of a mission, such as a

322 Fundamentals of the Shuttle GNC

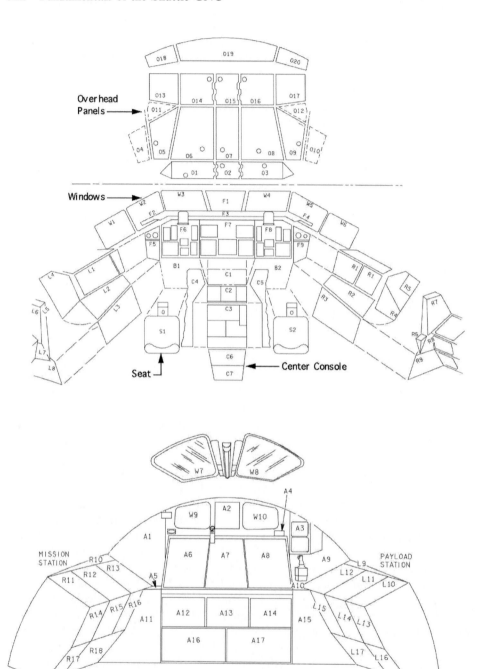

Layout of flight deck forward station cockpit (top) and aft station (bottom).

contingency two-engine-out abort maneuver in which they would be subjected to a high negative-g condition. The findings from this first rudimentary example of digital mockup were then validated using the foam mockup.

As the D&C requirements for the various subsystems continued to expand, this increased the workload of the astronauts and their understanding of each subsystem. A lot of effort was devoted to developing the operational nomenclature. The pilots suggested that schematic layouts be added to panels to assist in understanding their subsystems, and this was done for panels R1 (power distribution and control), L1 (environmental control), L2 (atmospheric pressure control), O7 and O8 (reaction control and orbital maneuvering systems), A1 (communications) and R12 (supply water).

A strong driver for the D&C design was maintainability of the control panels and their line replacement units. For this reason, many panels (not only those on the flight deck) had hinges to provide access to the volume behind. Each line replacement unit was mounted from the front of the panel in order to be readily installed and replaced without the need to remove the panel.

One early problem in developing the displays and controls was complying with the NASA design standards which prohibited the use of frangible materials. Most of the existing display devices, such a CRT displays and meters used a glass window as a mean of providing visual access and sealing the instrument case. The solution was to cover a display device that incorporated glass with Lexan, both to protect it and to contain glass fragments in the event of damage. These covers were then coated with antireflective material for correct optical properties. They were removable in order to simplify maintenance of the display devices.

In the weightless environment any contaminant or extraneous material poses a threat to the function of the controls and displays because if it is conductive and of sufficient size it can bridge across terminals on switches, circuit breakers, and meters. To prevent this, all internal and external electrical terminations on the panels were protected by a coating of resilient insulating material that also served as a humidity barrier.

Mindful of the fire in the Apollo 1 spacecraft that killed three astronauts during a test on the ground, all equipment and items in the crew compartment of the Orbiter had to satisfy stringent requirements for flammability and toxicity. All materials that were not within an environmentally sealed enclosure were reviewed by materials and processing specialists prior to use. Where existing hardware was used, special testing was carried out to define the flammability, toxicity, and outgassing characteristics of materials for which these were not already available. When an unacceptable material could not be changed, it was coated or otherwise protected. Sometimes waivers were granted after analysis indicated acceptability because of configuration, quantity, etc.

The controls and displays were divided between the forward flight station and aft flight station. The forward flight station was split into left, right, central and overhead areas. On the left side, panels housed circuit breakers, controls and instrumentation for the ECLSS, communications, heating controls, and the elevon trim and body flap controls. It also had the commander's speedbrake and thrust controller. Panels on the right side had more circuit breakers, controls for the fuel cells, hydraulic system, auxiliary power units, engines, the pilot's communication controls, electrical power distribution, and the pilot's speedbrake and thrust controller. A console between the two seats contained the flight control system channel selector, air data equipment, communication and navigation controls, circuit breakers for the fuel cell system, and the pilot's elevon trim and body flap controls. The overhead panel had the lighting controls, computer voting panel, and fuel cell purge controls. The centrally mounted forward panel had a number of screens to display software generated information, the C/W system, aerosurface position indicators, backup flight control displays, the fire protection system displays and controls, the all-important primary flight instruments, auxiliary power unit and hydraulic displays, and the controls for the landing gear.

The aft flight deck was split into three major zones called the mission specialist station, the payload specialist station, and the on-orbit station. Located on the right, the mission specialist station housed controls and displays for monitoring systems, communications management, payload operation management, and payload/ Orbiter interface management. On the other side, the payload specialist station incorporated a removable panel section to install controls for the specific payload of a mission. The on-orbit station was in the center, facing directly towards the rear of the payload bay. It was split into two work stations. On the right were controls for maneuvering the robotic arm. On the left were controls to enable the mission commander to maneuver the Orbiter during a rendezvous and/or docking, viewing out of the overhead or rear windows.

The launch of *Atlantis* on 19 May 2000 for the STS-101 mission may well have

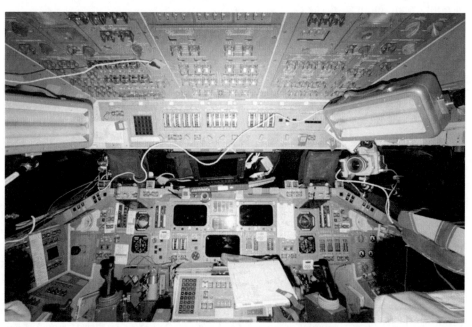

A view of the Orbiter analog cockpit.

A view of the Orbiter glass cockpit.

seemed to observers to be no different to her maiden flight in 1985, but for the crew on the flight deck there was a distinct difference. Hitherto, the forward flight station had resembled the cockpit of one of the jet airliners that initiated passenger service in the 1950s. Three CRT monitors on the central panel displayed information from the flight software, and there were analog flight instruments available to both pilots for use during re-entry and landing. But STS-101 introduced a "glass cockpit" in which all the analog and mechanical instruments had been replaced by advanced graphical representations on large LCD screens.

In NASA jargon, this was the multifunction electronic display system (MEDS). It involved four types of hardware: four integrated display processors (IDP), eleven multifunction display units (MDU), four analog-to-digital converters (ADC), and three keyboard units which together communicated with the GPCs over the DK data buses. The MEDS allowed monitoring of vehicle systems and computer software

326 Fundamentals of the Shuttle GNC

processing, and facilitated manual control of flight crew data and manipulation of the software. It provided timely response to crew enquiries by way of displays, graphs, trajectory plots, and predictions of flight progress. The pilots could alter the system configuration, change memory configurations, respond to error messages and alarms, request special programs to perform specific tasks, execute operational sequences for each mission phase, and request specific displays.

The most visible part of the new system were the eleven MDUs, full color, flat 6.7-inch square panels with active matrix liquid crystal displays. Their primary role was to drive the variety of MEDS color displays, and they were readily readable in direct sunlight. Nine MEDS were on the forward flight station. The others were on the aft station. Due to the digital nature of the new system, in order to display Orbiter system data the analog data provided by the on board sensors had to be converted to a digital format. This was the task of the ADCs, which converted analog subsystem data into 12-bit digital data. The IDPs processed this to generate the displays on the MDUs. The IDPs were the interface between the MEDS and the GPCs, formatting data from the GPCs and ADCs for display on the MDUs. They also accepted inputs from switches, edge-keys, and keyboards. The three keyboards remained unchanged from the previous analog flight deck configuration.

Primary flight display

Having the soul of an aircraft, some of the flight instruments present on the Shuttle flight deck were already familiar to its pilots, such as the attitude director indicator

A view the cockpit overhead panels.

The Orbiter flight deck 327

A view of the flight deck aft station.

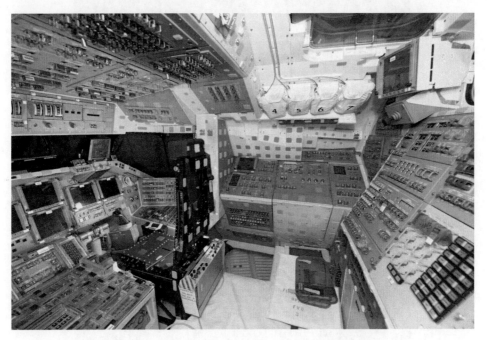
A view of the right-hand side of the flight deck.

328 Fundamentals of the Shuttle GNC

A view of the left-hand side of the flight deck.

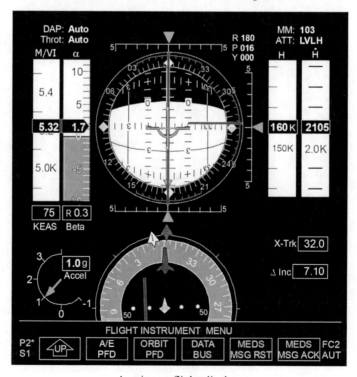

A primary flight display.

(ADI) and the horizontal situation indicator (HSI). For the glass cockpit these two instruments were replaced by a single flat screen on which this information could be graphically represented. This screen also displayed other flight instruments, such as the angle of attack tape, Mach/velocity tape, equivalent airspeed meter, altitude and altitude rate tapes, and a g-meter. Altogether, this graphical representation was called the primary flight display (PFD). It also displayed supplemental information such as status indicators for the current major mode, abort mode, ADI attitude settings, speedbrake, throttle, and digital autopilot status.

The positions of pointers and needles on the ADI displayed attitude information, attitude rates, and errors. The attitude of the Orbiter was displayed to the pilots by a software simulated ball (often referred to as the "eight ball") which was gimbaled to represent the three Cartesian degrees of freedom. Covered with numbers to indicate angular measurements,[2] the ball moved in response to software-generated commands to depict the current attitude in terms of pitch, yaw and roll. A white band (called the "belly band") indicated 0° of yaw. An artificial horizon indicated 0° of pitch, positive pitch angles (0° to 180°) were drawn on the white half of the ball and negative pitch angles (180° to 360°) were on the darker half. In addition to the graphical attitude representation of the ADI ball, there was a digital readout on the upper right corner indicating the current roll, pitch and yaw in degrees. Three rate pointers and three attitude error needles gave continuous readouts of rotational

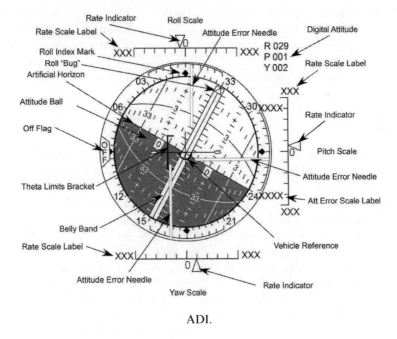

ADI.

---

[2]   An implied zero had to be added as the last digit of each number.

rates and attitude errors respectively. Both types of indicator were "fly-to", requiring the pilots to maneuver in the direction of the needle in order to null either the attitude rates or attitude errors. The attitude errors were relative to the body axis coordinate system, and hence were independent of the selected reference frame of the ADI. In this way, they showed the difference between the required and current vehicle attitude.

Each ADI also had a series of switches to select the mode or scale of the readout. For example, by using the ADI ATTITUDE switch the commander and pilot could select INRTL (inertial), LVLH (local vertical/local horizontal) or REF (reference) as the frame of reference. INRTL displayed the vehicle's attitude relative to the inertial reference frame. LVLH displayed it in an Orbiter-centered rotating reference frame with respect to Earth. REF was primarily used to display the attitude relative to an inertial reference frame defined at the last time that the switch called ATT REF was depressed. The REF position was useful when the crew wanted to resume an earlier attitude or to monitor attitude excursions during an OMS maneuver. The ADI RATE switch enabled the crew to select the magnitude indicated by the full-scale deflection of the rate pointers between HIGH (a coarse setting) and LOW (the finest setting). In a similar manner, the ADI ERROR switch selected one of three scales for the error needles.

The horizontal situation indicator (HSI) complemented the ADI by providing a pictorial view of the position of the vehicle relative to various navigation points, with a visual perspective of certain guidance, navigation and control parameters. This was available only during ascent and re-entry, and was used mainly to control or monitor heading/yaw performance. It served as an independent software source to compare against ascent and re-entry guidance to assess the health of individual navigation aids during re-entry, and of the information that the flight crew would require to fly these phases manually.

Control sticks

One of the most familiar pieces of apparatus on an aircraft's flight deck is the control stick used by pilots to direct commands to the aerosurfaces of the wings and tail. The Shuttle was no exception. In fact, it had two different control sticks.

The rotational hand controller (RHC) was more sophisticated than an aircraft's control stick because it enabled an astronaut to impart gimbal commands to the main engines and SRBs during the ascent, to command RCS and OMS firings on-orbit and during the early portion of atmospheric re-entry, and to command the aerosurfaces in the final portion of re-entry. In total, three RHCs were placed on the flight deck: one for each pilot on the forward flight station and one on the aft station. As on any aircraft, human factors dictated that an RHC deflection must produce a rotation in the same direction as the pilot's line of sight. For this reason, when using the RHC at the aft station, a switch called SENSE allowed the crew to select the line-of-sight reference, to accommodate the possibility of observing a payload looking towards the aft of the Orbiter (–X axis) or above the Orbiter (–Z axis).

A view of the THC (left) and RHC (right) on the commander's seat.

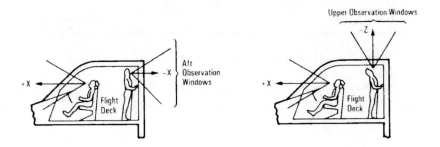

The SENSE switch concept.

There were a number of switches on the RHC. Pressing a red button on the right engaged the BFS. A switch on the left enabled the biasing of the stick's pitch and roll commands to be trimmed. And a push-to-talk switch controlled voice transmissions. With the exception of the final countdown, the RHC could be used in every major mode simply by moving it out of the so-called detent position which was defined by a deadband of 0.25 degrees in each axis. As the stick left the detent, the operative software downmoded the flight control system from the automatic mode to control stick steering (CSS) in the axis of deflection. This was also referred to as "hot-stick downmode".

It is worth noting that for CSS to be active during the ascent, either pilot had to depress a specific button to enable the CSS control mode axis by axis. The stick was also designed in such a manner that when the amount of deflection reached the "soft stop" there was a step increase in the force required to proceed further with that

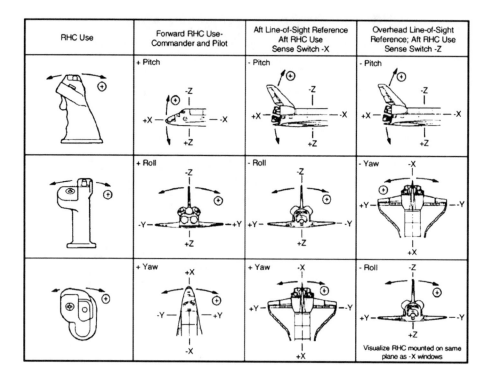

Positive deflection of the RHC by axis.

deflection. Internal control logic performed a self-check to exclude using the RHC in the event of failure, and for transmitting the commands to the GPCs for further processing to provide control responses and stability.

The second control stick was the translational hand controller (THC), and it was placed at the commander's station and aft station to allow for manual control of RCS jet firings to perform translations along the body axes. It was similar to the RHC in that when someone wanted to use the aft THC they would use the SENSE switch to select the direction of the line-of-sight. During a nominal mission, the commander's THC was active during orbit insertion, on-orbit, and deorbit. The aft THC was used only while on-orbit.

DIGITAL AUTOPILOT

In most of the workbooks and technical manuals concerning the Shuttle's GNC the digital autopilot (DAP) is referred to as the heart of the flight control system because it had to interpret maneuver requests, compare them with what the vehicle was doing, and generate commands for the relevant effectors. There were three main DAPs for handling the Orbiter, namely the transition DAP, the orbital DAP, and the aerojet DAP.

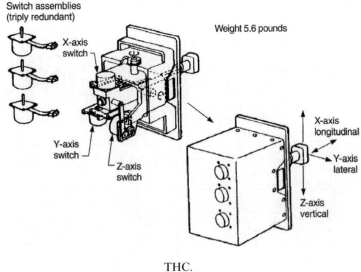

THC.

The transition DAP was used by the GNC for maneuvering during the delicate phase of orbital insertion; that is to say, from the moment at which the Orbiter had to safely separate from the ET through to insertion into the orbit desired for the mission. As its name implies, the orbital DAP was used during orbital flight to handle rotation and translation of the Orbiter as necessary to execute the mission. And the aerojet DAP assisted the Orbiter in its metamorphosis during re-entry from a spaceship to a hypersonic glider to conventional glider for landing on Earth. The digital autopilot managed how the vehicle was handled and maneuvered in the different flight phases. For example, during re-entry the aerojet DAP gradually allowed the aerosurfaces on the wings and tail to supersede the RCS jets, in order that in the densest part of the atmosphere the Orbiter would fly as a conventional aircraft based on its static and dynamic handling properties. As such, the digital autopilot of the Orbiter differed from the autopilot of a conventional airliner. For the Shuttle, ascent and re-entry were automatically flown by the GNC and orbital flight was driven by orbital mechanics. In other words, the DAP of the Orbiter was not used simply to go from A to B, but to control the vehicle in response to either automatic or manual commands during insertion, on-orbit and re-entry.

The crew could interface with the autopilot in several ways in order to command a maneuver and to set specific parameters for the DAP software module in operation during the current flight phase.

On the central console between the pilots' seats and on a panel on the aft flight deck there were 24-button keyboards to specify the characteristics of the Orbiter for a maneuver or to adopt and maintain an attitude. The keyboards were split into three distinct sections. The left uppermost section had only two buttons, labeled A and B, for selecting a particular family of values for the parameters controlled by the DAP module (transition, orbital or aerojet) currently in use. The DAP A

DAP pushbuttons.

configuration had wider deadbands and higher rates than those of DAP B. The wide deadbands were used to minimize fuel usage, while the tight deadbands allowed greater precision in executing maneuvers and in holding attitude. The configuration to use was specified by the flight plan and, if necessary, by Mission Control. For both DAP A and DAP B configurations, the flight crew had the option of reading and changing all parameters using the DAP CONFIG (SPEC 20) software page.

Four buttons in the right uppermost section of the keyboard allowed the crew to control the way in which the Orbiter would maintain its attitude. The AUTO mode maintained the vehicle at a specific attitude and attitude rate, set manually using the UNIV PTG software page.[3] On selecting this mode, the so-called universal pointing processor (a DAP software module) would rotate the Orbiter about the shortest angle to the desired attitude set by the crew. Once in attitude, RCS jets would automatically fire to hold the attitude and attitude rates within the deadband values on SPEC 20. The INTRL and LVLH buttons would also maintain the Orbiter at a fixed attitude, in the former case relative to an inertial reference frame and in the latter case relative to Earth. The FREE control mode left the Orbiter free to drift

---

[3] This software page will be described in detail in Chapter 12.

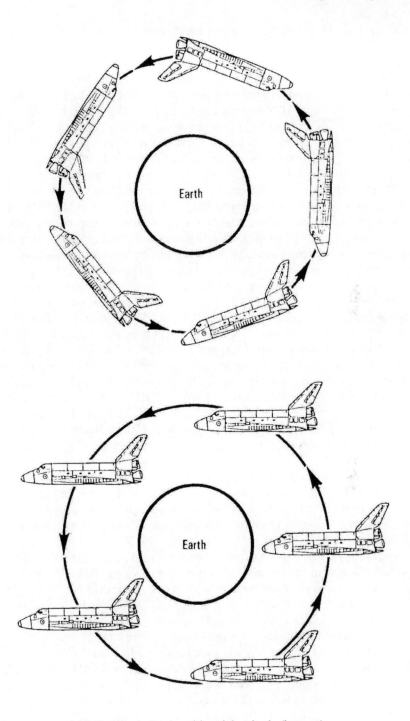

LVLH attitude (top) and inertial attitude (bottom).

around any axis. In this mode, attitude changes could be commanded only manually by the RHC.

Often it was necessary to manually maneuver the Orbiter, such as while docking with the International Space Station. The autopilot could be set to provide different types of control in response to the commands of the RHC and THC. For this reason, the lower part of the DAP keyboard contained two sets of nine buttons to control the behavior of the vehicle in translation and rotation. In terms of translation maneuvers, pressing the NORM button for any given axis caused the primary RCS jets in that axis to be fired for as long as the THC remained out of detent. This mode was used mostly for RCS translations and for propellant dumps. Depressing any of the PULSE buttons caused the appropriate primary RCS jets to fire to achieve the translational pulse rate specified by SPEC 20. In contrast to the NORM mode, the PULSE mode would fire the RCS jets only once. To fire them again, the THC had to be returned to the detent and then deflected again. It is worth remembering that both modes were independently available for all three directions of translation, so that a mix of modes was possible for manual control of translational. Owing to concern that up-firing jet plumes along the Z axis might damage either a payload or a spacewalker operating close to the Orbiter, a LOW Z translational mode was added to the autopilot. When selected, this inhibited all up-firing jets from firing. To achieve a +Z translation, the +X and +Y jets were commanded to fire because they each incorporated a small +Z thrust component. Operating this way increased propellant usage, but it was deemed more acceptable than causing damage to a payload or crew member working outside the Orbiter. If the final phase of a docking were to threaten a collision, then selecting the HIGH Z mode and making the appropriate deflection of the THC would cause all nine up-firing jets to fire to perform an emergency withdrawal. Although this blast of efflux might seriously damage a payload, the integrity of the Orbiter and the safety of the crew had higher priority.

For rotational maneuvers, the crew could select two different ways of imparting inputs using the RHC. On selecting DISC RATE on any axis, deflecting the RHC out of detent would cause the appropriate RCS jets to fire to attain the rotational rate specified by SPEC 20. Once the RHC was placed back in detent, the DAP would fire the jets to maintain that attitude in either INTRL or LVLH as appropriate. Depressing the PULSE button for any given axis would cause the RCS jets to fire for a specified increment in response to each deflection of the RHC in that axis. Once back in the detent position, the Orbiter was placed in free drift for the axis for which PULSE was selected. Irrespective of whether DISC RATE or PULSE were selected, moving the RHC past the soft stop would produce a continuous firing of the jets.

The DAP could operate the RCS jets in one of three control logic modes. From a hardware logic point of view, the jets were combined in several groups, each using a collection of jets located on the same pod (forward, left or right) pointing in the same direction. Within a group, each jet was given a priority ranking; that is to say, the jet with the highest ranking would be the one fired first for executing a maneuver using that group. If the jet that had the highest ranking was insufficient to accomplish the requested maneuver, then the second highest ranking jet would also be used, and so

Digital autopilot 337

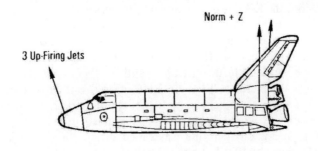

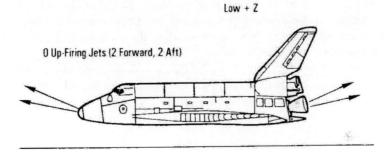

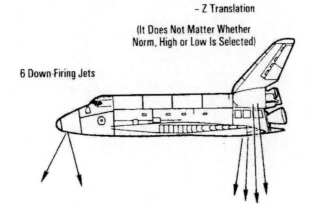

RCS firing for Z translations.

on. Prior to a mission, the DAP was loaded with a number of mission-based lookup tables to provide in a tabular manner a full description of the groups and the priority ranking assigned to each jet. Other tables provided a description of all the possible combinations of jet firing that the DAP would be able to use taking into account multiple jet failures, propellant feed constraints, and whether or not the jets were fed by OMS propellant. Whenever a THC or RHC moved out of detent, the DAP looked at its tables and, knowing the actual RCS configuration, could understand how many and which jets would have to be commanded in order to achieve the desired action. This means of operating was called the primary RCS jet selection logic and was used when the PRI button on the DAP keyboard was pressed.

A second logic available to the DAP with the RCS was the so-called alternate jet selection, activated by pressing the ALT button on the keyboard. If the vernier jets were not available, the DAP still used the primary RCS jets but tried to attain a finer degree of attitude control intended to permit almost unrestricted payload and RMS operations. In this mode, the DAP selected up to three jets whose rotational velocity increments were the closest to that required to satisfy maneuver request. For each of the eleven jet groups, the DAP looked up a table of the angular rates expected in each rotation axis as a result of a jet firing from each jet group. From this, the alternate jet selection logic chose the best jet group for the commanded maneuver in one of two ways. One method selected jets that minimized the difference between the angular rate obtained by firing the jets compared with that desired. The second method computed a dot-product to maximize the thrust component in the desired direction. If the RCS vernier jets were available, then this control mode could be selected by pressing the VERN button. This mode was preferred to ALT, since vernier control produced less stress on the RMS/payload, used less propellant, and achieved tighter deadbands than either PRI or ALT. Like the ALT mode, the best jet(s) to use were determined by the Orbiter/payload moments of inertia.

11

The art of reaching orbit

NOMINAL ASCENT: FIRST STAGE

A Shuttle launch is a spectacular sight for spectators. It is also a great moment for the crew setting out to put into practice the countless and often tiring hours of training. It was certainly a moment of intense concentration for the two astronauts who had the privilege of sitting in the front seats of the flight deck, without doubt the best place to be in a spaceship. In general terms, liftoff marked the culmination of a complex and lengthy procedure known as the countdown, in which the vehicle was configured for launch and checked for any sign of a problem. The real show started at T-6 seconds, when on board flight software started the sequence to ignite the three main engines and throttle them up to 100 per cent of their rated thrust. If any engine was delivering less than 90 per cent thrust after 4.6 seconds, or if any of the MAIN ENG STATUS lights was amber, there would be an immediate sequenced shutdown. Otherwise, at T=0, the solid rocket boosters would be commanded to ignite, marking the moment of liftoff. Thereafter, the nominal task of the crew was to monitor and perform those actions specified by the Ascent Checklist.[1]

Launch pad abort

On 26 June 1984 *Discovery* was on the pad attempting for the second time to begin her maiden voyage as mission STS-41D. At T-6.6 seconds the main engine sequence began. Rookie mission specialist Michael Mullane was on the flight deck, just behind the pilots. "I remember thinking, 'Well, we're going now. It's going to happen.' And you get that heavy vibration when the liquid engines start ... [and that] shakes you really good in the cockpit. And watching those numbers flicker off the countdown

---

[1] All procedures explained in the remainder of this chapter are taken from the Ascent Checklist of STS-114, the first return-to-flight mission after the *Columbia* disaster.

clock – and then to have the master caution light go on; and quiet." For the first time in the program, a launch pad abort had occurred. "Obviously, the engines shut down, or at least we thought they did because there were two shutdown lights on [and] one [that] wasn't. Mike Coats[2] kept pressing on the ... engine shutdown switch to get a light on, but nothing occurred. Whatever's going on, we wanted all the engines off." It was a perilous time. "You somehow sense, 'Oh, my gosh, we were down to within a couple of seconds of the SRBs igniting ... are they going to ignite?' In the back of your mind you're thinking, 'What happens if those ignite?' I mean, if they ignite, you're dead, because you have no liquid engines running. And that's one of those things you can talk yourself out of as an engineer. 'Oh come on, now. The safety system is working fine. Those things aren't going to ignite.' But it was disconcerting, to say the least, in the cockpit."

This was the first pad abort since an attempt to launch Gemini VI in December 1965. The launch control team was taken by surprise. For the crew of *Discovery* in the seconds immediately following the abort, the inability of the launch control team to provide clear instructions was of particular concern. As Mullane says, "You hear these calls, and it's obvious that it's not like, 'Oh, we really know what's going on here', like it had been in all the simulations in all of our training ... [and] they rapidly figured out what the heck the problem was. But there were those few seconds there where the tone of [their] voices ... and the things ... being said were like a dagger of fear right in the old heart. In the cockpit ... I will tell you, we were sitting there praying."

As soon as the pad abort occurred, a water shower was activated to spray water onto the main engines and aft fuselage of the Orbiter to prevent temperatures from rising and boiling the cyrogenic propellants, as gas inside the feedlines could all too easily make its way into the external tank and cause an explosion. For this reason, the procedures following a launch pad abort required shutting down all of the APUs which, being in the aft fuselage, were a source of great heat. Another hazard was the potential for a hydrogen fire. As Mullane says, "Then we got the word that somebody reported seeing a fire on the pad, not on the vehicle, but on the pad ... They reported it as a small fire on the pad. And I'll never forget thinking, 'There's no such thing as a small fire when you're sitting on four million pounds of propellant.' That was a real terror to hear the word fire."

On board *Discovery*, the crew did not yet have clear indications of what to do. Judy Resnik, who was on the middeck, unstrapped herself, awaiting word from her commander to open the cabin ingress/egress hatch and escape onto the access arm of the pad structure, but launch control told them to remain in the cabin. This proved to be the correct decision, since ultraviolet sensors on the pad structure were indicating the presence of burning hydrogen. As the naked eye is unable to see a hydrogen fire, if the crew had been told to perform an emergency egress they would have entered an invisible but deadly fire. To relieve the tension, rookie astronaut

---

[2] Pilot of the STS-41D mission.

Steve Hawley made a 'bad joke' that has become one of the most famous quotations of the program. As mission commander Henry W. "Hank" Hartsfield explains, "As soon as we looked at everything and everything was okay, Steve said, 'Gee, I kind of thought we'd be a little higher at MECO.'"

The rationale for instructing the crew not to evacuate the Orbiter is interesting. It was only partly due to concern about the hydrogen fire. The launch control team was also worried about something else. The pad had an emergency egress/escape system consisting of seven separate slide wires and seven multi-person basket assemblies. In the event of a potentially catastrophic situation during the countdown, the astronauts and other personnel working at the 195-foot level of the fixed service structure would be able to effect a rapid escape to a safe landing zone some 1,200 feet to the west of the pad where there was a bunker equipped with breathing air, first aid supplies, and communication equipment. The problem was that since the beginning of the Shuttle program, "No one had ever ridden the slide wire, and they were afraid to tell us to do it," Hartsfield remembers. "That bothered a lot of us, in that they were concerned enough about the fire that they really wanted us to do an emergency egress from the pad area, but since the slide wire had never been ridden by a real live person (they'd thrown sandbags in it and let it go down) they were afraid to use it, which was a bad situation." After this incident, procedures were revised and finally astronaut Charles F. Bolden took the ride and reported it to be "a piece of cake", thereby certifying the slide wire for human usage.

Charles D. Walker was the rookie STS-41D payload specialist, and he explains the incident, "What was happening, was that the sensors monitoring the engines had sensed that one of the valves controlling the thousands of pounds per second of liquid oxygen wasn't flowing. The valve was not running at the rate at which the computers were programmed to expect it to close. [So] the first engine starts, the second engine had started, and the third engine was just into the fuel flow process and hadn't ignited yet. And the second engine was shut down. All of them were shut down because of this valve on the second engine."

After STS-41D the Shuttle program suffered four further launch pad aborts. On 12 July 1985 the countdown for *Challenger* on mission STS-51F was halted at T-3 seconds when the GPCs detected a problem with a coolant valve on the left engine. The valve was replaced and the mission was launched on 29 July 1985. On 22 March 1993 the countdown for *Columbia*'s launch as STS-55 was halted by the GPCs at T-3 seconds owing to a problem with purge pressure readings in the oxidizer preburner of the left engine. All three main engines were replaced on the pad, and the mission got underway on 26 April 1993. On August 12, 1993 the countdown for *Discovery*'s third launch attempt ended at the T-3 second mark when the GPCs noted the failure of one of four sensors in the left engine which monitored the flow of fuel. Again all three main engines were replaced on the pad. STS-51 was successfully launched on 12 September 1993. Finally, the most spectacular Shuttle launch pad abort occurred on 18 August 1994 when the countdown for *Endeavour*'s first attempt for STS-68 ended at T-1.9 seconds when the GPCs detected higher than acceptable readings in one channel of a sensor that monitored the discharge temperature of the high pressure oxidizer turbopump in the right-hand engine. A subsequent test firing of this engine at the Stennis Space Center in Mississippi confirmed that a slight drift in a fuel flow meter in the engine caused an increase in the temperature of the turbopump. The test also confirmed that there had been a slightly slower start for this engine on the day of the pad abort, which could have contributed to the higher temperatures.

First-stage guidance

With SRB ignition, the first stage guidance was initiated to enable the Shuttle to fly through the densest portion of the atmosphere on a trajectory designed: (1) to avoid contact with the launch structure; (2) to avoid exceeding the maximum aerodynamic pressure (max-q) for the vehicle stack in nominal and in perturbed conditions; (3) to maximize performance; and (4) to avoid recontacting with the SRBs after they were jettisoned.

The moment of liftoff was always dramatic. Robert Crippen was the rookie pilot on STS-1 and he describes his feelings as follows, "I knew when the main engines lit off. It was obvious that they had, in the cockpit, not only from the instruments, but you could hear and essentially feel the vehicle start to shake a little bit. When the solids lit there was no doubt we were headed someplace; we were just hoping it was in the right direction. It's a nice kick in the pants; not violent. The thing that I liken it to, being a naval aviator, is that it's similar to a catapult shot coming off an aircraft carrier."

First-stage ascent guidance used an open-loop logic based on an attitude versus velocity data table to command the best attitude to maximize the performance of the vehicle within system constraints. This meant that guidance commands were based not on the performance of the vehicle but on its Earth-relative velocity, or V_{rel}. This was computed by the navigation software propagating the equations of motion using sensed accelerations from the IMUs, adjusted using the uplinked data on the winds and then scaled by the ratio of the speed of sound at sea level to the speed of sound at the current altitude. The value of V_{rel} was used to trigger important first-stage events based on I-load tables residing in the flight software.[3]

One of these tables provided a set of angle commands in yaw, pitch and roll that determined the precise attitude the Shuttle had to have during key ascent events. The first one was called the "vertical rise". Unlike all rockets previously built, the stack comprising the Orbiter, external tank and solid rocket boosters had a complex shape very different to the usual cylinder of a rocket. Furthermore, because the launch pad was a leftover from the Apollo program, it had not been designed with the Shuttle in mind and there was a significant risk of contact between the vehicle and the adjacent structure. To preclude this, the guidance software commanded the vertical rise so that the stack maintained its launch pad attitude up until V_{rel} exceeded 127 feet per second (I-loaded), by which time it would be clear of the launch structure. Then

---

[3] The term I-load, meaning "initialization load", referred to a variable in the GPC software that could be adjusted as mission design constraints changed from flight to flight.

Nominal ascent: first stage

	ASCENT PROCEDURES	HOOK
R180	LVLH	
.96M	√P_c → 72%	
1.24M	√P_c → 104%	
P_c < 50+5 s	√SRB SEP (Backup AUTO SEP 2:21)	
	√TMECO	
	*If <u>NOT STABLE</u> (10 sec):	
	NO COMM – CSS & MAN THROT	
MM103+10 s	√OMS assist	OMS 1 TGTING
	Close suit O2, open visor	
3:00	√EVAP OUT (T < 60)	
	* If Systems ABORT reqd:	
	RTLS at 3:40 or	
	TAL Select prior to ⟦23000⟧	
	Otherwise Manual MECO ⟦23700⟧	
V_I = ⟦13.2K⟧	√Roll Heads Up	
	* If Man Throttle (3 eng):	
	Man Shutdn at ⟦25700⟧	
	* If 1 eng:	
	⟦TRAJ⟧ √SERC ON	
	When MPS PRPLT = 2%:	
	MAN THROT, P_c → 67%	
	Man MECO @ CO mark **BFS**	
MECO	√√V_I = ⟦25819⟧	
MECO+20 s	√ET SEP	
	* If '<u>SEP INH</u>':	
	ET SEP – MAN	
	If Rates > .7,.7,.7:	
	MPS PRPLT DUMP SEQ – STOP	
	Null rates	
	ET SEP – SEP	
	Post ET Sep -Z xlation:	
	MPS PRPLT DUMP SEQ – GPC	
	If Rates < .7,.7,.7:	
	Assume Feedline Fail	
	If VI < ⟦25760⟧ or BFS Engaged:	
	OPS 104 – PRO (√BFS 104)	
	NOTE: Expect – 'Illegal Entry' (PASS)	
	'Illegal TIG' (BFS)	
▶ MM104+2 s	If ET Sep complete and HA > ⟦72⟧ :	
	+X xlation for 11 sec	
	√TGTS	
	√ASC PKT for failures	
	If OMS 1 not reqd:	
	OMS ENG (two) – OFF	
	Go to <u>POST OMS 1</u>	

FB 2-5   ASC/114/FIN A

Ascent Procedures cue card (from Ascent Checklist).

guidance began to issue commands for the "roll program", a maneuver around all three axes that was completed at T + 20 seconds. It left the vehicle in a "heads down" orientation. As Crippen recalls, "I think [the roll program] excited some of the spectators because they didn't know that we were going to roll, and truthfully, all of the previous launch vehicles had rolled also. It is just that it's not very obvious when you don't have wings sticking out there like the Shuttle does. All that was very comfortable." At the same time, control was passed from the Kennedy Space Center to the Johnson Space Center in Houston, Texas.

The roll program was very important, especially for the Orbiter. In particular, the pitch motion drove the angle of attack between the relative wind and the chord of the wing (an imaginary line drawn between the leading edge and the trailing edge) to a negative angle (negative q-alpha), in turn making the wings generate a force oriented towards the belly of the Orbiter, alleviating structural loads caused by flying in the densest part of the atmosphere. In addition, because ground-based radio antennas had a good line-of-sight, this improved transmissions using the S-band antennas on the Orbiter. The "heads down" attitude also enabled the crew to view the horizon, which for a pilot is always welcome. And last but not least, the roll program put the Shuttle on the correct azimuth to reach the desired orbit.

To appreciate this it is necessary to remember that one of the seven parameters

An Orbiter performing the roll program soon after liftoff.

that uniquely define an orbit is the inclination of the orbit, which is the angle between the plane of the spacecraft's orbit and the Earth's equatorial plane. Since the facilities were a carryover from the Apollo era, it happened that when a Shuttle was on the pad the tail of the Orbiter pointed south. The roll program would turn the vehicle so that its tail was aligned with the azimuth required to achieve an orbit that had the desired inclination. At the same time as the vehicle rolled onto the correct azimuth, it began a gradual pitch over to arc out across the Atlantic. Failure to start the roll program at the corresponding V_{rel} was a clear indication of a serious malfunction of the guidance software. Since the on board instrument displays could not provide the crew with the information that they would need in order to manually fly the roll program profile, in that situation they would immediately engage the BFS in order to preclude a loss of control.

At liftoff, both ADI ATT switches were in REF because LVLH was not able to monitor the initial roll maneuver due to a discontinuity at a pitch of 90 degrees that arose from a mathematical singularity in the LVLH reference frame transformation equations. Once the roll was completed, the ATT switches could be reset to LVLH to simplify monitoring and piloting tasks. As LVLH pinned the ADI yaw to zero, yaw could be read on the HSI as a delta from the NORTH. Throughout first-stage flight, the flight crew had the task of checking vehicle attitude and attitude rates against the ASCENT ADI cue card to identify any off-nominal guidance situation and take over manually if necessary.

As the Shuttle accelerated, the aerodynamic loads increased to max-q, when the dynamic pressure of the relative wind flowing around the vehicle reached a peak and subjected the structure to the greatest structural loads.

Applying adaptive guidance throttling, the software used an I-loaded table which provided main engine throttle commands as a function of V_{rel} in order to achieve the desired vehicle performance and to minimize max-q. In nominal conditions, the main engines were run at 100 per cent of their rated thrust at the moment of liftoff. This limit was imposed by vehicle acoustic and structural constraints that ruled out using a higher thrust on the pad. The immense power released at liftoff and its effects on the vehicle were not fully appreciated on the inaugural mission. As Crippen says, "We later learned that at liftoff from the solids, we had

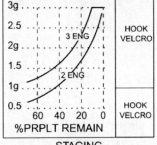

ASCENT ADI cue card (from Ascent Checklist).

At max-q.

broken a strut up in the nose of the vehicle that held the reaction control [system] tanks. It was caused by the shockwave of the solids against the Mobile Launch Platform ricocheting back up and hitting the vehicle. John and I, I don't think, realized that at that time; in fact, I know we didn't." The 100 per cent throttle was maintained until an I-loaded V_{rel} was achieved, then the engines were commanded to 104 per cent. This throttle setting would be held until another specific V_{rel} was attained, at which time the engines would go into the "thrust bucket" in order to limit max-q. The moment of max-q was readily recognized, as it coincided with going supersonic and the generation of a shockwave that produced a cloud of condensed water all around the vehicle. As Crippen recalls, "About the time we were approaching going supersonic, I've likened it to driving my pickup down an old country washboard road. It was that kind of shaking, but nothing too dramatic, and it didn't sound to me, or it didn't feel, as significant as what I'd heard John talk about on the Saturn V."[4]

The "thrust bucket" that was performed by adaptive guidance throttling prior to max-q gained its name from the shape of the line when the throttle commands were plotted as a function of V_{rel}. The objective was to keep the aerodynamic loads on the

---

[4] John Young, STS-1 commander, was also a veteran of two Gemini flights and two Apollo missions.

vehicle within structural limits by throttling down the main engines at the same time as adjusting the elevons to relieve the loads on the wing structure. Depending on the mission constraints, the main engines were throttled down to a value in the range 65 to 74 per cent. For example, the nominal ascent procedures for STS-114 called upon the flight crew to verify that the engines were throttled to 72 per cent at Mach 0.96 and up to 104 per cent at Mach 1.24. The variable geometry of the propellant grain in the solid rocket boosters enabled their thrust to vary in a programmed way to assist in minimizing max-q. But solid rocket motor performance is specified with reference to a nominal ambient temperature that is often not the ambient temperature at the time of launch, so this can cause an increase in thrust if the ambient temperature is higher than normal (a situation dubbed "hot") and a decrease if it is cooler (dubbed "cold"). In performing the thrust bucketing, the adaptive guidance had to take "hot SRBs" and "cold SRBs" into account in throttling the main engines. To do this, it compared the time at which a predetermined I-loaded V_{rel} would be reached in a nominal situation (neither "hot" nor "cold" SRBs) with the actual time that this velocity was reached. If it occurred early, then it was a "hot SRB" situation and the engines were commanded to throttle back more severely. If the velocity was achieved late, it was a "cold SRB" situation and the main engines would be throttled back less severely.

Changes in SRB performance were addressed not only by way of a biased throttle bucket, but also by varying the pitch commands to hold as closely as possible to the nominal trajectory. In the case of "hot SRBs" the increased acceleration would place the vehicle further up-range when a specific V_{rel} was attained. In order to prevent a "depressed trajectory", the nose was raised slightly. The opposite applied in the event of "cold SRBs". In this case the vehicle would be further down-range when a specific V_{rel} was attained. In order to prevent guidance from "lofting" the trajectory[5] (as in the case of a failed main engine) the nose was lowered. This also increased the down-range velocity so that the close-loop second-stage guidance would not have as large a deficit to work with in order to achieve the main engine cutoff (MECO) target. It is worth noting that these variations in the pitch profile were very difficult to observe by naked eye.

If one of the three main engines failed during the first-stage ascent, the guidance software would throttle the other two engines to compensate as best it could. If this occurred prior to aerodynamic load relief the thrust bucket would not be executed, in order to achieve the highest possible performance. In parallel with adaptive guidance, the flight control system also performed load relief in response to increased structural loads on the wings due to wind gusts and shears. First-stage guidance commanded the flight control system to steer the vehicle into the winds in order to alleviate the wing loading problem. But this had the direct consequence of making the Orbiter fly out of the planned trajectory plane, as yawing into the wind had the effect of rotating the thrust vector out of the current plane. To compensate, guidance

---

[5] For a brief explanation of what trajectory lofting is, refer to the paragraph below on a nominal second-stage ascent.

Elevons deflected for load relief during ascent.

would then issue roll commands to realign the thrust vector with the desired plane. As a further action during load relief, the flight control system used closed-loop logic to issue commands to deflect the elevons to null the hinge moment and minimize Nz loads on the wings.[6] Load relief and adaptive guidance would be terminated if an engine failure occurred during the load relief phase, but pitch correction would be made in order to prevent a depressed trajectory.

At T+100 seconds, the SRB separation sequence began to monitor the chamber pressures of both the left and right boosters. When the middle value of the three Pc transducers fell below 50 psi, physical separation occurred. If all Pc sensors on one booster failed "low" prematurely or "high", the act of separation was inhibited until a backup timer expired in order to ensure that physical separation was not attempted with excessive SRB thrust. The backup timer represented the latest time that a slow-burning booster would require in order to tail-off from combustion pressures which exceeded 50 psia. The auto separation sequence would also compare observed rate gyro roll, pitch and yaw rates and dynamic pressure with stored limits. If these limits were exceeded, then separation would be inhibited and the crew alerted. Only once the out-of-limit condition had been corrected could the auto separation sequence be terminated. Alternatively, the crew could manually command SRB separation.

Prior to flight software release OI-26,[7] as the vehicle approached SRB separation the pitch rate command trended to zero in the tail-off region, since the commanded pitch rate followed the slope of the pitch attitude as a function of velocity in a table that was uplinked before launch. This meant that the vehicle held attitude for several

---

[6] Nz is along the Z axis, and hence such loads are perpendicular to the wings.
[7] OI means "operational increment".

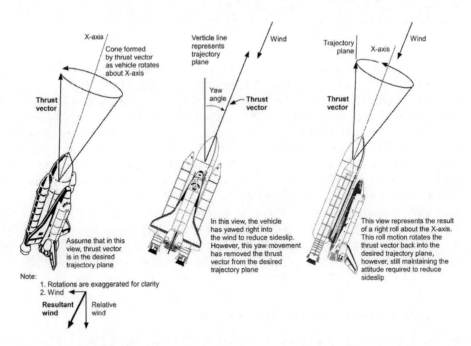

Load relief operation. (Courtesy of www.nasaspaceflight.com)

seconds prior to the software commanding attitude-hold for SRB separation. Second-stage guidance then had to compensate for the extended hold. However, starting with OI-26 the pitch steering during SRB tail-off reduced the pitch command after T+100 seconds at a constant rate so that the pitch command rate could stay at a constant 0.44 degrees per second for an I-loaded number of seconds beyond the moment when N_y feedback faded out.[8] This provided a more benign transient on initiating second-stage guidance, and increased the payload by 50 pounds. The only observable difference was that the angle of attack at separation was some 3.5 degrees more positive with this new procedure implemented.

SRB separation was another dramatic moment for the crew. Crippen recalls it as follows, "When we got to two minutes into flight, when the SRBs came off, that was enlightening in that the big separation motors that push the solids away, the ones up forward, actually you could see the fire come over the forward windows. I didn't know that I was going to see that, but it was there, and it actually put a thin coat across the windows that sort of obscured the view a little bit, but not bad. But the main thing when they came off, it was noisy."

Perhaps the most interesting part of SRB separation was the distinct change in the perceived acceleration. Terry J. Hart was mission specialist on STS-41C and was on the flight deck behind the pilots during launch on 6 April 1984. As he recalls, "Very

---

[8] The N_y feedback was what made the load relief possible.

quickly the solid rockets taper off and separate, and that was the first surprise I had because your G-loading builds up to, I guess, close to two and a half Gs or maybe a little bit more during that first two minutes as the solid rockets are reaching their peak thrust and the main engines are at 100 per cent. Then very quickly that thrust tapers off as the solid rockets burn off, and they separate. Well, the sensation that you have at that point, I wasn't quite prepared for. When you get used to two and a half, it feels pretty good. You're going somewhere. When you go back to one and half. That feels like about a half. So the sensation you have is that you're losing out, that you're falling back into the water. You don't think like you're accelerating as much as you should be to get going." Although Hart had worked on the SSME program and was familiar with their performance, he admits, "I think in the next minute I must have checked every five seconds ... the main engines to make sure they were running, because I [could have sworn that] we only had two working, because it just didn't feel like we had enough thrust to make it to orbit. But then gradually the external tank gets lighter, and ... of course, with the same thrust on the engines you begin to accelerate faster and faster. So after a couple of minutes I felt like, yes, I guess they're all working."

Shuttle day-of-launch trajectory design operations

A Shuttle mission was planned several years in advance, but the planning for one of the most crucial phases was done only a few hours prior to liftoff. The first priority of any launch vehicle is to safely insert as much mass into the desired orbit as possible. In particular, safety means that no structural or thermal constraints are violated at any point during the ascent, as they could rapidly cause a catastrophic failure. From the point of view of aerodynamic loads, the most dangerous part of the ascent occurred in climbing to an altitude of about 155,000 feet, because the vehicle had to endure the atmospheric conditions of the day, namely winds.

It is easy to appreciate that the stack comprised of the Orbiter (with its protruding wings and tail stabilizer), external tank and solid rocket boosters was not the ideal configuration for flying through the dense lower atmosphere. The structural limits for such a complex vehicle, with its elements held together by attachment points, required a large margin. As described earlier, first-stage guidance was based on a set of attitude commands as a function of V_{rel}. The commands in roll, pitch and yaw were specified a few hours prior to launch by the day-of-launch I-load update (DOLILU) carried out using ground software and personnel skilled in the design and verification of the trajectory.

This started about six hours before launch with the release of a number of weather balloons from a site near the pad to measure atmospheric conditions, particularly winds. The data from the balloons was relayed to the Johnson Space Center as initial input to an iterative computer simulation to calculate the initialization loads for the first-stage guidance. In this first step, the team processed the atmospheric conditions of the day to calculate the required roll, pitch, yaw steering commands and the engine throttle commands to satisfy the following conditions:

1. Hold the angle of attack (alpha) at –4 degrees to reduce wing loading during the ascent.
2. Hold the angle of sideslip (beta) equal to zero degrees to reduce side loads.
3. Keep the maximum dynamic pressure (max-q) near the design target. As the potential wind variations are greater in the winter than in the summer, this target was reduced for a winter launch.

As part of the output from this first step, a wind table was calculated for use by the flight software in determining the elevon schedule to minimize elevon hinge moment loading in flight. The simulation software used to calculate the I-load based its analyses on a filtered or smoothed wind profile in which wind features deemed unlikely to persist between the time of taking the wind measurements and the time of launch were removed.

To refine the steering and throttle commands a higher fidelity simulation checked the initial solution against an unfiltered wind profile, centering the trajectory on the induced alpha and beta spikes. This trajectory prediction was assessed to verify that it satisfied all of the trajectory, control, systems, structural loading, and performance constraints. Given the short time available in which to assess structural loading using complex algorithms, simplified structural load indicators (SLI) were developed from higher fidelity structural analytical limits to calculate element structural loadings for a trajectory designed on the day of launch. A total of forty-two SLI constraints were calculated for specific points on the stack, such as the Shuttle/ET attachments. Each SLI was evaluated in the high-q region of flight, namely in the range Mach 0.6 to 2.2, and in each case a limit (that could vary across the high-q region) was set, along with a minimum margin. During verification close to launch, if the SLI check against the latest balloon weather assessment reported a negative margin, the launch attempt was deemed NO-GO, scrubbed and rescheduled.

At about 90 minutes prior to launch the I-load tables were frozen and uplinked to the Orbiter. However, verification continued in order to be certain that the uplinked data would still be valid at launch. This was necessary because the DOLILU solution uplinked to the Orbiter was valid for the specific wind profile measured several hours earlier. A GO/NO-GO would be called by the loads and DOLILU officer (LDO) in the Launch Control Center, who collated reports from the prime team at the Johnson Space Center, a team of atmospheric specialists at the Marshall Space Flight Center, and an independent verification and validation (IV&V) team. The prime team was responsible for trajectory design, verification and integration of the DOLILU system. Their results were verified by the IV&V team at a contractor's location. The software tools used by the two teams were independently coded to the same requirements and same quality standard in order to ensure that there were no systematic programming flaws. The IV&V team independently received the wind profile data from the balloons. For DOLILU to be a GO, constraint margins between the two teams had to be consistent to within specific tolerances.

NOMINAL ASCENT: SECOND STAGE

Once the most powerful solid rocket boosters ever designed were jettisoned, only the main engines provided the necessary thrust to reach orbit. With the SRB separation there was an automatic and instantaneous transition of the flight software to MM103 for second-stage guidance.[9] Unlike first-stage, this made use of a closed-loop logic with each cycle computing where the vehicle ought to be in the sky in order to hit the I-loaded MECO target.[10] By navigation, the software knew where the vehicle was in relation to the nominal trajectory and processed the position and velocity errors to issue commands to steer the thrust vector so that the Orbiter would be in the specified position with the required velocity for MECO. In this way, any error (provided it was not excessive) inherited from the first stage could be corrected.

The task of the first stage of the Shuttle was to follow a trajectory designed to climb through the densest part of the atmosphere as directly as possible and thereby reduce aerodynamic drag at the expense of a reduction in performance. At SRB separation the velocity vector had a very large positive vertical component, meaning that the vehicle was climbing rapidly. But because by this point it was above most of the atmosphere it was free to accelerate without incurring serious aerodynamic loads. The second-stage guidance deflected the trajectory towards the horizon to build up a very large positive horizontal component and achieve the energy required for orbital flight. This is because achieving orbit is not merely a matter of reaching the proper altitude, it also requires gaining the necessary horizontal velocity. This "steerdown" (a sort of a spiral trajectory) by the second-stage guidance was performed through to MECO, at which time the velocity vector was nearly parallel to and almost the same magnitude as the target velocity vector. Cross-range steering was performed with the same logic in order to steer the Orbiter towards the orbital plane specified for the mission.

Second-stage guidance relied on the PEG 4 scheme[11] and was nominally done in two phases. The first phase computed throttle and attitude commands based on three main engines and a constant thrust requirement until achieving an acceleration of 3 g, which marked a structural loads constraint. At that point, typically one minute prior to MECO, the second phase would progressively throttle down the engines so as not to exceed that limit. Unlike the mighty Saturn V, which during ascent reached an acceleration peak of 8 g, the Shuttle could not withstand structural loads greater than 3.5 g. The strengthening needed for the Orbiter to have been able to withstand higher accelerations would have made the structure so heavy as to have severely reduced its operating altitude and payload. And the Shuttle program was intended to make space flight available not only to test pilots but also to "more common" people who did not have the training to withstand the high accelerations endured by jet

---

[9] MM means "major mode".
[10] Defined in terms of inertial velocity, flight path angle, radius from the center of the Earth and the desired orbital plane.
[11] PEG means "powered explicit guidance".

fighter pilots. For all these reasons, it was a requirement not to exceed 3 g during the ascent. Should an engine fail during this phase, the constant thrust phase would be resumed in an effort to limp into a low but stable orbit.

To eliminate or minimize "abort gaps", STS-1 in April 1981 and STS-26 (the first return-to-flight mission after the loss of *Challenger*) in September 1988, flew higher than normal trajectories. These "lofted" or "abort shaped" trajectories required a third PEG phase that ran from SRB separation to an I-loaded time and achieved lofting by assuming that an engine would fail and cause a loss of performance at that moment. After that PEG stopped assuming that an engine would fail, and continued to provide commands for a nominal trajectory. But this method created "black zones" where an unsurvivable entry/pullout condition would occur if two engines actually were to fail. For this reason, and because abort shaping reduced payload capacity, lofted trajectories were discontinued.

Soon after SRB separation, the crew had to verify that guidance convergence was achieved by no more than ten seconds after the software transition to MM103. This was done by visually checking that TMECO on the ASCENT TRAJ display became stable and that the ADI error needles were no longer in the stowed position. Upon initiating each major mode the convergence checks were usually satisfied within a couple of guidance cycles, which is to say between 4 and 6 seconds. An abrupt change in the actual acceleration after initial convergence, for example as a result of engine failure, usually caused the convergence to fail for no more than one guidance cycle. So the longest reasonable interval for PEG to fail to converge was ten seconds. Failure to achieve convergence for more than ten seconds indicated that an internal fault was preventing the PEG solution from satisfying the target conditions and the crew would fly the remainder of the ascent manually.

Once using flight software OI-26, the Shuttle introduced a maneuver called OMS-assist burn. This was normally carried out ten seconds after switching to MM103. It burned 4,000 pounds of propellant in 102 seconds to provide an increase in payload of 250 pounds.[12] Starting with STS-87 in November 1997, a "roll to heads up" was performed during a nominal ascent to facilitate an earlier communications capability with the TDRS geostationary relay satellites, as the Bermuda ground station had been closed to save money. This roll maneuver was begun at an I-loaded relative velocity that differed from mission to mission and was made at a rate of 5 degrees per second. It was accompanied by a change in pitch of between 20 and 30 degrees, changing the direction of the thrust vector.[13]

Forty seconds before MECO the position constraints were released, meaning that guidance no longer tried to achieve the targeted altitude and orbital plane, because a small change in position error so close to MECO would produce large changes in the thrust turning rate vector and result in over controlling. The so-called "fine

---

[12] With reference to STS-114 Ascent Checklist, this maneuver was specified by the line: *MM103 + 10s* $\sqrt{}$ *OMS assist*.

[13] With reference to STS-114 Ascent Checklist, this maneuver was specified by the line: $V_I = 13.2K$ $\sqrt{}$ *Roll Heads Up*.

counting" started ten seconds before a nominal MECO, terminating closed-loop guidance and commanding the engines to a power level of 67 per cent for three engines and 91 per cent otherwise. Then the flight path angle constraint was released, and the time-to-cut-off was calculated solely on the desired velocity change. When guidance saw that the Orbiter had achieved the correct inertial velocity, all of the operating main engines were commanded to shut down.

Brian Duffy, veteran of four Shuttle flights, two as commander, describes MECO as follows, "The engines cut off and you go from what [was] like the most violent place you've ever been in your life to the most peaceful place you've ever been in your life, in a couple of seconds. It's an interesting little transition, when the engines shut off and all the thrust stops, and all of sudden you're free-floating. Everything's quiet. It's interesting how it happens, and every time I was equally impressed. Even the fourth time, I just went, 'Wow, I forgot how much power there is.' It's awesome, it really is."

Once MECO was confirmed, the nominal ET separation sequence was initiated. This involved sequential acts to perform major functions such as umbilical unlatch and retract, commanding engine prevalves closed, electrical deadfacing, inhibit and separation mode checks, and structural separation. If the logic detected a problem in performing a safe separation, the sequence would be halted and the crew alerted. As listed in the ascent procedures, the first action to take would have been to set the ET SEP switch to the manual (MAN) position and then perform a check of the rates. If any rate exceed a preset value of 0.7 degrees per second the crew would set the MPS propellant dump sequence to STOP and null the rates. Normally, the MPS dump was performed after ET separation but certain failures could cause it to start early. In that case, the dump sequence had to be stopped because the dump would induce rates on the vehicle and make it difficult to null the existing rates. After nulling the rates, the crew would manually separate from the tank. The most likely reason for separation inhibition was failure of a feedline disconnect valve. Because there was no foolproof means of establishing whether a valve had really failed or it was only a sensor error, the crew had to assume that a valve had failed to close. The procedure

```
MECO              √VI = 25819
MECO+20 s         √ET SEP
              *  If 'SEP INH':                                    *
              *      ET SEP – MAN                                 *
                     If Rates > .7,.7,.7:
              *          MPS PRPLT DUMP SEQ – STOP                *
              *          Null rates                               *
              *          ET SEP – SEP                             *
              *          Post ET Sep -Z xlation:                  *
              *              MPS PRPLT DUMP SEQ – GPC             *
              *  If Rates < .7,.7,.7:                             *
              *      Assume Feedline Fail
              *      If VI < 25760 or BFS Engaged:                *
              *          OPS 104 – PRO (√BFS 104)                 *
              *          NOTE: Expect – 'Illegal Entry' (PASS)    *
              *                         'Illegal TIG' (BFS)       *
```

ET separation sequence (from Ascent Checklist).

was to wait six minutes to allow residual propellant pressures in the Orbiter and ET vents to fall to a level where the venting force was less than 1,800 pounds from an open valve. In this way recontact with the ET after separation due to the propulsive effect of escaping propellant would be minimized. Separation could then be manually performed.

After a separation (however it was achieved) the Orbiter would use its RCS jets to initiate a $-Z$ translation of 4 feet per second in order to establish an opening rate with the ET. If this maneuver could not be carried out automatically, the crew would do so manually via the THC.

In order to assess ET performance, many missions (especially after the *Columbia* accident) included a $+X$ axis translation of 11 feet per second to increase the opening rate, followed by a pitch up of 2 degrees per second to place the ET in the overhead windows of the flight deck to enable pictures to be taken. Two minutes after MECO was confirmed, a MPS dump was initiated. This took two minutes. During the dump, the main oxidizer valve was opened to dump residual oxygen and the fuel fill/drain valves and the fuel bleed valves were opened to dump residual hydrogen. If there had been any issues that limited maneuvering capability then the ET photography would be deleted. In a new ET pitch maneuver introduced by STS-114, the MPS dump was used to impart a pitch moment to the Orbiter, enabling the pitch to start two minutes earlier and the pictures to be taken from a closer range for improved resolution using handheld cameras.

Orbital insertion profiles

It would have been possible to take the external tank into orbit but the MECO target was selected to ensure that the tank re-entered the atmosphere over an unpopulated region. For this reason, the velocity at MECO was insufficient to put the Orbiter into a stable orbit. But it was close enough for the desired orbit to be achieved by making one or two OMS burns. At this point, two profiles were pursued during the program. The first one, called standard insertion (SI), required performing two OMS burns in sequence. Crippen describes this as follows, "We'd cut off the main engines so that we were somewhat short of the apogee, or the high point, of the orbit that we wanted. Then we did a burn [OMS 1] of the orbital maneuvering engines to get the apogee up at the right altitude. Then we flew around to the apogee and fired the engines again [OMS 2] to get the perigee up." This was undertaken by early missions, before full information became available on SSME performance and targeting precision.

As confidence in SSME performance grew, a new insertion profile was designed, called direct insertion (DI). This was first performed by STS-41C in April 1984 with Crippen as commander. "People looking at flight design came up with, 'Hey, we can do that by using the main engines to get the apogee at the right point, so you don't need that first burn.' It's a much easier task from a crew standpoint, because you're pretty busy right after main engine cutoff, and this took away some work, so it was a neat thing to try." The first OMS burn would be made at apogee to lift the perigee. Starting with STS-30 in May 1989 all missions employed direct insertion. Its main advantage was that it allowed higher orbital altitudes because more OMS propellant was available for orbital maneuvers.

Regardless of the insertion profile, on completing the –Z translation following ET separation the software automatically switched to MM104 and displayed the OMS 1 MNVR EXEC software page with an I-loaded target for the OMS 1 burn. For direct insertion, this represented a desired target with a delta-V of 11 feet per second, which was exactly the delta-V imparted to the Orbiter by the MPS dump two minutes after MECO. For a direct insertion mission the crew had nothing to do during this mode unless MECO had left an underspeed condition, in which case an OMS burn might be made as early as two minutes after MECO for a very large underspeed or left until two minutes before apogee for a small underspeed. For a nominal standard insertion the crew would perform the OMS 1 burn either using the I-loaded target or a new one provided by Mission Control. After this burn, the crew had to note any anomalies and use the THC to manually null out any residuals greater than 2 feet per second. OMS guidance remained active so that other contingency OMS burns could be made prior to transitioning to the next major mode.

The switch to MM105 was made manually, causing the OMS 2 MNVR EXEC page to appear with the I-loaded target for the OMS 2 burn. The crew verified the burn solution, and in particular the time of ignition (TIG) in order to center the burn at the moment of apogee for the most fuel optimal burn solution. Whilst coasting up to apogee, the crew also executed the MPS gaseous hydrogen inerting procedure and the ET umbilical doors were closed. The Orbiter adopted the calculated burn attitude, and at TIG the command was given to start the burn. If the burn was being controlled automatically, the crew were ready to take control at any moment using the THC and conclude it manually. After this burn, the crew took note of the post-burn condition, reconfigured the OMS/RCS for on-orbit operations, and performed the MPS liquid hydrogen vacuum inerting procedure. Finally, a manual switch to MM106 terminated the ascent phase and initiated "insertion coast".

NOMINAL ASCENT: DISPLAYS

The Shuttle could not be flown at all without its computers. If a problem disabled all the PASS computers, the crew would have no option but to engage the BFS. It was possible that a combination of software errors and/or hardware failures could prevent otherwise functional PASS computers from flying successfully. As an alternative to engaging the BFS, the pilots could manually fly the Shuttle during the ascent using a mode called control stick steering (CSS).[14] To do so, it was necessary to provide the flight crew with a means of monitoring the AUTO guidance system and performance of the vehicle so that they would know when to take over. And after manual control had been initiated, they would require a means of manually

---

[14] This was also referred to as "hot stick downmode".

doing whatever a healthy guidance system would have done. The information to perform these two tasks was provided by special computer-generated displays along with printed procedures.

Generally speaking, these displays provided a plot of the nominal trajectory with altitude on the vertical axis and relative velocity (V_{rel}) on the horizontal axis. The current state of the vehicle calculated by the navigation software was depicted as a moving triangle, and circles showed the state vector predicted for 20 seconds ahead on the first stage and 30 to 60 seconds ahead on the second stage. This enabled the pilots to monitor how well the guidance software was working. Such displays were generated by both the PASS and BFS, and although BFS was in "listening" mode its displays were more detailed and more accurate than those generated by PASS. At least one CRT (or MEDS later on) showed a PASS-generated trajectory display and another displayed the BFS-generated ascent trajectory information.

During the first stage of the ascent, both BFS and PASS showed a trajectory plot with its horizontal scale ranging from 0 to 5,000 feet per second and a vertical scale ranging from 0 to 170,000 feet. Ticks on the curved line corresponded to LVLH ADI ball angles of 70, 60, 50 and 40 degrees and were to be used in conjunction with the ADI and the Ascent Checklist to determine guidance and vehicle performance. The PASS display for the first stage (ASCENT TRAJ 1) provided information about the type of contingency abort available at the current point of the ascent, along with other details on special guidance actions that would be available if an abort were selected. The same display generated by BFS provided several other details. For example, the "Pc < 50" field would double overbright and flash if Pc sensors in the SRBs detected a combustion chamber of less than 50 psi. Flashing of this field would indicate that the SRB separation sequence had been initiated. If the dynamic pressure and/or vehicle body axes exceeded the I-loaded values, SRB separation would be inhibited and the crew informed by a double overbright and flashing SEP

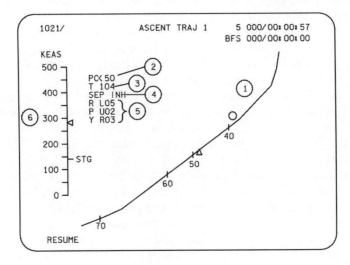

Ascent display for first stage.

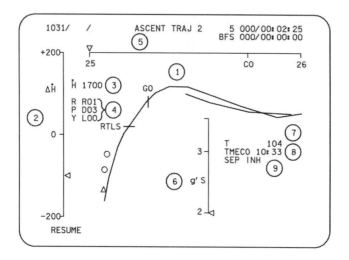

Ascent display for second stage.

INH on the ascent trajectory. The BFS display also showed its own roll (R), pitch (P) and yaw (Y) ADI errors for comparison with the PASS-driven ADI error needles. If the pilots engaged CSS then these errors would function as fly-to indicators. Finally, this display also showed the equivalent airspeed in knots (KEAS) scale with a moving triangle. A tick indicated the moment for safe staging (≈ 138.9 seconds), which also served as a good indicator of dynamic pressure.

After jettisoning the solid rocket boosters and switching to the second stage, the software would automatically display ASCENT TRAJ 2 information, and this would remain until completion of the $-Z$ maneuver for ET separation. As during the first stage, the PASS-generated display would show abort options and available functions and the BFS-generated display would provide more information about the trajectory. In particular, on the far left of this display a delta-H scale would indicate the actual radius rate (essentially the vertical velocity) from the navigation software minus the expected radius rate for the present velocity. The current delta-H would be shown by a moving left-pointing triangle, flashing once off-scale. The display would also show a digital readout of the current vertical velocity in feet per second, a digital readout of the current throttle command by the guidance software, and a digital readout of the predicted MECO time in minutes and seconds since liftoff. A horizontal scale at the top of the display would indicate the moment of MECO. This was an inertial velocity scale measured in thousands of feet per second, and since most MECO targets would be between 25,000 and 26,000 feet per second the scale was normally displayed with these limits, but since this was a flight-specific number the limits could be specified differently. A CO (for "cut off") on the scale would indicate when the main engines ought to be commanded off in order to achieve the nominal MECO inertial velocity. A third scale was located below the central plot to indicate the total load factor in g's. The current load factor would be shown by a moving left-pointing triangle, flashing on exceeding the 3-g limit. This scale would be used to monitor how well guidance was holding the acceleration within the limit during a nominal ascent and to indicate the progressive throttling required to maintain that load during a manual ascent.

CONTROL STICK STEERING

Riding in the front row of a Shuttle flight deck is one of the highlights in the career of a pilot, and being able to manually fly it into orbit would be the ultimate experience. The Shuttle required such extensive use of computers that without them it could not have been flown at all. Nevertheless it was essential to have available some sort of manual control during the ascent to overcome faulty on board navigation, erroneous guidance, impending loss of control, unexplained attitude/rate excursions, undamped oscillatory motions, winds, and SRB dispersions.

Early in the program it was determined that the crew could not manually control the vehicle during the roll and load relief programs, so control stick steering (CSS) could be selected only after 90 seconds of flight. It was selected by pressing either the PITCH or the ROLL/YAW CSS buttons on the commander's or pilot's eyebrow panel. This would give both the commander and the pilot manual control of all three axes. Pressing the button a second time would return control to the flight software.

After staging, the pilots would select CSS if the guidance remained uncoverged for more than ten seconds. This action had to be taken promptly to avoid the vehicle violating the range safety limits; that is to say, trajectory deviations that would take it close to populated areas. In this case (which fortunately never happened during the program) the ground would provide yaw steering commands to enable the pilots to veer away from the limit lines. To fly a nominal ascent to MECO in CSS mode, the crew would use the trajectory displays, the ADI and the HSI. If Mission Control was able to confirm that the NAV state was good, the crew would monitor the ASCENT TRAJ displays and impart pitch commands, small deflections of less than 5 degrees using the RHC to steer the vehicle and maintain the current "state bug" (triangle) and the two predictor bugs (circles) on the trajectory line. Roll would be monitored on the ADI and held at 180 degrees until it was time for the roll to heads-up maneuver, then it was held at 0 degrees. Finally, yaw would be monitored by watching the pointer on the HSI. If Mission Control reported that the NAV state was not good, then the pilots would fly using the ASCENT ADI cue card, which specified the profile in terms of pitch command, altitude, and altitude rate as functions of inertial velocity. That is, by using CSS and manual throttling they had to attempt to obtain the pitch command, altitude and altitude rate appropriate for the current inertial velocity.

In fact, the flight rules stated that on selecting CSS it was also necessary to select manual throttling, as unconverged guidance could potentially issue a spurious MECO command. Manual control of the throttle would prevent this. The selection of manual throttle would be done by the pilot pressing the red button on the speedbrake/thrust controller and then moving the lever to match the current commanded throttle setting within 4 per cent. At this point, manual throttling could begin. The rules also required manual throttling in the case of failure to initiate/

terminate thrust bucket commands for first-stage q-bar control, and in the case of failure to maintain the 3-g acceleration limit designed to ensure that the loads did not exceed the structural constraints.

INTACT ASCENT ABORT MODES

The 8.5-minute ride into orbit was a tiny fraction of a Shuttle mission lasting more than two weeks but the extremely dynamic nature of the ascent made it one of the most difficult phases to manage during an emergency. Prior to the Shuttle (and also afterwards) all spaceships were capsules mounted on top of the launch vehicle. The Mercury and Apollo capsules had a tower incorporating a solid rocket that would fire in the event of off-nominal performance and haul the capsule away from a failing rocket, to enable it to splash into the ocean as during a nominal re-entry. The Shuttle was completely different. Firstly, it was carried on the side of its launcher. And it was designed to land on a runway, not to ditch in water. A complex set of abort scenarios had to be devised to enable the Orbiter to fly to a suitable landing site in the event of an emergency during the ascent.

Two types of scenario were conceived, one for "intact aborts" and the other for "contingency aborts". An intact abort was intended to enable the Orbiter to fly to a planned landing site but a contingency abort focused on crew survival and accepted possible loss of the vehicle. Selecting the type of abort to perform would be based on the severity of the issue and the point in the ascent at which it occurred. If there was a single failure an intact abort would be appropriate, but if there were multiple failures a contingency abort might be the only option. There were four possible intact abort modes.

Abort to Orbit (ATO) was designed to attain a temporary orbit at a lower altitude

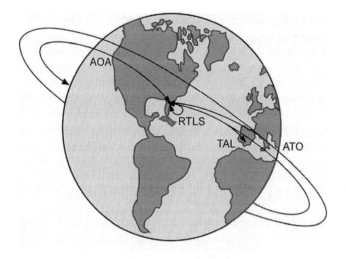

Intact abort modes.

than scheduled, to obtain time to evaluate the anomalous situation and choose either an early deorbit burn or an OMS maneuver to raise the orbit and continue with the mission. Abort Once Around (AOA) was designed for when it would not be possible to maintain orbit. It would use the remaining performance to make one revolution of the globe and land at either Edwards Air Force Base (EDW) in California, Northup Flight Strip (NOR) in New Mexico, or Kennedy Space Center (KSC) in Florida. The Transoceanic Abort Landing (TAL) would allow the vehicle to achieve a suborbital trajectory sufficient to enable the Orbiter to land on a runway in Canada, Europe or Africa depending on the vehicle performance and orbital inclination. And Return To Launch Site (RTLS) was designed to allow the vehicle to fly down-range to dissipate propellant and then turn around in powered flight in order to return and land on the runway near the launch site.

One of the factors taken into consideration in choosing the most appropriate abort mode was the so-called "performance capability" of the Shuttle, defined as a measure of its ability to achieve the desired orbit. If its performance was better than predicted then MECO would be attained with a surplus of propellant. But if the vehicle was not performing as well as predicted then it would not be able to attain the desired MECO velocity. The most probable cause of a MECO underspeed was a main engine failure. For underspeeds of several hundred feet per second, the situation could be recovered by increasing the delta-V of the OMS 1 and OMS 2 burns. If the underspeed was too large to be compensated using excess OMS propellant, the Orbiter would not be able to reach the desired orbit and either an ATO or AOA would be performed. An engine failure very early in the ascent would end in either a RTLS or TAL abort.

Selecting an abort mode in powered flight was done using a rotary switch located just in front of the pilots. In arm/fire fashion, this switch was turned to desired abort mode and an adjacent button pushed in order to initiate the abort. As a redundant line of communication, the abort button would illuminate in response to a command from Mission Control, with the light being extinguished when the crew executed the abort instruction. In the absence of a voice communication, this light would indicate to the crew that an abort was required, although it could not specify the desired mode. As a further level of redundancy, the crew could initiate an ATO or TAL by entering an item into the OVERRIDE display (SPEC 51). However, there was no RTLS option on this display. As a redundant method, the crew could request RTLS by selecting OPS 601 PRO. Upon selection of a given abort, the commander and pilot would start using the cue cards for that abort mode contained in the Ascent Checklist.

Return To Launch Site

The RTLS mode was the most aerobatic of all the abort procedures written for the Shuttle, and the least trusted by the pilots. Despite the extensive effort devoted to it in training, no one expected to return alive from such an abort. Fortunately, it was never required. It was feasible only after SRB separation and through to the point known as "negative return", where the vehicle was too far from the launch site and its down-range velocity too great to turn around and return. To execute this abort, a number of important actions had to be taken either automatically by the guidance software or manually by the crew.

On initiating a RTLS abort, an automatic OMS fuel dump would occur. The term "OMS dump" is a misnomer since it implies piping propellant overboard whereas in fact the dump was performed by burning propellant through the OMS engines and, if necessary, also through the aft RCS jets in a configuration called "interconnect dumps". The interconnect would be executed automatically by a software module called the "smart interconnect" as soon as the abort was initiated, but while the OMS engines would ignite immediately, the RCS jets would not be turned on until 3 seconds after the interconnect was complete. This interval was to enable the crew to verify correct positioning of the OMS and RCS valves in order to avoid consuming all of the RCS propellant and leaving the vehicle uncontrollable after MECO during the turnaround maneuver. A nominal interconnect would select all 24 aft RCS primary jets but if the smart interconnect detected a problem it would select only 10 jets (or indeed none in the most extreme case). The task of the smart interconnect was to ensure that only OMS propellant reached the RCS jets. In general, for any interconnected abort dump all 20 aft pitch and yaw jets would be fired together in a configuration called "null jets" that would impart no rotational or translational perturbations on the vehicle. In addition, the four +X jets that fired directly aft would also participate in the dump but on a timetable in the abort control sequence. The dump times for each mission were specified in the Abort Dump section of the Ascent Checklist.

If one RCS leg was unavailable for any reason, the corresponding leg on the other

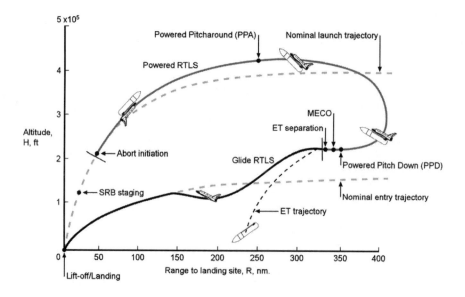

RTLS flight profile.

Intact ascent abort modes 363

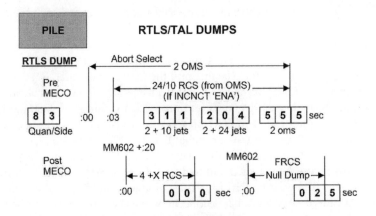

RTLS propellant dump.

side would be made unavailable in order to avoid an unsymmetrical dump. As each leg consisted of 10 jets, the interconnect dump would be limited to just 10 jets. This would prevent unwanted pitch moments from asymmetric jet firings. The affected manifolds would be reset to be fed from the RCS tanks only, and the OMS burn rate adjusted to provide an accurate measure of the propellant burned. After the required amount of propellant had been consumed, all the propellant feed paths would be reset to their normal configuration.

The OMS dump was necessary because when the Shuttle was launched it carried enough propellant to accomplish a nominal mission, including the deorbit maneuver. The fact that all of this propellant was in tanks in the aft fuselage placed the center of gravity much further to the rear of the Orbiter than it ought to be whilst flying in the atmosphere. On a nominal mission, all this propellant would have been consumed by the time the Orbiter performed re-entry. The only way to consume this propellant for a RTLS was by burning it.

In parallel, the guidance software would initiate "trajectory lofting", commanding the Orbiter to a more vertical attitude in order to minimize the loss of altitude whilst continuing to fly away from the launch site. This continued up to the point at which the vehicle would start to reverse its course by a somersault called the powered pitch around (PPA). As the Orbiter would be completely powerless after the main engines were shut down, guidance had to calculate the best way to attain the exact MECO in terms of altitude, flight path angle and velocity as functions of range to permit it to safely jettison the external tank and glide back to the launch site.

Soon after RTLS was initiated, guidance would switch from the PEG 4 scheme to PEG 5, specifically designed to deal with this kind of abort. Firstly, depending on the velocity at which the abort was initiated and the number of main engines running, the software would calculate the best pitch angle (theta) to perform the fuel wasting that was required for trajectory lofting. Given the assumption of an immediate PPA, the software would predict the trajectory to reach the PPA point and the trajectory up to MECO, computing the mass of the Orbiter at main engine cut off. If the predicted mass was still greater than the desired mass of 2 per cent,[15] then further fuel wasting would be necessary. The delta between the computed and desired mass was used to compute how much more fuel remained to be wasted, and this was used to compute the time of MECO. Thus the solution involved calculating the pitch angle at which to perform the fuel wasting and the moment at which to terminate the wasting and start the PPA. This ensured that from the start of the fuel wasting up to MECO, passing through the PPA, sufficient propellant would be wasted to allow the Orbiter to arrive at the desired MECO target in terms of altitude, flight path angle and range with only 2 per cent of propellant in the ET. Depending on the point during the ascent at which the abort was initiated, it might be that sufficient propellant had already been burned and there would be no need for fuel wasting.[16] In any case, once the fuel wasting was finished, guidance would command the PPA, a pitch turn at a rate of 10 degrees per second in which the attitude would be changed from heads-down going away from the launch site to heads-up pointing in the direction of the launch site.[17] Completion of the PPA marked the start of the flyback phase in which the Shuttle would thrust to cancel its motion away from the launch site and start to fly back towards it. By now PEG 5 would be computing the steering and throttle commands required to attain the MECO point.[18] The actual commands would not change much from cycle to cycle, but the accuracy of the PEG 5 prediction would improve as MECO approached.

Prior to MECO, the vehicle would execute a turnaround known as the powered pitch down (PPD). But first guidance would cease to calculate steering and throttle commands because near to MECO small deviations in the solution calculated at each cycle would result in excessively large commands that could cause loss of control of the vehicle. So about 6 second before PPD, guidance would lock out further attempts to correct the final MECO conditions thereby terminating the flyback phase. At this point the exact time of MECO would be determined. It would not be revised even if subsequent events made it non-optimum. To satisfy the altitude and flight path angle contraints at MECO, the vehicle would require a positive angle of attack (alpha) of

---

[15] Guidance had to ensure that the Orbiter would arrive at the desired MECO target with only 2 per cent of propellants remaining in the ET, as more might slosh, causing the tank to lurch and collide with the Orbiter at separation.

[16] It is important to note that OMS dumping and fuel wasting are two different actions. The former was carried out to control the center of gravity of the Orbiter and, as the name implies, it involved the OMS engines. The latter was carried out using the main engines in order to bring the vehicle to a specific altitude. The name "fuel wasting" derives from the fact that in this case the thrust of the main engines had a large vertical component that did not increase the down-range distance. During the second stage of a nominal ascent, the primary component of the thrust is horizontal. In this sense a large vertical component equates to wasting propellant.

[17] This rate is kept high in order to prevent the Shuttle from gaining too much altitude while passing through the vertical.

[18] Prior to PPA, PEG 5 only calculated the angle for the fuel wasting and the moment for PPA; it did not give commands for steering towards the MECO point.

about 30 degrees approaching MECO. But to safely separate from the ET the Orbiter would require an angle of attack of –2 degrees. The PPD would transition between these two angles. Because a leisurely transition would result in a sink rate larger than desired, resulting in overheating and/or overstressing of the vehicle, it would occur at a rate of 10 degrees per second and be executed as late as possible. This would leave only a few seconds to establish stability for MECO. The engines would be throttled back to their minimum thrust setting to reduce the effect of this change of the angle of attack on the trajectory.

The Orbiter would maintain an alpha of –2 degrees until the calculated time of MECO. In this "mated coast phase", the RCS jets would fire to maintain the Orbiter in the right attitude for ET separation. It was essential that there be no recontact. For a normal separation, the aerodynamics would cause the ET to "peel off" beneath the Orbiter, much like a fighter airplane dropping a belly tank. As this process would be too slow and unpredictable to be relied upon, especially because the attitude at ET separation might be far from ideal, the RCS jets would be fired to establish a greater opening rate. This would involve all of the down-firing RCS jets. With four forward down-firing jets and six aft down-firing jets, the simultaneous translation and rotation would carry the Orbiter up and away from the falling tank.[19] Should any pitch or roll be required during this separation maneuver this would be achieved by momentarily turning off some of the active jets, thus avoiding using any up-firing jets that might decrease the separation rate.

Separation firing would last at least 10 seconds and until an alpha of 10 degrees was attained. At this point the so-called powered RTLS (PRTLS) was over and the gliding RTLS (GRTLS) began. The energy of the Orbiter would determine how far and how long it was able to glide. The horizontal distance that an Orbiter can glide is a function of its lift-to-drag ratio (L/D), which can be adjusted by changing the basic flight parameters such as alpha and by moving aerodynamic control surfaces to affect drag and/or lift. It would have been unwise to plan to fly at the maximum L/D, since an extra headwind or unplanned drag would curtail the trajectory short of the runway. For this reason, the MECO target was designed to give the Orbiter more energy than was necessary to reach the runway. Should the Orbiter arrive at MECO with less than this energy but still be capable of a landing, it would be said to be "low on energy". If it was so low on energy as to be unable to reach the planned runway, downmoding would be necessary (a term which indicates an action is being undertaken that is less demanding than that planned). In this case, downmoding might mean changing the path to the runway from an overhead-turn approach to a straight-in approach. It could also mean changing the targeted runway to one with a shorter approach. Ironically, having considerably more energy than planned introduces problems too, because the Orbiter could conceivably fly past the runway and not be able to land. It had a very limited

---

[19] Rotation would occur during the translation since the forward down-firing RCS jets have a longer moment arm relative to the center of gravity than the aft ones, making them about twice as effective.

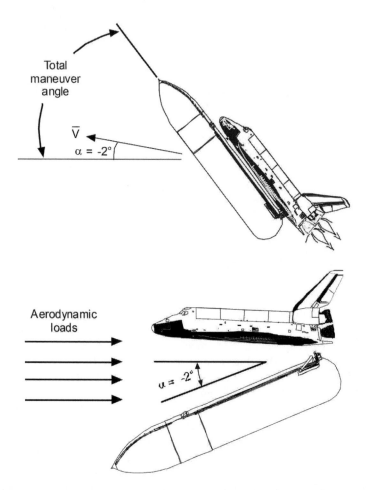

RTLS powered pitch down maneuver (top) and ET separation (bottom). (Courtesy of www.nasaspaceflight.com)

turning capability and could not just circle over the field, bleeding off energy prior to lining up for an approach. Excessive energy would be dissipated during the GRTLS phase in order to make the runway approach with an energy state as close as possible to that desired.

After jettisoning the ET, the thousands of pounds of propellants in the pipes of the MPS system would be dumped as during a nominal ascent. The liquid oxygen dump would be done by opening the main oxidizer valves of the three main engines. Owing to the dangers associated with liquid hydrogen, the fuel would be dumped overboard by special plumbing called the RTLS dump line, which exited on the left side of the aft fuselage and was designed to provide for the maximum dump under the accelerations experience during GRTLS. Due to the limited time available, both dumps would be carried out simultaneously. The software was also capable of a post-MECO dump of RCS propellant through the jets in order to help to adjust the center of gravity.

Due to the separation constraints, once clear of the ET the Orbiter would dive with an alpha of 10 degrees and start to pull up. To attain stable level flight, it would have to pullout before the aerodynamic forces became uncontrollable. This would involve maneuvering as soon as possible at a rate of 2 degrees per second using the RCS jets in order to achieve an alpha of 50 degrees. This phase was called "alpha recovery".

As the Orbiter fell deeper into the atmosphere, the increasing dynamic pressure would cause ever more lift to be generated by the wing, producing an acceleration Nz along the Z axis. The alpha of 50 degrees would be held until Nz reached 1.8 g, and the Nz-hold phase started. Then guidance would start to reduce the angle of attack (continuing to hold Nz at 1.8 g) until the rate of descent was less than 250 feet per second. At this point, the pullout was essentially complete and the transition to stable flight would begin. In this phase the angle of attack would be adjusted to conform to a predetermined alpha profile as a function of velocity that would optimize control of the vehicle until it was time to start the approach and landing.

Transoceanic Abort Landing

The Shuttle Landing Facility at the Kennedy Space Center in Florida was not the only place where an Orbiter could land in an emergency. If the down-range distance and energy were such that a RTLS abort was not possible, then if the conditions were right a landing could be made on the other side of the Atlantic on a certain number of runways in Europe and Africa.

On initiating this TAL abort mode the flight software would immediately start an OMS dump in order to manage the center of gravity and to meet Orbiter gross weight constraints. Its duration would depend on whether RCS interconnection was enabled and, if so, by the number (24, 10 or 4) of RCS jets used. The OMS dump would be continued after MECO if the dumping prior to MECO had left more than 36 per cent of the OMS propellant remaining. Post-MECO it would also assist in separating from the ET, in reducing wing leading edge temperatures, and in meeting center of gravity and landing weight constraints. The start/stop times for each dump were specified in the section of the Ascent Checklist for TAL procedures.

For example, with reference to the Ascent Checklist for STS-114 the pre-MECO OMS dump would start immediately with the two OMS engines and be followed, if enabled, by activation of the RCS interconnection (*"if INCNCT 'ENA'"*) with either 24 or 10 jets. The duration of the burn would depend upon the configuration. A dump with only the two OMS engines would last 495 seconds but a dump with both OMS engines and all 24 RCS jets would last only 278 seconds. For the post-MECO dump, the procedure set the duration in terms of a percentage of the remaining fuel and the acceleration along the +Z axis. For a normal load on the vehicle equal to 0.05 times the acceleration of gravity, the dump would be terminated irrespective of whether the minimum fuel condition was met since the OMS tanks were not designed to support dump operations under accelerations along axes other than the X axis.

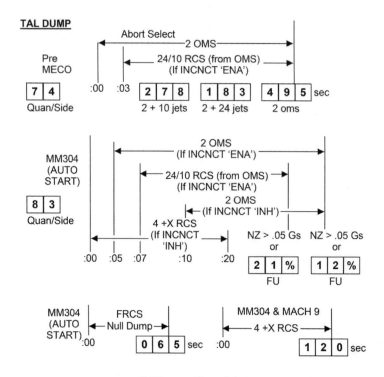

TAL propellant dump.

After STS-87 in November 1997 the nominal ascent program included the "roll to head up" maneuver and this was performed at a velocity of 12,200 feet per second. If a TAL abort was selected after this velocity, the roll would already have taken place, but if the abort occurred prior to this point then the roll would be carried out not at 12,200 feet per second but at 14,000 feet per second. During the powered portion of a TAL abort, guidance would also use the so-called "variable IY steering" to reduce the performance requirements and attain earliest abort capability.[20] Regardless of the mission inclination, yaw steering was a function enabled prior to liftoff and served as a precaution against suffering two engine failures. As the term implies, the maneuver would yaw the Orbiter in order to swing the trajectory towards the TAL landing site and ensure that the cross-range limits were not violated.

There were a number of TAL sites available to insure against weather and landing/navigation aid issues. The "prime" TAL site was specified for each mission on the day of launch, but several others were also available and were referred to as augmented contingency landing site (ACLS). The adjective "augmented" derived from the fact that although these sites were not prime, there were NASA personnel, hardware and software support available. Landing site selection was based on

---

[20] IY refers to a unit vector perpendicular to the desired orbital plane.

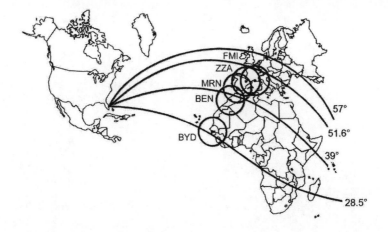

Ascent ground tracks for various inclinations and primary TAL site available for each inclination.

Inclination	Available TAL Sites in *geographic* order of preference			
28.45°	BYD Banjul, The Gambia	BEN Ben Guerir, Morocco	MRN Moron, Spain	
39.00°	MRN Moron, Spain	ZZA Zaragoza, Spain	BEN Ben Guerir, Morocco	
51.60°	FMI Le Tube, France	ZZA Zaragoza, Spain	MRN Moron, Spain	BEN Ben Guerir, Morocco
57.00°	FMI Le Tube, France	ZZA Zaragoza, Spain	MRN Moron, Spain	BEN Ben Guerir, Morocco

TAL sites.

several factors, including weather, navigation and landing aids, and the dimensions and load bearing capacity of the runway. From a flight design and ground rules standpoint, the closest TAL site to the ground track of the Orbiter would be chosen in order to minimize cross-range when flying the atmospheric re-entry portion of the abort. On launch day the best choice would be the designated prime site. If the status of that site changed during the powered flight, another site would be selected from the available ACLSs. In the improbable case of both the prime and augmented sites becoming unavailable during powered flight, it would be necessary to try to reach an emergency landing site (ELS).

Because the Orbiter had more cross-range capability during powered flight than in a powerless glide, all TAL *targets* (both prime and augmented) were defined as a point 500 nautical miles from the actual landing site. From that point on, the Orbiter would have enough energy to glide in for a landing. On a map showing the locations of the TAL sites, it is often possible to see circles drawn around each site, each with a

370 The art of reaching orbit

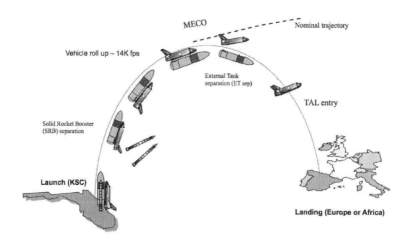

TAL flight profile.

radius of 500 nautical miles. The TAL target towards which guidance yaw steering would bring the Orbiter represented the tangent of the ascent trajectory ground track with the landing site cross-range circle.

As in a nominal ascent, guidance would compute the MECO point which would provide the desired relationship of range and velocity, cross-range, flight path angle, and radius/altitude. Specifically, guidance aimed the Orbiter for a MECO flight path angle and altitude that would provide a flight path angle of 0 degrees and an altitude of 360,000 feet at ET separation. Once all the powered flight guidance targets had been achieved, MECO would be commanded and separation 20 seconds later would be followed by the –Z translation of 11 feet per second. As during a nominal ascent, the act of ET separation could be halted in the event of rate violations or the failure of a feedline disconnect valve. If so, the crew would null the rates. If the automatic sequence did not resume after 15 seconds, the crew would initiate the separation manually. This procedure had to be completed before the dynamic pressure increased to a level that would impede separation.

Owing to the criticality of the period after ET separation, the commander would have three minutes to verify that the attitude and rates of the Orbiter were valid prior to switching to OPS 3. If necessary, he would activate CSS and maneuver to the proper attitude as soon as possible. Since manually re-establishing the attitude could take as long as 45 seconds, this would delay the OMS dump and result in the Orbiter being 4,000 feet nearer the ET when that ruptured upon re-entry. After the attitude check, the automatic switch to OPS 3 would be monitored and if it failed to occur it would be performed manually. The software was designed to automatically switch to MM104 after MECO. While in this major mode, the transition DAP, which was not designed for atmospheric flight, would control the vehicle. Within three minutes after ET separation (less in the case of a significant underspeed), the dynamic pressure would begin to build up, requiring guidance using the aerojet DAP and hence the switch to OPS 3. This was because the aerojet

DAP could command more RCS jets and use the aerosurfaces. In procedural terms, a nominal TAL re-entry was very similar to a nominal end of mission re-entry, the principal difference being that the post-MECO OMS dump had to provide a delta-V of 100 feet per second between the Orbiter and the ET and the initial angle of attack for the pullout maneuver would be 43 degrees instead of 40 degrees when starting a nominal re-entry.

Abort To Orbit

On 29 September 1988 *Discovery* lifted off for STS-26, the first mission since the *Challenger* disaster. Former astronaut and Shuttle pilot Jack Lousma had been hired as a commentator by ITN, a British television news station. During the ascent he described the abort modes as they became available. The host, Alastair Burnet, asked Lousma his preferred abort mode. Lousma immediately replied, "Abort To Orbit." ATO is indeed the best of the abort modes because, as the name implies, it allows the Orbiter to achieve an orbit at an altitude of about 105 nautical miles. Although much lower than for a nominal ascent, this temporary orbit provides time to assess the problem prior to choosing either an early deorbit burn or an OMS maneuver to a higher orbit in order to continue the mission.

Upon this abort mode being selected, guidance would select an ATO MECO target designed to prevent the ET from falling on a critical land mass, and then it would work to ensure minimum performance capability, minimum altitude constraint, and minimum SSME net positive suction pressure (NPSP) constraints upon MECO. In parallel with the determination of the new MECO target, guidance would also have started the usual OMS dump for center of gravity control and to reduce the overall weight in order to enhance performance in powered flight. As was so for all OMS dumps, the ATO dump would be controlled by the abort control sequence with a flight dependent configuration of OMS engines, with or without RCS interconnect. Finally, guidance would initiate variable IY steering with the objective of changing the orbital plane target at MECO from the nominal ascent value to a current in-plane performance-optimum value for an ATO abort. Variable steering has already been described in relation to the TAL abort, but it is worth expanding on the concept.

Generally speaking, IY refers to a unit vector perpendicular to the desired orbital plane and implicitly specifies the inclination and node constraints. The most efficient way to launch a space vehicle is due east, into an orbit with an inclination matching the latitude of the launch site (28.5 degrees for the Kennedy Space Center). But often the mission requires a different inclination, and this obliges the vehicle to gradually steer its velocity vector towards that plane. In the case of ATO abort, guidance would use IY steering to change the IY target from the nominal mission inclination to the in-plane value at the time that the abort was initiated. In other words, the inclination would be frozen and all the thrust directed in-plane to optimize propellant usage and performance. However, IY steering would be set to attain a minimum IY value if the inclination at the time of the abort was less than the minimum acceptable inclination defined prior to launch. In this case, out-of-plane steering would continue until this minimum inclination was achieved.

372 The art of reaching orbit

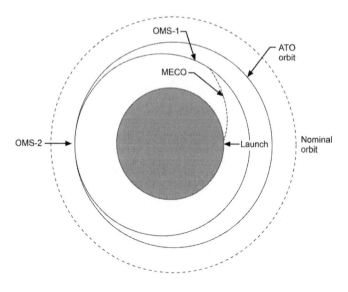

ATO flight profile.

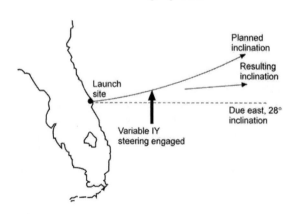

Result of variable IY steering.

After MECO during ATO, an OMS burn would be made just as for a standard insertion ascent. This would be to raise the orbit apogee (HA) to 105 nautical miles. On reaching apogee, a second burn would circularize the orbit at that altitude. Based on OMS fuel reserves, this second burn might be canceled if the attained perigee was greater than the minimum required perigee (HP) of 90 nautical miles. A circular orbit at 105 nautical miles would allow 24 hours to assess the condition of the Orbiter and decide whether to perform further OMS maneuvers to raise the orbit and accomplish most of the mission objectives or instead to return home. An ATO minimum-perigee orbit would enable the Orbiter to remain in space until the first opportunity to land at the primary landing site. Deorbit and re-entry from an ATO orbit would be similar to concluding a nominal mission, and very likely return to the

Intact ascent abort modes

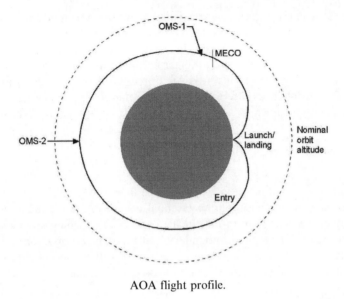

AOA flight profile.

same landing site. But an off-nominal ATO insertion might necessitate landing at an off-nominal site.

Abort Once Around

As the name implies, the AOA abort mode involved reaching orbit and performing a nominal re-entry after a single revolution. This option was available only following a nominal MECO or an ATO abort, and the decision would be made based on complex criteria that took into consideration any underspeed and malfunctioning systems.

An AOA would be chosen only after the OMS 1 burn of a standard insertion, or prior to the OMS 2 burn of a standard insertion or a direct insertion. In the case of a significant underspeed and a MECO apogee of less than 95 nautical miles, an OMS 1 burn would be made to raise the apogee height in the same manner as for a standard insertion profile. For both standard and direct insertions, the OMS 2 burn would include a radial component to shift the line of apsides and thereby rotate the orbit so that the perigee would occur at the entry interface appropriate for the return to Earth. Although not retrograde like a nominal deorbit, in this case the OMS 2 burn would be carried out by OPS 3. If excess propellant had to be burned in order to adjust the center of gravity or to satisfy the OMS tank landing weight constraints, the requisite propellant wasting would be included in the burn. If the underspeed was so great that the perigee after the OMS 1 burn was lower than 75 nautical miles, a special angle of attack management procedure would be applied because in this case the orbital drag between the OMS 1 and OMS 2 burns would be significant. The objective would be to minimize the frontal profile of the vehicle and hence the drag that it endured.

Two types of re-entry trajectory were available in this abort mode, a "steep" AOA and a "shallow" AOA. The steep AOA would be very similar to a nominal re-entry and landing but a shallow AOA would fly a flatter trajectory that was less desirable since it would expose the Orbiter to a longer period of atmospheric heating and less predictable aerodynamic drag. Edwards Air Force Base would be the prime landing site because its dry lakebeds offered considerable margin in an emergency resulting from vehicle energy problems.

Intact abort modes: selection criteria

If one reads the transcripts of the ground-to-air communications of a Shuttle ascent, it is evident that there is not much talk between the flight crew and Mission Control. It was policy to limit communications to essential calls, which were made in technical jargon. Many of the calls were to inform the crew of the opening and closing of the various abort modes. These "abort boundary" calls were based on information from a software system known as the abort region determinator that was able to calculate the best abort option available based on real-time ascent data. These calls included:

- *"Two Engine TAL"*
  With this first abort boundary call by Mission Control, the crew was advised that they had reached the earliest inertial velocity at which AUTO guidance would be able to achieve the desired MECO target for the designated prime TAL site in the event that one of the main engines failed. A RTLS abort was still viable, but TAL would be the preferred option for engine failure since single engine completion capability could be reached much earlier for TAL than for RTLS. For flights at a high inclination, such as to the International Space Station, the prime TAL in-plane site was Zaragoza air base in Spain (ZZA), but since the capability to reach out-of-plane TAL sites at Morón air base in Spain (MRN) and Ben Guerir air base in Morocco (BEN) could be achieved several seconds earlier than for the prime TAL site, the call "Go to TAL" would indicate to pursue this earlier capability rather than wait for the in-plane site. If an engine failed right at the "Two engine TAL" call, then the guidance software would have to perform a landing site redesignation.
- *"Negative Return"*
  This second call indicated the vehicle had too much down-range energy to be able to return to the Kennedy Space Center for a RTLS abort. Since no one ever had much faith in the RTLS mode, being informed that it was no longer feasible was a relief for the crew.
- *"Press to ATO"*
  This third call told the crew that if one main engine failed then the Orbiter had enough down-range energy to reach the "design underspeed" at MECO, this being the largest underspeed from the nominal MECO that achieved the desired ET impact area and the Orbiter achieving minimum performance, minimum altitude, contingency center of gravity envelope, and minimum net positive suction pressure for the main engines. Executing an ATO at this

stage would also require an OMS dump that could be varied in amount using the OVERRIDE (SPEC 51) display as directed by Mission Control. In the meantime, a TAL abort remained a viable option.

- *"Press to MECO"*
This call marked the boundary beyond which it would be possible to achieve the design underspeed MECO without necessitating an OMS dump.
- *"Single Engine OPS 3 (109)"*
Beyond this mark, the Orbiter had enough velocity to perform a TAL abort with two main engines out. A second engine failure prior to this point would have required performing a contingency two-engine-out maneuver for an emergency landing site or perhaps even a bailout. To understand this, it is necessary to introduce the concept of "droop" boundary. When two engines fail, the thrust-to-weight ratio of the vehicle becomes less than one and the trajectory starts to fall (or droop). As propellant is consumed with the thrust constant, the thrust-to-weight ratio exceeds one and the Orbiter can resume its climb. Single Engine OPS 3 (109) represented the first point in the ascent at which, in the case of running on a single engine, the vehicle would not fall below 265,000 feet. Prior to this point, a second engine failure would cause the vehicle to descend below this threshold, with the high dynamic pressure rapidly violating the thermal constraints of the ET. If a TAL occurred after this call, one of the augmented out-of-plane TAL sites would be used rather than the primary in-plane TAL site. In this situation, the remaining engine would be run at 109 per cent of rated thrust.
- *"Single Engine Zaragoza"*
This informed the crew that from now on the in-plane prime TAL site (here Zaragoza) could be achieved with a second engine failure.

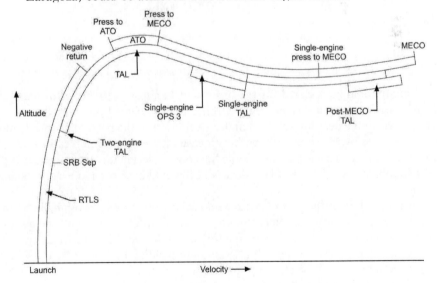

Abort calls as a function of velocity and altitude.

- *"Single Engine Press to MECO"*
  From this point, the AUTO guidance could achieve the design underspeed at MECO using only one engine throttled at 104 per cent of its rated thrust.

Given the requirement for the crew to be constantly aware of which abort options were available at any given moment in the ascent, the Ascent Checklist included the NO COMM MODE BOUNDARIES cue card which listed the abort boundaries and their inertial velocities in feet per second. As the name implies, this card served as a backup in case of loss of communication with Mission Control and the crew would have it handy in order to be ready for a loss of communication. The boundaries in bold type were those typically called by Mission Control. Note that the boundaries on this cue card did not reflect off-nominal performance, such as a stuck throttle. Only the abort region determinator could calculate the abort boundaries based on the actual performance. Furthermore, the cue card was specific to a given launch, as the times could change significantly owing to differences in mass properties, environmental modeling, and performance characteristics. Looking at the cue card, it is interesting to observe the presence of three additional TAL-related boundaries which were *not* called out by Mission Control.

- *NEG [TAL site]*
  This boundary represented the last opportunity to abort to the prime TAL site selected for that mission and still be able to achieve the desired cross-range performance.
- *LAST PRE MECO TAL*
  Beyond this point, the AUTO guidance would not be able to achieve the designated TAL/ACLS landing sites. If a TAL was required, the crew would have to manually command MECO prior to a certain (mission dependent) velocity and then execute a post-MECO TAL.
- *LAST TAL*
  This represented the highest MECO velocity at which a TAL or emergency landing site was viable. For example, "LAST TAL MRN 24300" meant that after 24,300 feet per second a TAL to Morón was no longer a viable option.

An abort could be declared for reasons other than loss of flight performance, for example the failure of a system on the Orbiter that precluded pursuing the nominal mission. Again the decision to abort and the selection of the most appropriate mode would be made by Mission Control, but in case of loss of communication the SYS FLIGHT RULES cue card would enable the crew to decide for themselves. If RTLS and TAL aborts were both available, the RTLS would be chosen in order to achieve the earliest landing.

A critical failure of the OMS system could arise from a loss of both helium tanks, a loss of one propellant tank on each pod, a loss of both oxidizer tanks or both fuel tanks. In any of these situations a TAL abort would be required since there would be insufficient propellant to reach orbit. With an impending loss of all APU/hydraulic systems (APU/HYD) a trajectory option with the minimum time to reach the ground would be selected in order to maximize the likelihood of landing prior to total loss of

NO COMM MODE BOUNDARIES

NEG RETURN (104)	8100	2 ENG ZZA (104)	5900
PRESS TO ATO (104)	10600	ABORT TAL ZZA (4)	
SE OPS 3 (109)	12100	EO VI	
SE ZZA (104)	14300	SE OPS 3 ZZA (109)	(4)
PRESS TO MECO (104)	17600	SE ZZA (104)	(4)
SE PRESS (104)	19000		
NEG MRN (2 @ 67)	19800	2 ENG MRN (104)	5800
LAST PRE MECO TAL	23000	ABORT TAL MRN (3)	
LAST TAL		EO VI	
YJT	20100		
YYT	20200	SE OPS 3 MRN (109)	(3)
YQX	21900	SE MRN (104)	(3)
IKF	23600		
INN	24100	2 ENG FMI (104)	6200
FFA	24200		
MRN,BEJ	24300	ABORT TAL FMI (29)	
KBO	24400	EO VI	
ESN	24800		
ZZA,KKI	25000	SE OPS 3 FMI (109)	(29)
FMI	25100	SE FMI (104)	(29)
JDG	25200		

NO COMM MODE BOUNDARIES.

PILE   SYS FLIGHT RULES

	RTLS	TAL
OMS – 2 He TKs		X
– 1 OX & 1 FU TKs (diff pods)		X
– 2 OX or 2 FU TKs		X
APU/HYD – Impending loss of all capability	X	X
CABIN LEAK – (-EQ dP/dT > .15)	X	X
CRYO – All O2(H2)	X	X
2 FREON LOOPS $\downarrow$ [Accum Qty ($\downarrow$ and decr) and/or Flow ($\downarrow$)]	X	X
2 MAIN BUSES $\downarrow$	X	
THERMAL WINDOW PANE	X	

SYS FLIGHT RULES.

hydraulics. Loss of cabin pressure integrity (CABIN LEAK) would impact both crew safety and equipment cooling, making a RTLS abort preferable. A loss of cryogenics (CRYO) would cause a loss of fuel cells and all electrical power, necessitating either a RTLS or TAL abort. The loss of both Freon coolant loops (TWO FREON LOOPS DOWN) would eventually cripple all three fuel cells. Since time would be the critical parameter, a RTLS abort would be preferable. If two main buses failed (TWO MAIN BUSES DOWN) then closure of the ET doors in the belly of the Orbiter would be compromised. Because TAL trajectories are thermally severe a RTLS profile would be preferable. If that mode was no longer available, it would

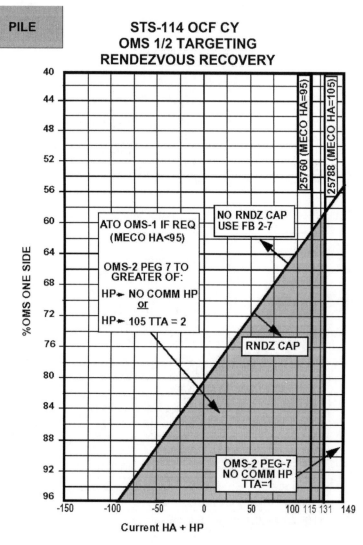

OMS 1/2 TARGETING RENDEZVOUS RECOVERY.

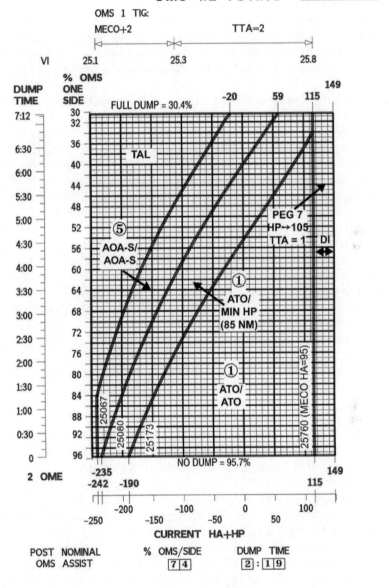

OMS 1/2 TARGETING.

be necessary to continue uphill. Finally, for a damaged outer window pane (THERMAL WINDOW PANE) a RTLS abort would be preferable for its more benign heating characteristics.

The Ascent Checklist had cue cards to be consulted if communications were lost during the orbital insertion phase. For example, the RENDEZVOUS RECOVERY card would enable the crew to determine whether the MECO velocity attained was sufficient for a rendezvous with the mission's target, such as the International Space Station.

The card specified that on achieving MECO with no communications, the crew should compare the sum of the apogee (HA) and perigee (HP) altitudes of the orbit at MECO[21] with the quantity of OMS propellant still available. If the OMS propellant quantity was below the diagonal line then there was sufficient for the OMS 2 burn to initiate the rendezvous. For an OMS quantity above the diagonal line there was no chance of a rendezvous and the crew would pass on to the OMS 1/2 TGTING card, a diagram that indicated the most appropriate abort. Despite being divided into regions, this card was used in the same manner as the previous one. For example, for the case of STS-114 with HA + HP = −50 and 44 per cent of the OMS propellant remaining, a shallow AOA abort would be required. An ATO would have been possible with 76 per cent of the OMS propellant remaining. For a mission not involving a rendezvous, the Ascent Checklist would not have included a RENDEZVOUS RECOVERY card.

It is interesting to note that in case of HA + HP $\geq$ 115 a direct insertion would have been possible irrespective of the quantity of OMS propellant remaining (a minimum of 30.4 per cent). Once it was established that an AOA or ATO abort was required, the crew would turn their attention to the OMS TARGETS – DIRECT INSERTION cards where the target data for AOA and ATO aborts was provided in flowchart style as a backup to that already loaded in the software memory. This card would also be used for a nominal ascent to provide the target parameters for a nominal OMS 2 burn after loss of communications.

For system failures occurring prior to the OMS burn of a nominal direct insertion the OMS 2 TARGETING cue card provided information about which abort to perform on the assumption that there was no underspeed at MECO (that is, that no OMS burn was required immediately after MECO) and that there were no other performance issues. If both OMS nitrogen tanks failed, a nominal OMS 2 burn could still be made but using only one engine in order to preserve the start capability of the other one for the deorbit burn. For loss of one propellant tank or one helium tank, the OMS 2 would be targeted to attain a mimimum perigee of between 85 and 90 nautical miles (cutoff at minimum perigee). If both OMS engines were lost, this burn would be performed by feeding OMS propellant to the RCS jets. Because the RCS jets were much less efficient than the OMS engines the burn would be minimized in order to provide no more delta-V than needed to raise the perigee to the lowest

---

[21] Remember that this value was a good indicator of the velocity attained at MECO, and hence an easy means of determining whether an underspeed MECO condition had occurred.

OMS 2 TARGETING
(DIR INSERTION)

FAILURE	OMS 2 TARGET
OMS – 2 N2 TKs (Perform burn Single OMS Eng)	NOM
OMS – 2 OMS ENGs – 1 He TK – 1 PRPLT TK EPS – MNA & B, MNB & C MNA & C ET SEP	PEG 7; TTA = 1 CUTOFF HP = 85
OMS – 2 OX or 2 FU TKs – 1 OX & 1 FU TK diff PODs – 2 He TKs	AOA-S
2 FREON LOOPS [Accum Qty ($\downarrow$ and decr) and/or Flow ($\downarrow$)] 2 H2O LOOPS CABIN LEAK (-EQ dP/dT $\geq$ .08)	AOA
APU/HYD – impending loss of all capability	AOA
CRYO – All O2(H2)	AOA

OMS 2 TARGETING.

acceptable altitude. In this way the deorbit burn performed using the RCS would be easier. A minimum perigee altitude would also be targeted in the event of the failure of two main buses, since it would give time either to recover the lost buses or to reconfigure the electrical loads. This procedure would permit the orbit to be raised if the problem was resolved, and minimize the deorbit if it was not. Loss of all OMS oxidizer or fuel, loss of all OMS helium, or loss of oxidizer and fuel in different pods would eliminate future usage of OMS propellant, thus requiring the RCS to draw upon its own supplies to provide the deorbit capability for an AOA return. If any of these failures occurred, it would be necessary to perform a shallow AOA since the RCS would be unable to complete a steep AOA deorbit burn. Finally, failures such as loss of both Freon coolant loops, loss of both water coolant loops, a cabin leak above limits, a third APU about to fail, and loss of all cryos would require an AOA as the fastest way to return to Earth.

Of course, these scenarios also apply to missions flown with a standard insertion profile. The same cue cards were provided but they specified different values, and in the case of the OMS 2 TARGETING card there was an additional column with abort indications for a failure prior to the OMS 1 burn.

CONTINGENCY ABORTS

Along with procedures for the intact abort scenarios, the Shuttle program also created a series of contingency abort (CA) modes to be executed if none of the intact aborts were feasible. Generally speaking, all aborts other than auto-completion RTLS, auto-completion TAL, a steep AOA, and ATO were listed as contingencies. For example, a late TAL to a contingency site involving manual procedures, or a TAL using low-energy guidance were contingencies even though a safe landing might be achieved. A contingency abort would also be required in the event of a single engine failure in which the other two engines were not operating at full performance, perhaps because a throttle failure had left them stuck in the "bucket".

Owing to the number of possible multiple failures that could occur during ascent, several CA procedures were developed to enable the vehicle to achieve a safe gliding flight condition. Several runways apart from the Kennedy Space Center and the TAL (prime and ACLS) sites were made available. These emergency landing sites (ELS) included Bermuda (BDA), east-coast abort landing (ECAL) in the continental USA, and several other sites across the Atlantic. If there was no prospect of landing, the only option was for the crew to bail out of the gliding Orbiter, leaving it to plunge into the ocean. It should be noted that while the intact abort scenarios were well defined and confirmed by simulations, the many unknowns involved in the contingency abort modes meant the applicability of these procedures was debatable. It was widely believed by the crews that contingency aborts were added merely to suggest to the public that a crew would have a means of saving their skins from even the most absurd failure scenarios. Many astronauts considered a contingency abort to be unsurvivable.

CA procedures were part of the Ascent Checklist and would be executed by the pilots assisted by the mission specialist seated behind them on the flight deck. In the main, they were presented on two-sided cue cards which had color-coded references and were divided into "procedure blocks" that gave instructions on how to properly activate the contingency abort software, and "monitor blocks" that described step-by-step the maneuver in progress. Each mode could be divided into three major phases: (1) powered contingency, (2) ET separation and mated coast, and (3) re-entry.

Powered contingency with two engines out

Selecting the most appropriate powered CA maneuver would be based on the abort after the first SSME failure (non-RTLS versus RTLS) and trajectory parameters such as altitude rate, inertial velocity, and equivalent speed. A color-code was assigned to each type of CA maneuver and this was displayed on the ASCENT TRAJ as a cue to the crew when referring to the Ascent Checklist for the information on the maneuver selected by the flight software.

For example, in the case of losing two main engines during a nominal ascent, two maneuvers were available coded blue and green. But a two-engine-out contingency during a RTLS abort presented five maneuvers coded blue, yellow, orange, green

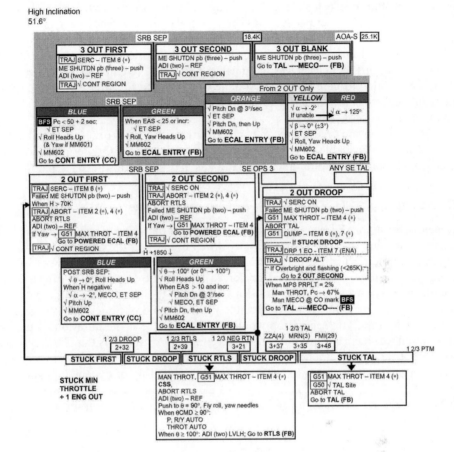

Cue card from Ascent Checklist with contingency abort instructions.

and red. Generally speaking, for a two-engine failure the remaining engine would have to be used after solid rocket booster staging to reduce the sink rate during re-entry. This could be done by directing the thrust vector in order to lower the altitude and/or bring the flight path angle[22] to zero prior to jettisoning the ET.

For failures before the velocity exceeded approximately 7,000 feet per second the trajectory was still very steep, so the vehicle would have to pitch towards the horizon to increase the horizontal component of its velocity. This would reduce the apogee, shallow the re-entry flight path angle, reduce the severity of the pullout, and limit the thermal loads. Without this maneuver the flight path angle would have remained steep and resulted in a ballistic trajectory with relatively intense re-entry thermal and structural loads. This was the basis for the two-engine-out blue CA mode. The green

---

[22] The flight path angle is known as gamma.

CA mode would be selected if the second engine failure occurred at a velocity in the range 7,000 to 12,000 feet per second. The trajectory would already have leveled off and the flight path angle would be close to zero, so vertical thrust vectoring would be applied in order to minimize the downward trend due to loss of thrust. Like in the previous case, this maneuver was designed to reduce the re-entry pullout rates.

It is interesting to note that no CA mode was designed for after SE OPS 3, since by then the Orbiter would have sufficient energy either to reach orbit or to execute an intact TAL abort.

If the Shuttle were to lose a second engine during a RTLS abort after the blue CA mode, the other available modes were designed to maintain the Orbiter in an attitude that would reduce structural and thermal loads and lead to a safe ET separation.

ET separation procedures and the three-engine-out CA mode

Generally speaking, ET separation was a function of the angle of attack (alpha) and dynamic pressure (q-bar). Analyses by Rockwell found that the Orbiter could safely separate from the ET at a q-bar value of up to 10 pounds per square foot. For higher dynamic pressures, the ET would be forced against the Orbiter in a catastrophic recontact. Analyses also found that for a q-bar in the range 2 to 10 pounds per square foot a safe separation was feasible with an angle of attack of –2 degrees.[23] Below –2 degrees, safe separation would still be possible if the duration of the –Z translation was increased to prevent contact. But in CA situations this ET procedure would not have been applicable. In fact, in a CA the high sink rate would have caused a rapid re-entry into the atmosphere and a build-up of dynamic pressure. In most of the CA scenarios, there simply would not have been time to undertake a RTLS-type powered pitch down and a normal mated coast separation sequence.

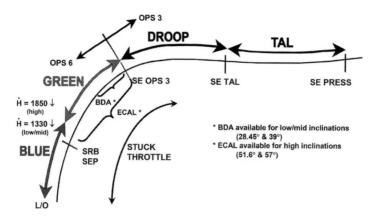

Two-engine out contingency abort.

---

[23] This separation technique was the one chosen for "normal" RTLS.

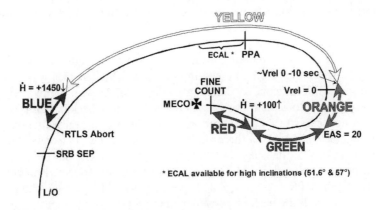

RTLS two-engine out contingency abort.

In order to improve these recontact problems, a "fast separation" sequence lasting only five seconds was devised. It could be executed in one of seven ways, depending on vehicle attitude and q-bar. The first case was an immediate, manually initiated fast SEP where there was no time available to maneuver into a better attitude. The second was a fast SEP with enough time to achieve an alpha of –2 degrees before the value of q-bar exceeded 10 pounds per square foot. The third method was the so-called "rate" fast SEP, in which the Orbiter already had the high angle of attack desired for re-entry but there was insufficient time to lower it to –2 degrees before reaching the maximum limit on q-bar. In this case, instead of pitching to a negative angle of attack, which might take as long as 30 seconds, the separation would be commanded while at a high angle of attack but with a pitch rate of –3 degrees per second so that the ET was essentially "pushed" clear. This means of separation offered distinct advantages in many CA scenarios. Firstly, it allowed separation to be delayed as long as possible in order to prolong the time available for powered flight. Secondly, in many cases the Orbiter would already be near the angle of attack required for re-entry, and the pitch rate of –3 degrees per second would require only to be slowed as the proper angle of attack was reached. If the Orbiter were taken to a negative alpha for separation, it would have required two pitch maneuvers, a negative pitch to the separation angle followed immediately by a pitch up to the angle for re-entry. A pitch rate of –3 degrees per second was selected because it offered a compromise between the need for a rapid pitch down to perform ET separation and the subsequent need to maintain control and slow the pitch rate.

The fourth type of fast SEP was the "(SRB) PC < 50", which would be used only in the event of losing all three main engines during the first stage of the ascent. In this case the Orbiter would separate from the ET and the still-firing SRBs. The fifth mode was the so-called attitude-independent SEP. In this case the separation could occur in any attitude so long as there were zero rates and q-bar was low enough to prevent recontact. The sixth mode applied if a high q-bar would prevent the pitch down to an alpha of –2 degrees. In this case the vehicle would be pitched *up*, to achieve an alpha of 125 degrees for ET separation followed by a pitch down to the re-entry alpha. The final fast separation type would apply if there was a failure to achieve separation at the intended attitude, in which case it would be performed at the minimum speed of 77 knots.

In the case of a three-engine-out contingency abort the Orbiter would no longer be providing any thrust, so the ET represented dead weight. The task was to separate as soon as possible and attempt to glide to a landing site. As with the two-engine-out scenario there were color-coded procedures in the Ascent Checklist with each color referring to a particular fast ET separation mode based upon the value of q-bar at the time the engines failed. For a three-engine-out case during the second stage of the ascent with a velocity in the range 14,000 to 22,000 feet per second, the contingency logic would be invoked and would depend on the predicted apogee. For example, if the loads during the OPS 3 re-entry pullout were predicted to be below 3.5 g, an OPS TAL completion would be performed rather than an OPS 6 fast separation.[24] If the velocity was greater than 22,000 feet per second, the contingency logic would not be required.

Re-entry

The powered flight, mated coast, and ET separation portions of a contingency abort were designed to deliver the Orbiter to acceptable re-entry conditions without loss of control. In general, reducing the ET separation altitude and sink rate would improve re-entry conditions. On the other hand, any error that occurred during powered flight or ET separation would result in a more stressful re-entry.

On completing ET separation, the Orbiter would be maneuvered to an alpha that optimized the L/D ratio in a manner similar to GRTLS (alpha recovery). This angle was a function of velocity, and could range from 20 to 58 degrees. A ramp down was required for flight control (pitching moment and stability) constraints at low Mach numbers, and a maximum of 58 degrees was chosen since the alpha/Mach indicator showed a maximum of 60 degrees and because beyond 60 degrees the total vertical force that the Shuttle could generate started to decrease and make alpha recovery less effective. In re-entry, the pullout would be managed by controlling the acceleration normal to the Z axis and the equivalent airspeed in what was known as "Nz-hold". Guidance would calculate a target Nz to balance the Nz and EAS constraints, then pitch down to maintain that g-level. Following Nz-hold, as in a "normal" RTLS, the contingency abort guidance would enter the alpha transition flight phase and fly a Mach-alpha profile that would ensure a safe attitude in hypersonic and supersonic flight, and would usually be good for subsonic flight. When transitioning to subsonic flight, the commander would also confirm that the Orbiter was pitching down to the horizon and that the airspeed was not so slow as to cause a stall to occur.

---

[24] This prediction was made by comparing the predicted apogee altitude with a 3.5 g apogee altitude, in terms of a velocity profile. If the predicted apogee altitude was below this profile, then the OPS 3 re-entry pullout loads would not exceed 3.5 g and a fast separation would not been required.

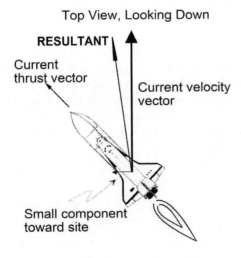

Yaw maneuver for contingency abort.

Depending upon the intended orbital inclination, most uphill and some RTLS and TAL trajectories offered a landing on the east coast of North America (ECAL) or on the Caribbean island of Bermuda (BDA) after losing two or three main engines. A common aspect for all two-engine-out contingencies was a "yaw" in which guidance commanded an unguided 45 degree yaw.

This maneuver was designed to reduce gliding cross-range by turning the velocity vector towards the chosen landing site. Once the equivalent airspeed was greater than 4 knots, the vehicle would yaw back into the velocity vector in readiness for MECO and ET separation. This deflection merely "assisted", since most of the cross-range would have to be performed during the gliding phase. It should be noted that this yaw steering was different from that during a TAL abort. In a TAL abort the steering was to achieve a position from which the Orbiter *would* have the cross-range capability to glide to the landing site. In this case, the steering was unguided, meaning that it was a turn in a direction that would *assist* the Orbiter in aiming for one of these emergency landing sites but *could not guarantee* the conditions required for successfully gliding the remaining cross-range. Generally, for high inclination missions, ECAL sites were available for two-engine-out scenarios from a velocity of about 6,000 feet per second (just past staging) until about 12,500 feet per second (just past single-engine droop). They were also available on RTLS profiles for a second engine failure during fuel dissipation between 5,700 feet per second up to the initiation of PPA. Sites in Canada were available for high inclination three-engine-out scenarios from 14,200 feet per second to about 19,000 feet per second (until "3 EO TAL" capability) for in-plane launches. For launches due east, Bermuda was generally available during a MRN or a BEN TAL between 9,000 and 12,000 feet per second. The site selection could be done by the guidance software based on a table of I-loaded velocities that defined the windows for ECAL and BDA. Yaw steering would lead the crew to the procedures for attempting a landing on the selected site. The indication of yaw steering was the primary cue to the crew that ECAL/BDA were available. The crew could also choose the landing site (especially in the case of guidance failure) using tables in the Ascent Checklist that listed the landing opportunities based on the velocity at the time of the second engine failure, catering for cases where the engines failed at different times or simultaneously.

If a second engine failed outside these landing windows, then there was no reason to undertake yaw steering. In this situation, the crew would continue on the current heading. Studies showed that the Orbiter could not survive a gear-up water landing, so in the absence of an accessible runway the crew would wait until the vehicle was subsonic and then activate attitude-hold for stability so that they could bail out. The attitude-hold mode was activated by rotating the abort switch to ATO and pressing the abort button, then switching from CSS to AUTO. Guidance would take snapshots of the current airspeed and roll angle, and try to hold both of these values constant. Once the altitude had reduced to 30,000 feet, the crew would depressurize the cabin, blow the side hatch, and make their escape.

Survivability

It is easy to appreciate that the contingency abort procedures had many limitations, most notably the existence of "black zones" for some Orbiter configurations; that is to say, regions where structural failure or loss of control would be unavoidable. And for those scenarios in which simulations showed that the CA *should* work, there were several limiting assumptions made in the analysis that cast doubt on the efficiency of these procedures. Fortunately, none of them were ever required during the Shuttle program.

12

Orbital dancing

MANEUVERING THE SHUTTLE

Science fiction movies have accustomed us to spaceships that fly in space as if they were advanced jet fighters. But the reality is very different. Flying in space requires careful planning and coordination between the crew of the vehicle and controllers on the ground or aboard a space station towards which it is heading or from which it is departing. Due to the limited amount of propellant available on board, as well as the performance of rocket engines, each maneuver must first be analyzed and optimized to make the most of the resources available. In the case of rendezvous missions with a space station or a satellite, saving fuel means taking advantage of orbital dynamics that require maneuvers to be made at precise times and positions in space.

The two powerful OMS engines and the array of RCS thrusters gave the Orbiter great agility in performing the maneuvers for orbit and attitude control. Following the orbital insertion phase, the digital autopilot was set to orbital DAP mode. This used a scheme called PEG 7 (or "external delta-V") that differed from PEG 4 by using an open-loop logic. That is, the OMS burns were computed to achieve a given delta-V without placing any constraints on the position of the Orbiter at the conclusion of the burn. Orbital dynamics is less demanding than either ascent or re-entry, and transfers are a matter of changes in velocity rather than of position. It is only necessary to start the maneuver at a predetermined time and execute it at the designed thrust and for the calculated duration in order to be sure of reaching the rendezvous target at a certain range, after which small corrections can be performed to close in either to undertake proximity operations or to dock.

During the mission, the orbital DAP was always either in RCS mode (also called RCS DAP mode), which was the default mode to perform attitude-hold or attitude maneuvers using the RCS thrusters, or in TVC mode (TVC DAP mode), which was only used for executing OMS burns.

RCS DAP mode

Several missions required the Orbiter to maintain a specific attitude in order to point scientific experiments in a particular direction. If this was not a requirement, then the attitude of the vehicle still had to be managed in order to uphold thermal, lighting and communications constraints. To do this, the RCS DAP mode operated a logic known as "error correction".

A spacecraft in orbit cannot hold the same attitude indefinitely, since atmospheric drag, the gravity gradient, solar pressure, payload operations, and astronauts moving inside all contribute perturbations. For example, if the spacecraft is required to hold a specific attitude in order to aim a telescope at a star, this involves nulling the attitude rates. On the other hand, to uphold thermal constraints it might be necessary for the spacecraft to rotate about one axis at a given rate to even out the temperature of the structure. Perturbations can impart motion around any one or even all of the axes. To return the vehicle to the desired attitude and attitude rate, the error must be nullified.

The RCS DAP computed these errors on an axis-by-axis basis as the difference between the desired attitude and attitude rate and the current attitude and attitude rate measured by the IMUs. These differences were then processed by a software routine called the "phase plane module". A phase plane may be visualized as a graph which plots the rate errors against the attitude errors for one axis, with a "box" being drawn around the center whose limits represent equations whose coefficients (attitude and attitude rate deadbands) are set by the SPEC 20 display. The box is used to determine when, if, and in what direction commands must be generated to null the errors. If the Orbiter was within the deadbands, then the plotted point would be inside the box. If the point left the box, jets would be commanded to fire to drive it back into the box, reducing the errors once more to an acceptable level. In an attitude-hold situation the plotted point cycled around the zero error point with the RCS jets remaining inactive until the limits were exceeded. This means of operating was called "limit cycling", with the cycling rate depending upon the deadband limits and the jets selected by the crew. The same logic applied when the crew manually

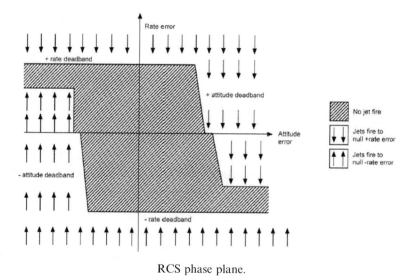

RCS phase plane.

commanded a rate using the THC/RHC. In this case, the commanded rate would become the new attitude to reach and the error to null with respect to the current attitude.

TVC DAP mode

OMS burns were performed using a completely different logic known as the "rate command system". In this case, the guidance software issued commands to the OMS both to achieve the desired delta-V and also to gimbal the engines to point the thrust vector in the desired direction. In order not to produce a turning rate, the gimbals had to be pointed to deliver the net thrust through the center of gravity of the vehicle. The TVC mode was activated any time that an ON command was sent to either or both of the engines, so long as there was at least one good IMU to provide attitude data. In addition the DAP had to be put into AUTO control mode using the DAP keyboard, and the RHC placed into its detent. If these conditions were satisfied, the burn could commence and the TVC mode remained active until either engine cut-off occurred, the OMS was turned off, or attitude data was lost. If the TVC failed to maintain the OMS properly oriented during the burn, the guidance software would command a so-called "RCS wraparound" in which it fired RCS jets to hold the vehicle in the proper attitude for the burn.

Display and control for maneuvers

To control the attitude of the Orbiter to precisely point instruments or to initiate OMS burns at the proper times to perform a rendezvous, the crew had two displays linked to the GNC software.

To set up, execute, and monitor translational maneuvers using either the OMS or the RCS the crew would employ the MNVR EXEC display which was available for post-insertion orbital flight, rendezvous, and de-orbit. Firstly, they would select one of four engine configurations: (1) both OMS, (2) left OMS only, (3) right OMS only, and (4) RCS. Irrespective of the configuration chosen, every OMS burn had to be targeted and enabled using this display. As RCS burns did not have automatic flight control, these could be performed only by deflecting the THC out of its detent. In this case the display allowed the pilot to monitor the progress of the burn and determine when the desired effect had been achieved. On choosing the engine configuration, the crew had to provide the necessary parameters for computing a solution using either the PEG 4 (only available for the post-insertion and de-orbit phases) or PEG 7 (only available for orbital flight) guidance schemes. For PEG 4 they had to provide two parameters to enable the software to establish the relationship between the horizontal and vertical velocities,[1] the height of the target

---

[1] Remember that PEG 4 guidance, also called linear terminal velocity constraint (LTVC) guidance, achieved a specified relationship between the horizontal and vertical velocity components at a certain target point in orbit.

point to be attained, and the angle from the current position to the target point. For PEG 7 they had to provide a delta-V in each of the X, Y and Z axes of the LVLH frame of reference.

Once these parameters were entered, the crew would load the data and a time of ignition (TIG) to enable guidance to make its calculations and display the attitude in inertial roll, pitch and yaw coordinates that the vehicle should adopt to carry out the burn. Achieving this attitude could be accomplished either automatically by selecting the AUTO control mode of the DAP, or manually, in which case the crew would steer to the ADI error needles. Pressing EXEC on the DAP keyboard enabled OMS ignition, which would occur at the specified time. If EXEC was pressed after the scheduled TIG, the burn would begin immediately. The display also showed data for use in monitoring the progress of a burn. In particular, when a target set was loaded, the guidance software computed and displayed the total change in velocity (in feet per second) needed to satisfy this requirement, along with its three components in the body axis coordinate system and the predicted apogee and perigee at the end of the burn. All these parameters were dynamically displayed to provide real-time feedback for either monitoring an automatic burn or controlling a manual burn.

Joseph P. Allen was a rookie mission specialist on STS-5 in November 1982. He describes the firing of an OMS engine as follows, "The first time we did an OMS burn . . . .to my astonishment, it looked like the back of the Orbiter blew off. [There was] this enormous flash of light, totally unexpected. You hear kind of a "whump" of the engine starting, and see a flash of light. It just is there and then it's gone, even though the engine continues to burn." As he later learned, there was a reason for this

OMS burn.

Maneuvering the Shuttle 393

dramatic spectacle. "The engines are started rich, more fuel than oxidizer, in order to make sure a clean burn starts, and then the mixture is made lean again, such that everything gets burned and there's no light at all. You would think there would be light from a rocket; there's none, at least looking out the back."

UNIV PTG (universal pointing) was used for automatic attitude maneuvers. By entering data on this display, the crew could command automatic maneuvers to point a given body axis at a specified target, to maneuver to a predetermined attitude, to set up a rotation around a given body axis, or to maintain an attitude. The display also showed attitude information and completion time for the current maneuver. They could select one of four options: maneuver (MNVR), track (TRK), rotation (ROT), and cancel (CNCL). As long as the DAP AUTO control mode was

1. Start with a body vector of pitch = 0°, yaw = 0°, which is pointing along the +X axis of the vehicle (Figure 5-3).

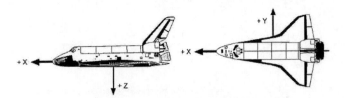

Determining the body vector - step 1

2. Pitch the body vector up (in a positive direction about the +Y axis) to the desired pitch angle. In Figure 5-4, P = 120°.

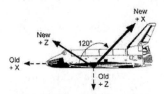

Determining the body vector - step 2

3. Rotate the body vector to the desired yaw angle in the new X-Y plane (in a positive direction about the new +Z axis). In Figure 5-5, Y = 45°.

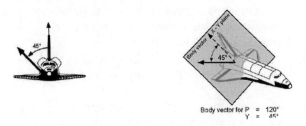

Determining the body vector - step 3

UNIV PTG pointing vector determination.

394 Orbital dancing

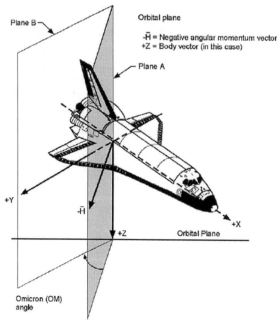

Plane A – Formed by the negative orbital angular momentum vector (always perpendicular to the orbit plane) and the body vector.

Plane B – Formed by the +Y axis of the orbiter and the body vector. The body vector forms the "hinge line" between the two planes.

Definition of pointing vector Omicron angle.

selected, all these options could be carried out automatically, otherwise the crew had to manually fly to the desired attitude.

The maneuver (MNVR) option would adopt the inertial attitude specified by the display. Once the flight control software had determined that this had been achieved, it would engage the attitude-hold function of the RCS DAP mode logic. Track (TRK) was by far the most complex functions of the universal pointing software, because it enabled the vehicle to track a specified target. To do so the crew had first to define a body pointing vector that was based on the center of gravity of the Orbiter and was defined in terms of pitch and yaw angles relative to the +X axis of the vehicle and the omicron angle defined the Orbiter's rotation relative to the pointing vector. Upon selecting TRK, the Orbiter would perform a three-axis maneuver at the rate specified by the current DAP to align the body vector with the selected target, and thus permit active tracking of the target. The software had six submodes.

In the orbital object tracking submode the body vector was oriented towards an orbiting vehicle in order to maintain a line of sight. To do so, GNC, computed a so-called relative target vector as the algebraic sum of the state vector of the Orbiter and the state vector of the target. This was the most important piece of data necessary for making a rendezvous. In the Earth-centered submode, the body vector was aimed at

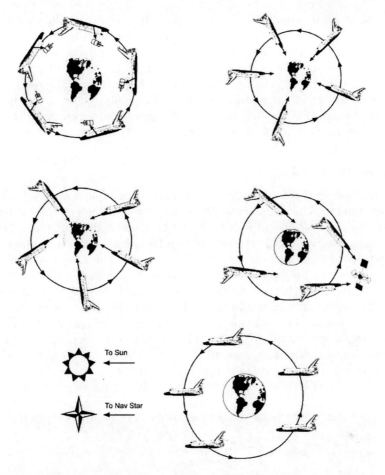

UNIV PTG tracking modes: from top left corner: orbital object, Earth center, Earth surface, tracking geosynchronous satellite, Sun and star.

the planet's center. To maintain this orientation the Orbiter had to establish a rate of rotation to match its orbital rate, resulting in a LVLH attitude-hold. This was often used when no specific attitude was required. Similarly, for the Earth-surface submode the body vector pointed towards Earth but at a point on the surface specified in terms of latitude, longitude and altitude. Once the aim point had been selected, GNC would constantly maneuver the Orbiter in such a way as to track the surface point regardless of whether it was above the horizon or not. Care had to be taken in choosing such a target point, because if the Orbiter were to pass directly over it the increased rate of rotation to maintain tracking might exceed the maneuver rates set in the DAP. With a trick, this submode could be used for tracking a point that was actually farther away from Earth than the Orbiter was (i.e., a geostationary satellite). To do this it sufficed to specify the altitude of the target above the surface of the Earth. The Sun-centered submode oriented the body vector towards the Sun

396 Orbital dancing

for thermal control purposes, such as to provide illumination or shading during payload deployment. As the Sun is so remote, this essentially placed the Orbiter in an inertial attitude. The celestial target submode pointed the body vector at a celestial object for astronomical observations. The navigation star target submode facilitated tracking of one of 100 navigation stars known to the software. The rotation (ROT) option was used for initiating a rate about a given body vector as specified by the DAP CONFIG display. CNCL canceled all current and future MNVR, TRK or ROT options and then invoked attitude-hold.

Manual attitude control

Despite the high level of automation built into the execution of maneuvers, the crew could perform translations and rotation by purely manual inputs. For example, with the DAP mode in INRTL and with DISC RATE selected, deflecting the RHC out of detent fired the relevant RCS jets until an angular rate was achieve in that axis equal to the rate specified in the DAP CONFIG display. Once this rate was achieved, the jets were turned off until the RHC was returned to detent or deflected in another axis. Deflecting the RHC beyond the soft stop made the jets fire continuously to produce a constant angular acceleration. Once the RHC was moved back inside the soft stop, the jets would be turned off but the rates would continue until the RHC was returned to the detent. With PULSE selected, RHC deflections fired jets to achieve the angular rate change specified on the DAP CONFIG display. High rate maneuvers could also be made to change attitude rapidly in an emergency. To do so with the DAP ROTATION set to DISC, the RHC would be pushed past the soft

RCS jet firing.

stop and back inside the soft stop several times until the desired rate was achieved, taking care not to let the controller return to its detent. To stop the maneuver, the RHC had to be returned to the detent position.

Generally speaking, firing the RCS jets was quite an interesting experience. For example, Bob Crippen says, "On the first flight we learned that the reaction control jets really are loud. It sounds like a Howitzer gun going off outside the window, and of course, no sound is transmitted through space; you're getting it back through the structure of the spacecraft." Vance D. Brand commanded three Shuttle missions including STS-5, the first operational flight, and he remembers that the RCS firings were quite visually striking, "At night it looked like a Fourth of July display, because you could look out over the nose and see these tubes of fire going up from the RCS jets. There were fantastic visual effects."

ORBITAL RENDEZVOUS MANEUVERS: DEVELOPMENT

When on 25 May 1961 President Kennedy set his nation the challenge of *"achieving the goal, before this decade is out, of landing a man on the Moon and returning him safely to the Earth"* many aspects of orbital flight mechanics were unknown. One of these was how one spacecraft could rendezvous in orbit with another and then dock with it. The Gemini program enabled NASA to learn how to do this, and thus contributed directly to the success of Apollo 11 in making the first human landing on the Moon in 1969. With this experience, mission planners presumed that rendezvous involving the Shuttle would be child's play, but reality soon proved otherwise.

The Shuttle was a spacecraft completely different from anything built previously, and this held true for its orbital maneuvering. In terms of rendezvous and proximity operations the Shuttle represented a significant point of departure from Gemini and Apollo. Most of its rendezvous targets did not possess active navigation aids such as transponders or lights, and nor were they originally designed to support rendezvous, retrieval and on-orbit servicing. Many Shuttle missions also involved deploying and retrieving the same or different satellites on a single mission, whilst others involved more than one rendezvous. Whereas for Gemini and Apollo the rendezvous targets were similar to or smaller than the active spacecraft, prior to missions to the Russian Mir space station and to the International Space Station the Orbiter was much larger than its targets. And whilst dockings by Gemini and Apollo spacecraft were done at a closing rate of about 1 foot per second, satellite retrieval using the robotic arm of the Shuttle required essentially zero relative velocity between the vehicles. Docking with Mir or the ISS required a contact rate an order of magnitude lower than for Gemini or Apollo, and tighter positional tolerances. Whilst Gemini and Apollo dockings were axial, along the crew's line of sight, with the Shuttle most of the dockings would be carried out using cameras to provide crew visibility and the cues for final control. All of these factors complicated the task of Shuttle mission planning, demanding the definition of new rendezvous techniques.

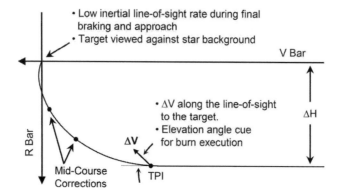

Terminal phase for coelliptic rendezvous.

Developing a Shuttle rendezvous profile

The first rendezvous approach profile investigated for the Shuttle was drawn from the experience gained with the highly successful Gemini and Apollo programs. Both employed a "coelliptic profile" in which the active vehicle entered an orbit that was coelliptic[2] with that of its target and at a lower altitude. To perform the rendezvous, the pilot of the active vehicle would visually observe the target against the stars and perform a burn for a delta-V along the line of sight to initiate an intercept trajectory. A number of corrections along the way would compensate for maneuver execution errors and targeting dispersions.

This profile worked well for targets which had active relative navigation aids, but these were not presumed when Shuttle rendezvous missions were baselined. On the other hand, one of the many missions for the Orbiter was to retrieve satellites whose designers had not envisaged them being retrieved and serviced. Using the previous profile, skin-tracking radar would have had only a few minutes to acquire and track a passive target prior to the active spacecraft making the interception burn. This would have also meant precise timing of launch and the orbital maneuvers leading up to the interception. In fact, being coelliptic does not mean that the two spacecraft travel on parallel paths; the fact that they are at different altitudes means the lower one travels more rapidly, with the result that only at certain times is one able to view the other. As a result, passive relative navigation was deemed to be a major challenge.

For this reason, the coelliptic profile was improved with the addition of a second coelliptic maneuver (NSR 2) to bring the active vehicle nearer the target's profile in order to reduce the difference in orbital velocities and provide more time to acquire and track the target.[3] This also provided flexibility in selecting the level of ground

---

[2] Two orbits are said to be coelliptic when their semi-axes coincide and when they are at different altitudes.

[3] In terms of a coelliptic rendezvous, NSR means "nominal, slow rate".

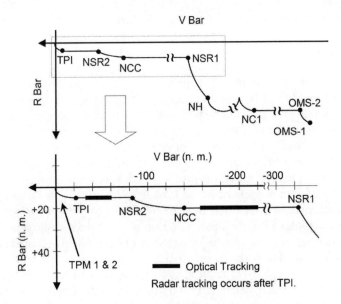

Relative view of dual coelliptic rendezvous.

tracking required, as well as in the selection of on board relative navigation sensors. And this second coelliptic phase would permit all subsequent maneuver points to be chosen to maximally exploit reflected sunlight for optical tracking of passive targets. Furthermore, this new segment ensured the same relative geometry from the onset of optical tracking through to intercept for variations in liftoff time and target orbital altitude, meaning that a wide variation in liftoff time was feasible without resulting in an excessively long phasing period. To better explain this concept, let us return for a moment to the simple coelliptic profile in which liftoff had to occur in such a manner as to place the active vehicle in view of the target in favorable illumination. Knowing orbital mechanics and the orbital altitudes of the vehicle and its target, it is possible to calculate when the vehicle will be able to see its target in reflected sunlight against the background of stars. For this to happen, the active spacecraft has to be launched at a specific time. If the launch window is missed, it is necessary to wait for the next one and this, combined with other constraints such as adverse weather, could result in significant delays. Adding the second coelliptic segment made the rendezvous almost independent of the time of liftoff. Once on the first coelliptic segment, the active vehicle would have several opportunities to start the second segment at the required phase angle and then acquire and track the target with favorable illumination.[4]

---

[4] The phase angle (theta) is defined as the angle between the target position vector and the projection of the active spacecraft's position vector onto the target's orbital plane. It is positive when the active spacecraft is trailing.

One important issue remained unresolved. The target would probably not possess strobes, as targets had in previous programs. The lighting requirements for the pre-TPI[5] optical pass and for initiating manual piloting at sunrise (several thousand feet from the target) in order to make the most of the daylight period meant that the TPI burn had to be made after sunset. If the tracking device failed, a manual backup by optical tracking would not be possible because the target was invisible in darkness! Doubts about whether this "dual elliptical profile" would be sufficient persisted into early 1980s. Of particular concern was the capability to obtain sufficient on board optical tracking of a target by reflected sunlight when in the presence of the bright limb of the Earth or a bright celestial object. In addition, due to the high relative approach velocity inherent in such a profile, there was some concern about early depletion of the limited propellant in the forward RCS pod. This latter aspect was highlighted when planning the Solar Max mission.[6] All of the simulations indicated that due to the complexity of the mission, which also involved deploying the LDEF satellite in a different orbit earlier on,[7] it would not be possible to have good lighting conditions without depleting the forward RCS propellant.

Drawing on the Gemini experience, mission planners tried to adapt the so-called "stable orbit" profile used by Gemini XI. This involved initiating the intercept from a station-keeping[8] point on the –V-Bar (that is to say, from some distance behind the target) instead of from a coelliptic orbit. The advantage of this approach was that the active vehicle and its target would have the same velocity. Station-keeping at a safe distance from the target would allow the crew to verify that their vehicle was ready to start the final approach, and would allow the time at which to start the final approach to be chosen to provide the required lighting conditions. Generally speaking, station-keeping can be achieved both on the V-Bar and on the R-Bar. However for a given range, V-Bar station-keeping uses less propellant than station-keeping on the R-Bar because both vehicles are in almost the same orbit. This provides the most efficient type of station-keeping because minimal propellant is required to maintain the active spacecraft at the same altitude and coplanar with the target. On the other hand, R-Bar station-keeping represents an unstable configuration because the two vehicles are

---

[5] The terminal phase initiation (TPI) maneuver was the moment at which to make the burn that would put the spacecraft on a direct intercept course with the target.

[6] The Solar Max satellite was launched into orbit around Earth on 14 February 1980, primarily to study the Sun during the intensive part of its activity cycle. A malfunction in January 1981 prompted the planning of a Shuttle mission to recover the satellite in order to restore it to full operation. This was achieved by the crew of STS-41C in April 1984.

[7] The Long Duration Exposure Facility (LDEF) was a passive rack comparable in size to a school bus, to expose various materials samples to the space environment. Deployed by STS-41C as planned, it was to be retrieved in 1986 but the *Challenger* disaster meant that it remained in space until retrieved by STS-32 January 1990 shortly before its orbit would have decayed.

[8] Station-keeping is the technique used to maintain the Orbiter at a desired relative position, attitude and attitude rate with respect to a target vehicle.

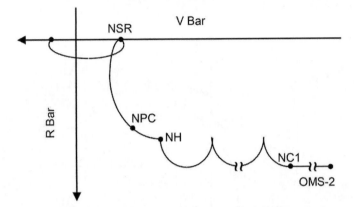

A stable orbit rendezvous profile proposed in March 1982.

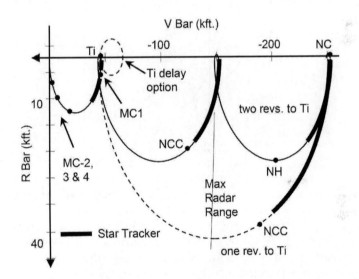

Stable orbit rendezvous profile improved (1983-1997).

in different orbits and over an extended period their differential velocities will cause the separation to vary. R-Bar station-keeping is therefore more complex, requiring more propellant. A number of stable orbit profiles were proposed, differing in the manner in which the final approach was flown.

The advantages of a stable orbit profile were: (1) lower propellant consumption, (2) simpler procedures for the crew and Mission Control, (3) stable station-keeping points on the –V-Bar in the event of either a system anomaly or a change in mission planning, and (4) no need to undertake optical tracking using star trackers so long as the radar functioned. It potentially offered more straightforward trajectory design for missions involving rendezvous from in front of or from above the target. Several

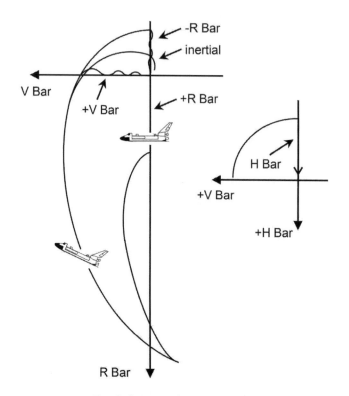

Proximity operations approaches.

weaknesses exposed by simulations were fixed by revising the profile. Firstly, the station-keeping point on the –V-Bar was eliminated in favor of performing a target interception (Ti) maneuver at the previous point of station-keeping. In the event of a systems anomaly, a station-keeping maneuver would be made until it was possible to resume. The purpose of Ti was to bring the Orbiter to a specific position relative to the target preparatory to starting the terminal phase of the approach.

It is important to understand that the Ti burn would not be along the line-of-sight to the target. On the Gemini and Apollo spacecraft it was possible to proceed with a TPI burn after a GNC malfunction by visually sighting the target to orient the vehicle and then firing along the line-of-sight. This was not the case for the Ti burn, but the greater confidence in the GNC of the Orbiter made it acceptable to move the Ti point out of the line-of-sight to the target. To further optimize the approach and minimize propellant expenditure, four midcourse correction (MC) burns were placed between Ti and the target interception point. In addition, to reduce the out-of-plane velocity component after the first midcourse correction maneuver, the on board tracking was extended prior to Ti to include one or two star tracker passes beginning at a range of 40 nautical miles for an overlap of ground and on board tracking for cross-checking purposes in advance of committing to an intercept trajectory. Finally, a "corrective combination maneuver" (MCC) was added to ensure that the Ti point would be in the same orbital plane as the target, and also at a point above the –V-Bar so as to make trajectory dispersions more manageable when near-continuous manual piloting was initiated at a range of about 2,000 feet. From there onwards, several different types of approach were studied. In one, Ti was targeted to place the Orbiter several miles in front of the target on the +V-Bar, after which it would move in along the +V-Bar. In another, Ti targeted the Orbiter for a point 5,000 feet ahead of the target and 1,500 feet above it, from which it would fly a "glideslope approach" that would avoid firing the RCS and prevent impingement on the target. But analysis showed that a direct (inertial) approach could be flown with a transition to the +V-Bar at a range of about 500 feet, combining an acceptable propellant consumption and a low risk of plume impingement. In this way, the glideslope approach was discarded. R-Bar approaches were also studied because the "natural braking" (slowing of the Orbiter) of orbital mechanics would reduce the need for thrusting. In addition, an approach on the H-Bar (out-of-plane) was extensively tested in order to bring it up to the same degree of maturity as the R-Bar and V-Bar approaches. The advantages of the H-Bar approach were consistently good lighting conditions for piloting, and Y motion (in the LVLH reference frame) uncoupled from motion in the other two axes. But unlike R-Bar, an H-Bar approach could not exploit the natural braking effect of orbital mechanics, since both vehicles would be traveling parallel to each other and hence have the same velocity. It would require frequent thrusting at the target during the approach, and out-of-plane motion would continue after the relative translational rates were nulled. It was discarded due to safety, station-keeping, propellant consumption, and plume impingement issues.

Despite their similarity, the inertial and –R-Bar approach necessitated holding the Orbiter in two completely different attitudes. With an inertial approach the Orbiter maintained its attitude fixed relative to the target all the way to the target grappling or docking. In a –R-Bar approach, the Orbiter constantly altered its attitude until arrival on the R-Bar, where the payload bay would be facing the target and perpendicular to the R-Bar, after which the Orbiter would translate towards the target along the R-Bar.

The stable orbit profile was the one flown by every mission involving rendezvous and proximity operations, starting in June 1984 when it was used by STS-41C for the Solar Max rendezvous through to the first Shuttle-Mir missions in the mid-1990s.

The plume impingement problem

While mission planners were studying the best rendezvous profile for the Shuttle, another problem was lurking ahead. It was known that the jets of the RCS could have a significant effect on the target when in close proximity. This plume impingement issue was first observed when Gemini XI spacewalker Richard Gordon hooked up a tether between his spacecraft and a docked Agena target vehicle. The experiment was to create some "artificial gravity" by setting the linked vehicles rotating around their common center of gravity. Analysis of the 16-mm footage showing tether dynamics in response to thruster firings established plume impingement to be an issue. The first manned mission to Skylab provided confirmation. During launch of the space station, one of its solar panels became jammed and was unable to deploy in space. When the first crew reached the orbital outpost the Apollo spacecraft was maneuvered in such a way that a crewman standing in the hatch could reach the array using a deployment tool. The plumes from the Apollo thrusters as it nulled the closing rate were powerful enough to alter the attitude of the much larger space laboratory, in response causing Skylab to fire its own thrusters to maintain its attitude. The overall result was to increase the separation. Once the astronauts were on board the station, they deployed a large parasol through a small airlock on one side of the main compartment to compensate for the loss of the thermal shield during launch. In performing a fly-around prior to heading home, RCS plumes significantly disturbed the makeshift thermal shield. The issue was considered so serious that when Soviet engineers were planning the first docking in space between an American spacecraft and a Soviet spacecraft scheduled for 1975, they demanded that four of the RCS jets on the Apollo be inhibited within two seconds of contact in order to prevent plume impingement on the solar arrays of their Soyuz spacecraft.

As early as 1973 the payload community which was investigating missions and payloads to be flown by the Shuttle expressed concern that RCS plume impingement could contaminate sensitive instrumentation on a satellite that was being approached by the Orbiter. The presumption was that because they had not been designed to be serviced, many of these satellites would not have any means of protecting their most sensitive parts and wouldn't be able to adjust their attitude to prevent contamination. For this reason, an approach trajectory was proposed that minimized the expulsion of combustion byproducts towards the target.

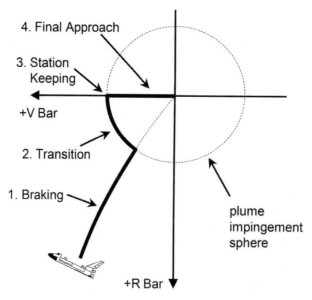

Terminal approach to minimize plume impingement on target (October 1973).

This introduced the idea of a "plume impingement sphere", a border within which Orbiter RCS jet firings in the direction of a target would be prevented so as not to pollute it. On reaching the border of this sphere (whose radius would be defined on a target-by-target basis) the Orbiter would switch from the direct approach trajectory to a station-keeping point on the velocity vector to the target. Once the robotic arm was verified ready to grapple the target, the Orbiter would initiate the final approach. As effective as this might seem, in 1975 when work got underway to define the plan for routine servicing of Solar Max and for retrieving LDEF, it was soon realized that the plume impingement sphere was inadequate.

Studies showed that due to the relatively small sizes of the Solar Max and LDEF satellites compared to the Shuttle, the plumes from the RCS jets could induce attitude rates on the target, or even push it away. Solar Max had an attitude control system but it was incapable of maintaining attitude against such powerful plumes. LDEF had no active attitude control system, it stabilized itself in the gravity gradient.[9] The root cause of the problem was that the sizing, placement, and orientation of the thrusters on the Orbiter had been chosen in order to provide the desired degree of flight control authority throughout the Shuttle's flight envelope and to avoid impingement on its aerosurfaces; impingement on satellites or the robotic arm was not a factor, and when this was realized it was too late. By June 1976 several simulations had produced the conclusion that the plume impingement would induce dynamics at RMS release or grapple ranges which might render deploying and retrieving a satellite like LDEF impossible. Work was ordered to develop improved models of the RCS jets and their plumes. In addition, several means of avoiding plume impingement were proposed. Some envisaged alternative recovery techniques using new hardware (such as stand-off berthing using a mast or a tether), a payload bay mounted cold-gas propulsion system, and "hardened" payloads. Owing to complexity and cost these were all soon discarded in favor of solutions requiring new piloting techniques and modifications to the flight control system. But this was easier said than done, since it would mean increasing propellant usage and complexity of the software and crew procedures. The propellant usage was particularly critical as, unlike Gemini and Apollo, the Orbiter was limited in terms of its forward RCS propellant. The concern was that it could run out of forward RCS propellant during the terminal phase, under dispersed trajectory conditions and in the event of a radar failure. Simulations showed that one technique that worked quite well for approaches along all the three axes of the LVLH frame of reference (V-Bar, R-Bar and H-Bar) used the Orbiter $+/-X$ body axis RCS jets for braking;[10] therefore the "Low Z" mode was added to the flight control system.

---

[9] Since an orbiting body is a three-dimensional object, parts of it are marginally closer to the Earth while others are farther away and this subjects the body to different values of gravitational attraction that cause it to rotate until a stable attitude is reached. This stable attitude is such that the different gravitational forces acting on the body are in equilibrium and no further torque is applied.

[10] See Chapter 10 for an explanation of the Orbiter body axes.

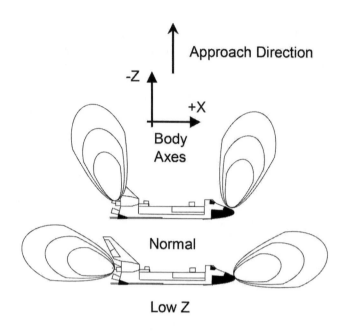

Comparison of plumes.

The X axis RCS jets had a thrust component that was primarily along the X body axis. But the serendipitous canting of the X axis RCS jets also created a small thrust component in the –Z direction, thereby providing a minimum braking capability. When in Low Z mode all the up-firing RCS jets would be inhibited, but this meant a very high usage of propellant in performing a braking maneuver. If it was done using the up-firing jets it would use one-twelfth the propellant. Since it did not demand extensive modification of the Orbiter or the target satellites, the Low Z mode proved to be a good solution to the plume impingement problem.

The development of new flight control techniques was fed into specific mission planning, and plume impingement became a manageable issue for the wide variety of targets associated with the program. In March 1977 the term "proximity operations" (shortened to "prox ops") was coined. It was defined as operations that occur close to a target (within 2,000 feet) and characterized by almost continuous trajectory control, as distinct from rendezvous control maneuvers which typically occur at intervals of hours or tens of minutes.

Rendezvous with Mir and the ISS: a new approach

On 17 July 1975 history was made when for the first time Americans and Soviets shook hands in space. Twenty years later, history repeated itself when on 29 June 1995 *Atlantis* accomplished the first docking of a Shuttle with the Russian Mir space station. Despite the Shuttle program having already seen several successful missions involving rendezvous and proximity operations, new challenges arose in docking an

Orbiter with a massive station. The stable orbit rendezvous profile was designed to optimize approaches along the V-Bar with the option of transition to the –R-Bar. The Low Z flight control mode that was instrumental in preventing damage to a payload during deployment and retrieval used a lot of propellant during braking maneuvers. Plume impingement was a concern for docking with Mir because the station was so large, had so many solar panels, and so much exposed apparatus. Planning missions to Mir (and later to the ISS) prompted renewed studies of the direct approach along the +R-Bar which would exploit orbital mechanics to minimize the Low Z thrusting during braking and also to accelerate separation after undocking. Studies indicated that this was feasible without changing the on board computer targeting constants for stable orbit profile. The availability of new laser sensors to measure the range and range rate provided a form of redundancy that was not available when the direct +R-Bar approach was considered for the first time for the Skylab reboost mission.[11] After extensive analysis, procedures development overcame programmatic resistance to the idea and the direct +R-Bar approach was approved by the program in April 1994. All nine docking missions to Mir were conducted in this manner. Nevertheless, there was room for improvement. In particular, it was necessary to find a way to optimally set up the initial conditions for a low-energy coast along the +R-Bar. The solution was found in targeting the Ti burn and the first three corrections (MC) not for interception with the target but for the manual takeover point at 2,000 feet. The MC-4 burn would then be targeted to place the Orbiter on the +R-Bar, 600 feet below the target. Then all that the Orbiter was required to do was to perform several thrusting maneuvers to translate up the +R-Bar axis against the gravity gradient that was tending to draw it down, thereby arriving with almost zero relative velocity after only a few braking maneuvers in Low Z mode.

This new profile was called the optimized R-Bar targeted rendezvous (ORBT), and although similar to the classic stable orbit profile it had one key difference. In the stable orbit profile the Ti burn aimed to place the Orbiter on an intercept course with the target and then four corrective maneuvers would put it on the terminal approach along either the +V-Bar or the –R-Bar. In ORBT, the Ti and MC burns placed the Orbiter on a trajectory to intercept not the target but a point on the +R-Bar *below* the target. Therefore the Orbiter did not have to do anything to place itself onto the R-Bar approach; it was already there! The stable orbit profile required further braking maneuvers after the MC-4 burn in order to slow the approach, and if this braking was done in Low Z mode it would use a lot of propellant. ORBT

---

[11] Because Skylab's orbit was decaying more rapidly than expected, in 1977 a reboost was assigned to an early Shuttle mission. The crew would fly by remote control a device that was basically a propulsion module with a docking system. Once docked, it would raise the station's orbit to preserve it for possible future occupation. But delays in the development of the Shuttle made this impossible. The reboost was canceled in December 1978, and on 11 July 1979 America's first orbital outpost fell into the atmosphere and burned up over the Pacific Ocean.

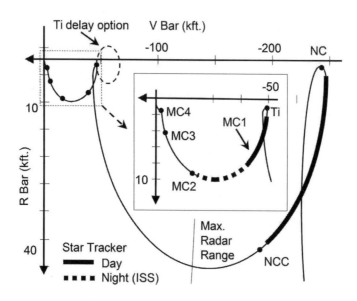

Optimized R-Bar targeted rendezvous (1997 to end of program).

involved fewer braking maneuvers and required less propellant. Another difference was that in ORBT the Ti point was placed below the V-Bar in order to allow for an optimized MC-4 burn. For the first six missions to Mir, the Shuttle flew a +R-Bar approach from a stable orbit profile but the final three missions used the ORBT profile. Docking missions to the International Space Station used the ORBT profile even when it was necessary to approach along the +V-Bar. In this latter case, once the Orbiter was on the R-Bar, a maneuver would be initiated to swing it up onto the +V-Bar preparatory to initiating the final approach.

After the *Columbia* disaster, one of the many recommendations provided by the CAIB was to "*develop a practicable capability to inspect ... the Thermal Protection System ... taking advantage of the additional capabilities available when near ... the International Space Station*" [R6.4-1]. A low-cost solution that did not involve the creation of new tools was to have the Orbiter halt beneath the ISS and carry out a sort of a somersault to expose its belly to the station so that detailed pictures could be sent to Mission Control for analysis. This R-Bar pitch maneuver (RPM) was performed 600 feet below the ISS. This position was selected because the Orbiter was already scheduled to halt at 600 feet as part of the ORBT profile prior to departing to make the V-Bar approach. The maneuver started by firing a pair of symmetric down-facing forward RCS jets to impart a positive pitch. As soon as the belly of the Orbiter was in sight of the ISS, that crew began to shoot pictures through the nadir windows in order to determine the status of the tiles on the underside. When the Orbiter was back with its payload bay facing the station, down-facing aft RCS jets were fired to cancel the pitch motion. Simple though it might seem, this maneuver was a serious violation of one of the most important ground rules for rendezvous introduced at the beginning of the Shuttle program: namely that during proximity

Orbital rendezvous maneuvers: development 409

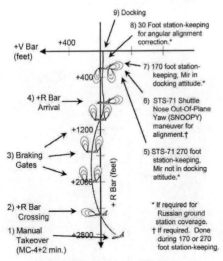

Plus R Bar approach to Mir from Stable Orbit profile, STS-71 (June 1995) & STS-74 (Nov. 1995).

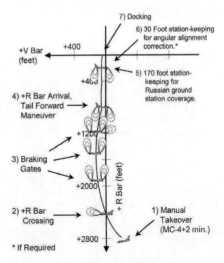

Plus R Bar approach to Mir from Stable Orbit profile, STS-76 (March 1996), STS-79 (Sept. 1996), STS-81 (Jan. 1997), & STS-84 (May 1997).

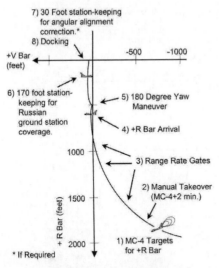

Plus R Bar approach to Mir from ORBT profile, tail forward docking, STS-86 (Sept./Oct. 1997) & STS-91 (June 1998).

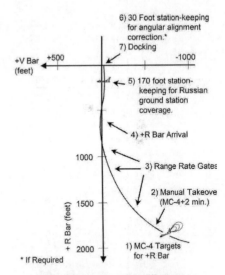

Plus R Bar approach to Mir from ORBT profile, nose forward docking, STS-89 (Jan. 1998).

Approaches to Mir.

410 Orbital dancing

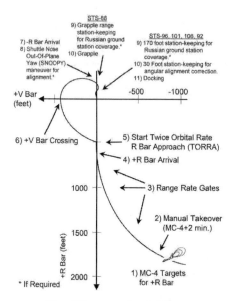

Minus R Bar approaches to ISS,
STS-88 (Dec. 1998) to STS-92 (Oct. 2000).

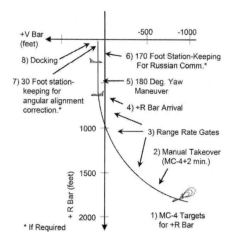

Plus R Bar approaches to ISS,
STS-97 (Nov./Dec. 2000) & STS-98 (Feb. 2001).

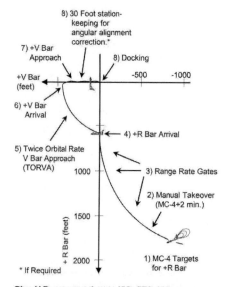

Plus V Bar approaches to ISS, STS-102
(March 2001) to STS-113 (Nov./Dec. 2002).

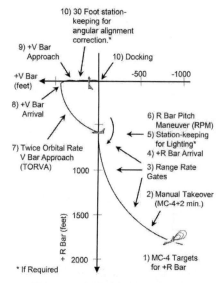

RPM and plus V Bar approaches
to ISS, STS-114 (July/Aug. 2005) to end of
program (2011).

Approaches to the ISS.

operations the crew must maintain visual or sensor contact with the target. This rule ensured that the crew would always be aware of their situation and capable of taking action to prevent a collision. But the Shuttle was not provided with sensors capable of tracking a target without the crew having a line of sight. The RPM put them in a blind situation, and for this reason it required careful planning to avoid any unsafe situation occurring while the crew were unable to see the station. On the other hand, during that period the Orbiter was being observed by the station crew.

Hubble servicing missions

Most Shuttle missions could have passed almost unnoticed to the general public, but not the highly successful Hubble Space Telescope (HST) servicing missions. Since it was deployed by *Discovery* during the STS-31 mission in April 1990, five servicing missions enabled the telescope to provide unprecedented astronomical observations for more than two decades.

The planning for this kind of mission involved trade-off studies, simulations, and technical discussions covering both nominal and contingency plans and procedures. The rendezvous with the HST and its grappling by the robotic arm were nominally scheduled for flight day three. The proximity operation phase would begin once the Orbiter was within 2,000 feet (i.e. after MC-4). The relative motion trajectory was designed to accommodate both Orbiter and HST constraints such as RCS jet plume impingement, HST thermal control and power generation – with the latter being a delicate issue because the attitude of the telescope had to be carefully managed to ensure that the solar arrays would generate adequate power despite the constraints imposed on its attitude control. Before the Shuttle initiated the final approach, the – V3 high gain antenna of the HST was stowed and latched and the two solar arrays were rotated to be parallel with the V1 axis. The HST then performed a roll maneuver to place the grapple fixture on the north side of the orbital plane. It then held this inertial attitude throughout the final approach. As the nominal grapple attitude of the HST was not optimal for power generation by the solar arrays, upon finishing this roll maneuver a 180-minute timer was started. If by that time the Orbiter had not grappled the HST then the telescope would start a low rate maneuver to an attitude for optimal power. However, this contingency was never invoked on any of the servicing missions because the grappling was accomplished without difficulty.

The servicing missions were quite demanding because, despite the HST having been designed to be serviced in space,[12] its design was completed before the plume impingement issue was fully understood. Mission planners had to take into account

---

[12] The original idea was to carry out on-orbit servicing every 2.5 years and return the telescope to Earth every 5 years for a more intense refurbishment, after which it would be relaunched. But by the late 1970s concerns about contamination of the telescope and the structural loads that it would suffer during ascent and re-entry prompted NASA to limit it to on-orbit servicing.

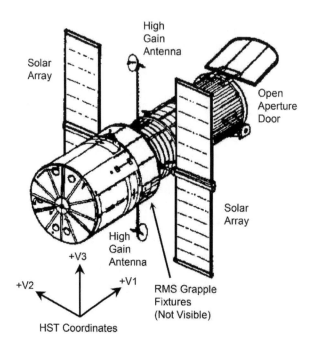

Hubble Space Telescope.

the fact that the Shuttle needed to climb to a much higher altitude in order to deploy the HST than any other payload and (as if this was not enough) the telescope did not have a propulsion system for orbit maintenance. As a result, the HST had always to rely on the Shuttle servicing missions to raise its orbit to counteract the natural orbital decay due to atmospheric drag. At the end of each servicing mission the orbit of the HST was raised to the maximum altitude that the Orbiter could reach. A further issue was that using the Low Z mode during proximity operation was expensive in terms of RCS propellant. The servicing missions therefore stretched the Orbiter's operating envelope to its limits.

The deployment of the HST was originally scheduled for STS-61J to be flown in August 1986 but the loss of *Challenger* in January of that year grounded the entire fleet. Although the Shuttle resumed flying in September 1988 a number of payloads had higher priority than the HST. It was assigned to STS-31 and finally launched on 24 April 1990. The next day the telescope was lifted from the payload bay using the robotic arm. When solar array #2 failed to deploy, Bruce McCandless and Kathryn Sullivan prepared to make a contingency spacewalk to try to free it but engineers on the ground were able to coax the panel to deploy using a different procedure and the spacewalk was canceled. Once the HST was released, *Discovery* separated to a safe distance and the Space Telescope Operations Control Center at the Goddard Space Flight Center set about verifying the correct functioning of the telescope. Meanwhile, *Discovery* withdrew to conduct station-keeping at a point 40 nautical miles behind HST on the −V-Bar. If the large protective door on the front of the telescope failed to open, the Orbiter would return and astronauts would make a contingency spacewalk to manually open it. Long range station-keeping was maintained until the activation of the HST was complete and the door verified open. About 1 day and 19 hours after deployment *Discovery* departed, using an orbit that was coelliptic to that of the HST in order to ensure safe separation.

The first servicing mission was by *Endeavour* as STS-61 in February 1993. This was of extreme importance, as in addition to several replacement parts there was an instrument that would enable the telescope to recover from a crippling optical flaw. In planning STS-61, an important issue regarding the attitudes of the Orbiter and the HST during the grappling operation had to be addressed. It was proposed that to ease this task the telescope should perform a roll maneuver about the V1 axis to place the grapple fixture in the proper orientation for grappling by the robotic arm. Doing so would have avoided having the Orbiter perform a large yaw maneuver which would have been executed at a safe distance and in Low Z mode. The disadvantages of such a maneuver included exacerbating orbital mechanics effects that could cause the two vehicles to separate as a result of the larger than usual station-keeping distance, cross-coupling from the Low Z mode increasing the difficulty of piloting, and difficulty in performing the yaw maneuver while station-keeping. On the other hand, having the Orbiter yaw into the correct attitude for grappling was highly desirable from an HST point of view because it would avoid placing the telescope in a non-optimal attitude for solar power generation, minimizing both depletion of its batteries and the risks of increased thermal heating and outgassing. Electrical load reduction was undesirable, and the solar array orientation in the grapple attitude was untested and could expose the arrays to increased plume impingement. Despite the good reasons for deleting the HST roll maneuver, concern for the safety of the Orbiter, its increased propellant consumption, and the greater plume impingement and contamination of the HST led to the decision to have the telescope perform the maneuver. This procedure became standard for every servicing mission.

Endeavour reached the HST by flying a stable orbit profile with an inertial final approach. After the Ti burn by the Orbiter, the HST was configured to minimize its electrical power requirements in order to accommodate the roll to grapple-attitude

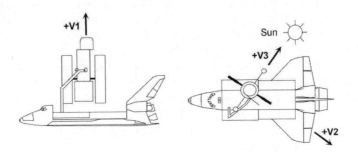

HST deployment.

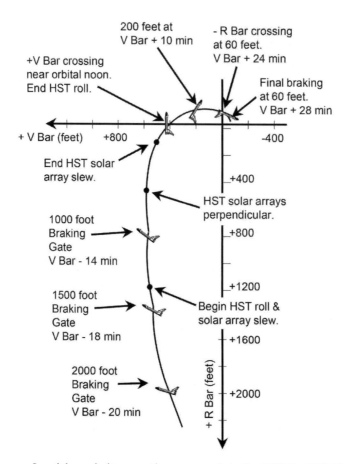

Inertial proximity operations approach to the HST for STS-61.

during proximity operations. At the start of the terminal phase (after MC-4) the +V3 axis of the HST was aimed at the Sun, the solar arrays were aligned with the V1 axis, and the –V3 high gain antenna was stowed in order to maximize clearance for the robotic arm. Approximately 20 minutes before the Orbiter reached the +V-Bar, the HST began the roll maneuver to place the RMS grapple fixture on the north side of the orbital plane. At a range of 400 feet the flight control system of the Orbiter was switched to the Low Z mode to prevent plume overpressure on the solar arrays of the HST. The roll maneuver by the HST was finished by the time the Orbiter arrived on the +V-Bar at a range of about 350 feet. At this time the –V1 axis of the HST (i.e. the end opposite the aperture door) was pointing at the Orbiter. *Endeavour* continued the inertial approach until reaching the station-keeping range of 35 feet, where the arm successfully grappled the HST on schedule, 10 minutes after orbital sunset in order to minimize Orbiter camera blooming and to permit completion of photography of solar array deflection during sunset.

The second servicing mission, STS-82 in February 1997, again pursued a stable

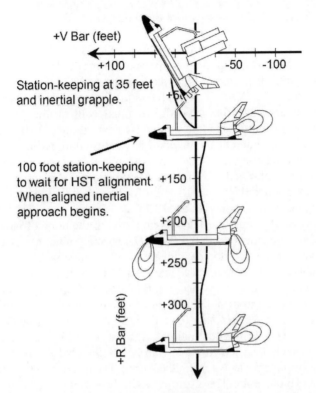

Final +R-Bar approach and inertial grapple.

orbit rendezvous profile but approached on the +R-Bar. This mission included a new deployment requirement. It had been found that ultraviolet light reflected by Earth could enter the telescope when the aperture door was open and cause contamination that might accumulate on the mirror during the servicing mission and permanently adhere to the mirror. For this reason, the deployment was done with the Orbiter in a different attitude. The new requirement was to point the +V1 axis of the HST away from the bright limb of Earth when being released in daylight prior to sunset in order to allow the Sun sensor of the HST sufficient acquisition time, and with the +V3 axis pointed sunward. Two deployment and separation profiles were prepared, depending on which side of the orbital plane the Sun was located. The first separation burn was made by two forward-firing –X RCS jets with the flight control system operating in free drift. As the Orbiter moved away from the HST, the +Z thrust component of the forward jets caused the vehicle to pitch down until commanded to stop a short while later. This nose-down rotation gave adequate clearance to the cabin without the crew losing their direct line of sight to the telescope. The –X jet separation also used less propellant than a Low Z separation and posed a lower risk of plume impingement. This separation procedure would become standard for subsequent servicing missions.

Beginning with STS-82, a contingency rendezvous profile was added in case the aperture door of the HST failed to close before the rendezvous. The procedure called for following a stable orbit rendezvous profile with a final inertial approach since the +R-Bar approach would expose the telescope optics to RCS jet plume contamination. With the proper timing the failed-open aperture door could be pointed away from the approaching Orbiter throughout an inertial approach and thereby minimize the risk of contamination. The −V1 axis would be pointed at the Orbiter two minutes after the MC-4 maneuver, when the crew took manual control and placed their spacecraft in an inertial attitude. When the Orbiter arrived on the +V-Bar, the −V1 axis of the HST would be pointed at the payload bay. This combination of Orbiter and HST attitudes ensured that the failed-open aperture door would always be pointed away from the approaching Orbiter. Starting with the third servicing mission, STS-103 in December 1999, an ORBT profile was used and the robotic arm grappled the HST in the same manner as STS-82.

STS-400 is the only Shuttle mission that was prepared in the hope of never being flown. After the *Columbia* incident, a final, fifth HST servicing mission was added to the manifest. However, since the Orbiter designated for the mission would not have had the capability to use the International Space Station as a safe haven in the event of irreparable damage to its thermal protection, STS-400 was prepared as a rescue mission for the crew stranded in orbit. The rescue concept would have required the pre-launch parallel processing of *Atlantis* for STS-125 and of the rescue Orbiter at the Kennedy Space Center.[13] In the nominal plan the rescue Orbiter would perform all of the ORBT maneuvers to reach *Atlantis* from below. Just before the newcomer performed the MC-4 burn, *Atlantis* was to maneuver to face its payload bay to Earth and its nose towards orbital south. This attitude would be maintained by the vernier RCS jets and, if necessary, by using the Low Z mode to avoid damaging the rescue Orbiter. The +R-Bar approach of the proximity operation profile would start with the crew taking manual control after the MC-4 maneuver. Unlike ISS-bounded missions, the RPM would not be performed. From a range of 1,000 feet through to grappling, the rescue Orbiter would use the Low Z mode. This range was chosen because crews of missions to the ISS were familiar with initiating Low Z operations at this range. Capture would be performed by the robotic arm of the rescue Orbiter grappling the forward grapple fixture of the berthed OBSS on *Atlantis*. Once grappled, the OBSS would roll out and the arm of the rescue Orbiter would align *Atlantis* so that both vehicles were nose-to-nose for effective attitude control in the mated configuration. Finally, the rescue Orbiter would maneuver the stack into a gravity gradient attitude for stability.

Separation would have involved the rescue Orbiter leaving from the −V-Bar. Contingency plans were also developed in case the rescue Orbiter proved unable to execute the nominal rendezvous profile because of either an underspeed at MECO or

---

[13] After several switches in the flight manifest due to technical delays, *Endeavour* was eventually chosen as the rescue vehicle and it would have been flown by the four flight deck crew members of the STS-123 mission.

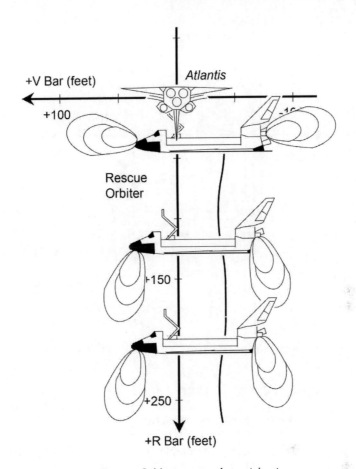

Rescue Orbiter approach to *Atlantis*.

a propellant failure. While it was preferable to fly the mission with the rescue Orbiter approaching from below and behind *Atlantis*, provisions were made to approach it from above and ahead. In this case *Atlantis* might have been called upon to perform some altitude and phasing adjustments to enable the rescue Orbiter to complete the rendezvous. Propellant margins on both vehicles would have been carefully managed to ensure that the rescue Orbiter had sufficient propellant for a safe deorbit.

ORBITAL RENDEZVOUS MANEUVERS: OPERATIONS

Generally speaking, during the 30 years of the Shuttle program three different types of rendezvous were available. For example, profiles which involved the interception of a satellite already in orbit were defined as "ground-up" missions. Another profile was the so-called "deploy/retrieve" mission in which the Shuttle would be launched with the satellite in its payload bay, deploy it, and later retrieve it for return to Earth.

418 Orbital dancing

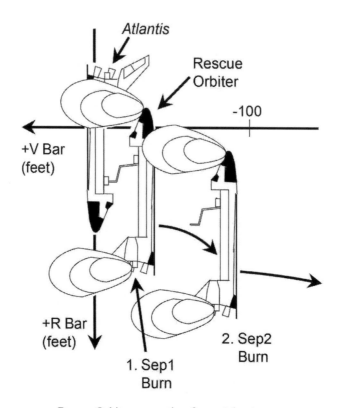

Rescue Orbiter separation from *Atlantis*.

Finally, a "control-box" profile was available for satellites that were capable of active translation control. In this case the satellite would be responsible for placing itself in an orbit that was within the rendezvous capability of the Shuttle. Of these three types of mission the ground-up profile was the most commonly flown, as illustrated by the missions to the Mir space station, to assemble the International Space Station, and to deploy commercial and scientific satellites. Deploy/retrieve profiles were also flown many times and are well represented by missions involving deployment and retrieval of the Spartan, SPAS, and Wake Shield Facility platforms. The control-box profile was flown only once, when the Intelsat VI F-3 commercial satellite maneuvered into a lower orbit so that *Endeavour* could retrieve it and install a new propulsive stage during the STS-49 mission.

Irrespective of the profile flown, each rendezvous mission consisted of a series of maneuvers to bring the two vehicles together in space. Rendezvous missions required the use of many different translational maneuvers whose procedures were detailed in the Rendezvous Checklist. Among other things, this document, which was unique to each mission, specified a rendezvous timeline in terms of the "phase elapsed time" (PET) relative to the time of ignition (TIG) of the target interception (Ti) maneuver. In the ensuing discussion, a typical rendezvous mission to the ISS will be described, as this was the target for most of the rendezvous missions of the program. Typically, a profile is plotted on a graph whose axes lie in the orbital plane, with the horizontal axis representing the delta-range between the active vehicle and the target and the vertical axis being the delta-height between their individual orbits. The direction of orbital motion is towards the left and each "U-shape" represents one revolution around the globe. This representation is called a "view along the $+Y$ axis" and is the most common means of analyzing such a mission. The rendezvous profile typically began with the OMS 2 burn that established the Orbiter in a stable orbit, and was followed by four phasing maneuvers (designated NC for "n^{th} closing burn") that enabled it to chase its target in a choreographed manner. With some basics of orbital mechanics it is easy to appreciate the purpose of these maneuvers. Because the Orbiter was always inserted into an orbit that was lower than its target, its greater orbital velocity resulted in a closing rate which, if not properly managed, would cause the Orbiter to overtake its target and thus make the rendezvous more difficult and more expensive in terms of propellant. With each posigrade NC burn the semi-major axis (or energy) of the orbit was progressively drawn out closer to the orbit of the target, reducing the range and slowing the closing rate.[14] The X axis distance, or down-track, was most affected by perturbations from water dumps, attitude maneuvers, inadequately modeled drag, and propagation errors, so daily phasing burns were performed in order to control the closing rate. Typically Mission Control would schedule two phasing maneuvers per flight day, one in the crew's morning and one in the evening. If necessary, the flight plan could also accommodate height adjustment maneuvers (designated NH for "n^{th} height correction") to control the Orbiter's height with respect to the target's V-Bar at the Ti maneuver. In fact, an NC burn could also change the Orbiter's height with respect to the target's V-Bar. The difference was that an NC burn was done mainly to change the phase angle between the Orbiter and the target with any change in height being a secondary effect, whilst an NH burn changed the height without affecting the phase.

In some cases, the mission could require a coelliptic maneuver (designated NSR for "n^{th} slow rendezvous burn") in order to divorce the illumination from the phasing. For a target in an approximately circular orbit it might be desirable to go coelliptic with the target so that the phasing (NC) burns could be initiated to achieve specific orbital lighting requirements for the terminal phase rather than to achieve placement of the line of apsides. This maneuver aligned the line of apsides of the active vehicle with that of the target to make the two orbits run practically in parallel and maintain an essentially constant differential height. An NSR would typically be executed at the apogee of the orbit to minimize the radial delta-V component, but it could be made anywhere along the orbit. Later in the program this kind of maneuver was eliminated because proper NC burn timing was able to guarantee the desired lighting conditions.

For ground-up rendezvous, the flight plan included one plane change maneuver

---

[14] As the Orbiter adjusted its orbit ever closer to that of the target, its orbital velocity would more closely match that of the target, thereby reducing the closing rate.

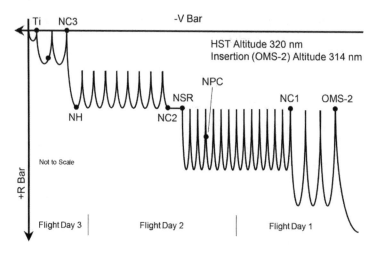

Example of a ground-up rendezvous profile (flown on STS-82).

(designated NPC for "n^{th} plane change") to put the Orbiter into the same orbital plane as the target at a specified time. The need for this maneuver arose from the fact that dispersions during the ascent could result in significant planar errors. In a ground-up rendezvous mission the Orbiter was not inserted directly into the plane of its target but into a so-called "phantom plane" that was designed such that the regression of the line of nodes would rotate the plane of the Orbiter's orbit to match that of its target at the time of intercept. The NPC burn was targeted in such a manner that out-of-plane errors would be completely nulled at some future time (usually Ti) since differential nodal regression ceases to be a factor when the altitudes of the two orbits are similar. Plane changes are extremely expensive in terms of propellant. For example, a change of only one degree requires a delta-V of 400 feet per second. For the Shuttle no more than 17 feet per second of aft propellant was budgeted for the NPC maneuver, so it was essential that the ascent and insertion errors be tightly constrained.

The rendezvous maneuvers described thus far were targeted by Mission Control[15] since the on board navigation software could not determine the parameters for these burns. Prior to a maneuver, the ground would read up the requisite delta-V and the crew would write this data onto a preliminary advisory data (PAD) and feed it into the MNVR EXEC display to be automatically performed at the proper TIG. The first maneuver to be targeted by the on board software was the NCC (for "n^{th} corrective combination") whose role was to locate the Ti burn in the correct place. The Orbiter would compute all subsequent burns because its relative navigation system could provide more accurate state vectors for the Orbiter and the target than those supplied

---

[15] Targeting a burn means defining the changes in velocity (delta-V) required to bring the spacecraft to a given point along its rendezvous trajectory at a given time.

Orbital rendezvous maneuvers: operations

by Mission Control. This was due to ground tracking uncertainties and the relatively close proximity of the two vehicles. It was just too dynamic a period for the ground to compute all the required burns and inform the crew, particularly during the final phase in which midcourse corrections were as close as 10 minutes apart.

To begin computing on board rendezvous solutions, the crew would first enable the REL NAV (SPEC 33) and ORBIT TGT (SPEC 34) displays, both of which were

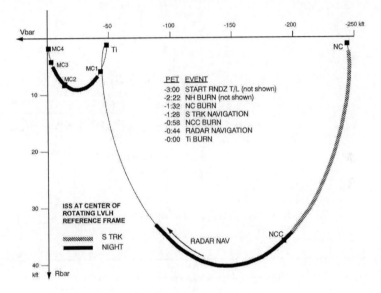

Orbit rendezvous profile.

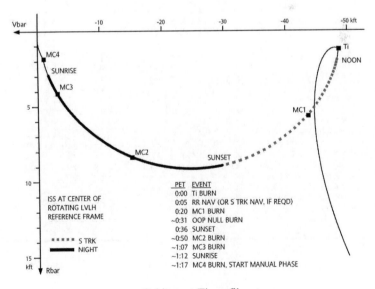

Orbit post-Ti profile.

422 Orbital dancing

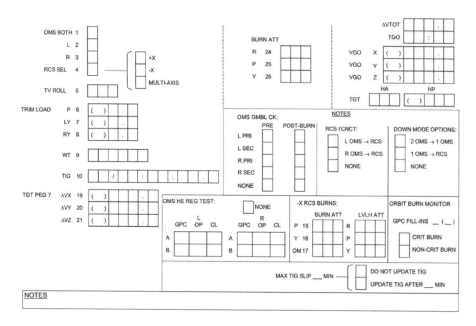

PAD for an NC rendezvous maneuver.

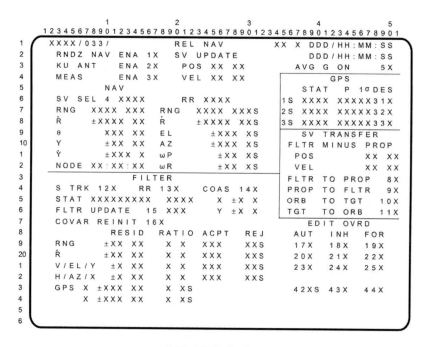

REL NAV display.

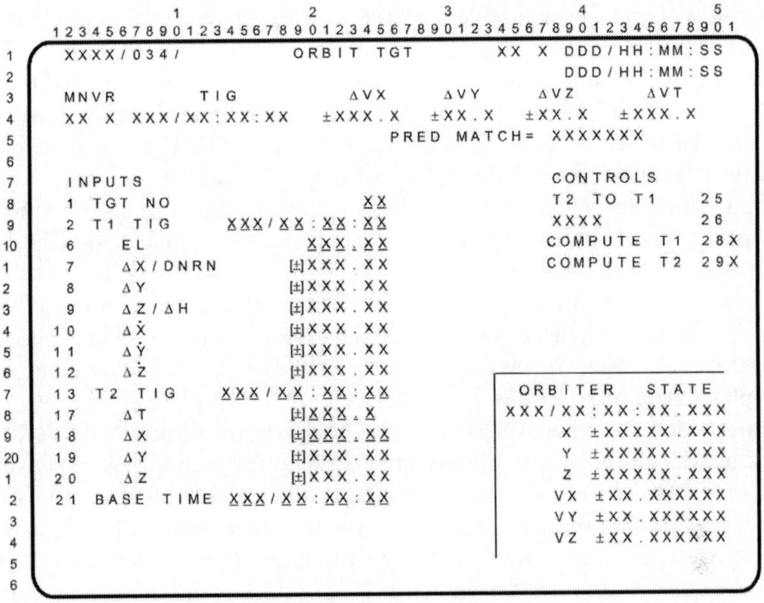

Orbital rendezvous maneuvers: operations   423

```
                      1         2         3         4         5
             1234567890123456789012345678901234567890123456789 0 1
          1    XXXX/034/           ORBIT TGT     XX X DDD/HH:MM:SS
          2                                         DDD/HH:MM:SS
          3    MNVR       TIG        ΔVX     ΔVY     ΔVZ       ΔVT
          4    XX X XXX/XX:XX:XX   ±XXX.X  ±XX.X   ±XX.X   ±XXX.X
          5                        PRED MATCH= XXXXXXX
          6
          7    INPUTS                             CONTROLS
          8    1 TGT NO              XX           T2 TO T1      25
          9    2 T1 TIG      XXX/XX:XX:XX         XXXX          26
         10    6  EL              XXX.XX          COMPUTE T1   28X
          1    7  ΔX/DNRN       [±]XXX.XX         COMPUTE T2   29X
          2    8  ΔY            [±]XXX.XX
          3    9  ΔZ/ΔH         [±]XXX.XX
          4   10  ΔẊ            [±]XXX.XX
          5   11  ΔẎ            [±]XXX.XX
          6   12  ΔŻ            [±]XXX.XX
          7   13 T2 TIG     XXX/XX:XX:XX         ORBITER   STATE
          8   17  ΔT           [±]XXX.X          XXX/XX:XX:XX.XXX
          9   18  ΔX           [±]XXX.XX          X   ±XXXXX.XXX
         20   19  ΔY           [±]XXX.XX          Y   ±XXXXX.XXX
          1   20  ΔZ           [±]XXX.XX          Z   ±XXXXX.XXX
          2   21 BASE TIME   XXX/XX:XX:XX         VX  ±XX.XXXXXX
          3                                       VY  ±XX.XXXXXX
          4                                       VZ  ±XX.XXXXXX
          5
          6
```

ORBIT TGT display.

TARGET NCC BURN [11A]

> **FINAL SOLUTION**
> OPS 202 PRO
> [GNC ORBIT MNVR EXEC]
> √Eng Sel CORRECT

CRT   √SV SEL correct
      [GNC 34 ORBIT TGT]
      TGT NO - ITEM 1 + 9 EXEC
      √TGT Set data:
          T1 TIG = NCC BURN SOLUTION TIG
          EL   + 0
          ΔT   + 57.7
          ΔX   − 48.6
          ΔY   + 0.0
          ΔZ   + 1.2
      COMPUTE T1 - ITEM 28 EXEC
      Record solution in PAD

> **FINAL SOLUTION**
> If > 40 marks in current sensor pass and
> SV UPDATE POS < 0.5 for the last 4 marks:
> | Burn FLTR soln
> If FLTR within ground solution limits:
> | Burn FLTR soln
> If PROP within ground solution limits:
> | Burn PROP soln
> If none of the above:
>   Burn ground soln EXT ΔVs

Instructions for NCC burn from Rendezvous Checklist (STS-135).

GNC specialist functions available for OPS 2. The relative navigation system was used to provide data and controls for rendezvous navigation operations, state vector management, thrust monitoring, and Ku-band antenna enabling. Orbital targeting (ORBIT TGT) was used to predict and select a two-impulse maneuver sequence. At initialization of the display, a parameter called BASE TIME was initialized at the value of the nominal Ti burn TIG as provided by the checklist. The next step was to enable and initiate target tracking, selecting the appropriate option from the UNIV PTG display. At this point, the crew started to use the Rendezvous Checklist for their actions.

The NCC burn was normally scheduled to occur an hour prior to the Ti burn but its computation began half an hour earlier. The Rendezvous Checklist called for the pilot to set it up by using keyboard strokes to recall the targeting data on the ORBIT TGT display (TGT NO – ITEM 1 + 9 EXEC). TGT NO representing the target set number on which the maneuver was based. For crew convenience, the Rendezvous Checklist provided a table with all the targeting data for each target maneuver based on the target altitude. The same data were then reported in the Rendezvous Checklist instructions. Each data set was composed of the time of ignition (T1 TIG, in minutes) for the burn, the elevation angle relative to the target (EL, in degrees), the time of reaching the point for the NCC burn (delta-T, in minutes), and the spatial coordinates (delta-X, delta-Y and delta-Z, in thousands of feet) of the target point (in this case Ti). Once the data had been loaded the computation was executed (COMPUTE T1 – ITEM 28 EXEC) to release the burn solution, which was automatically displayed in the MNVR section of the display and output to guidance for display on the ORBIT MNVR EXEC page in the section for PEG 7 input. It is important to understand that the solution calculated by ORBIT TGT represented the delta velocities in the LVLH frame of reference to reach the target point (in this case Ti) described by the loaded data set. However, this solution was not the definitive one since it was based on the state vector of the Orbiter at the moment of the computation. With about 30 minutes remaining until the burn, the state vector would change considerably with consequent refinement of the burn solution. For this reason, a second solution (intermediate) was calculated in the same way with the same data set. The third (definitive) solution was computed several minutes before the burn was due. The reason for calculating three solutions over a period of about half an hour was to monitor the trend and ensure that the trajectory was not being influenced by dispersions.

The commander would record every burn solution in the PAD for the corrective maneuver to be performed. Prior to execution, the solution calculated on board was compared with a targeted solution calculated by Mission Control. The flight rules stated that the on board solution would be used if at least 40 "navigation marks" had been taken by the present sensor in acquisition (i.e. rendezvous radar or star tracker) and processed by the navigation software with a state vector position update of less than 500 feet for the final marks. The reason for so many sensor marks was to have at least that many updates factored into the state vector in order to ensure the on board solution was more accurate than the ground one. The rule that the final few updated state vectors differ by less than 500 feet was to eliminate transient effects and ensure stability in the filtered state. If neither the rendezvous radar nor the star

Orbital rendezvous maneuvers: operations 425

SPEC 34 ITEM NO	1			6	17	18	19	20	
TGT ALTITUDE	TGT NO	DESCRIPTION	T1 REL TO BASETIME	EL (DEG)	DT (MIN)	DX (KFT)	DY (KFT)	DZ (KFT)	NOTES
130	9	NCC	-0/00:55:48	0	55.8	-48.6	0	+1.2	BASETIME = Ti TIG
	10	Ti	0/00:00:00	0	74.4	-0.9	0	+1.8	
	11	MC1	0/00:20:00	0	54.4	-0.9	0	+1.8	
	12	MC2	0/00:47:24	28.45	27.0	-0.9	0	+1.8	BASETIME = MC2 TIG
	13	MC3	0/00:17:00	0	10.0	-0.9	0	+1.8	
	14	MC4	0/00:27:00	0	13.0	0	0	+0.6	
	19	MC2 ON TIME	0/00:00:00	0	27.0	-0.9	0	+1.8	
150	9	NCC	-0/00:56:18	0	56.3	-48.6	0	+1.2	BASETIME = Ti TIG
	10	Ti	0/00:00:00	0	75.1	-0.9	0	+1.8	
	11	MC1	0/00:20:00	0	55.1	-0.9	0	+1.8	
	12	MC2	0/00:48:06	28.46	27.0	-0.9	0	+1.8	BASETIME = MC2 TIG
	13	MC3	0/00:17:00	0	10.0	-0.9	0	+1.8	
	14	MC4	0/00:27:00	0	13.0	0	0	+0.6	
	19	MC2 ON TIME	0/00:00:00	0	27.0	-0.9	0	+1.8	
170	9	NCC	-0/00:56:48	0	56.8	-48.6	0	+1.2	BASETIME = Ti TIG
	10	Ti	0/00:00:00	0	75.7	-0.9	0	+1.8	
	11	MC1	0/00:20:00	0	55.7	-0.9	0	+1.8	
	12	MC2	0/00:48:42	28.66	27.0	-0.9	0	+1.8	BASETIME = MC2 TIG
	13	MC3	0/00:17:00	0	10.0	-0.9	0	+1.8	
	14	MC4	0/00:27:00	0	13.0	0	0	+0.6	
	19	MC2 ON TIME	0/00:00:00	0	27.0	-0.9	0	+1.8	
190	9	NCC	-0/00:57:12	0	57.2	-48.6	0	+1.2	BASETIME = Ti TIG
	10	Ti	0/00:00:00	0	76.3	-0.9	0	+1.8	
	11	MC1	0/00:20:00	0	56.3	-0.9	0	+1.8	
	12	MC2	0/00:49:18	28.85	27.0	-0.9	0	+1.8	BASETIME = MC2 TIG
	13	MC3	0/00:17:00	0	10.0	-0.9	0	+1.8	
	14	MC4	0/00:27:00	0	13.0	0	0	+0.6	
	19	MC2 ON TIME	0/00:00:00	0	27.0	-0.9	0	+1.8	
210	9	NCC	-0/00:57:42	0	57.7	-48.6	0	+1.2	BASETIME = Ti TIG
	10	Ti	0/00:00:00	0	76.9	-0.9	0	+1.8	
	11	MC1	0/00:20:00	0	56.9	-0.9	0	+1.8	
	12	MC2	0/00:49:54	29.07	27.0	-0.9	0	+1.8	BASETIME = MC2 TIG
	13	MC3	0/00:17:00	0	10.0	-0.9	0	+1.8	
	14	MC4	0/00:27:00	0	13.0	0	0	+0.6	
	19	MC2 ON TIME	0/00:00:00	0	27.0	-0.9	0	+1.8	

Example of targeting data from Rendezvous Checklist (STS-135).

tracker were available to the navigation software to update the state vector this rule would not be applied since manually taking and incorporating that many COAS navigation marks would have been unrealistic. In this case, the burn solution would be calculated by means of filtering (FLTR) or propagation (PROP)[16] using the ground initialization state vector plus any IMU-sensed delta velocities. If either of these solutions agreed with that provided by Mission Control then the on board solution would be accepted, otherwise the maneuver would be performed using the ground solution. With the chosen solution entered into the PEG 7 input section of the MNVR EXEC display the pilot would execute ITEM 22 to compute the attitude of the Orbiter for the burn and the desired change of velocity in the body X axis (VGOx). As dictated by flight procedures, no action would be taken for a VGOx of less than 0.2 feet per second. As no automatic guidance was provided for RCS burns the commander would execute it using the THC, monitoring his progress on the MNVR display and terminating the burn when the maneuver was complete.

Several minutes later, work got underway for the next step in the rendezvous, the Ti burn. This was prepared in the same manner with three solutions computed and the third one selected for execution. The Ti burn initiated the final (transition) phase of the rendezvous. It was a posigrade phasing burn, and after a fuel-optimal trajectory would place the Orbiter at the MC-4 position some 900 feet aft and 1,800 feet below the target. If the need arose, the timeline could be delayed at any time prior to Ti. The most propellant efficient way of delaying a rendezvous was to slow the catch-up rate such that the Orbiter would rendezvous with the target an integral number of orbits later, maintaining the same lighting conditions. This could be achieved by any of the preplanned NC maneuvers, typically without a propellant penalty, or by a maneuver inserted into the profile for that purpose. The best point to efficiently make a phasing adjustment was at apogee so that the perigee could be raised and the rate slowed. The last opportunity to undertake a delay was the Ti burn itself. A "Ti delay" was a viable option if it proved necessary to recover some essential lost systems capability prior to committing to the intercept. For a delay of no more than two orbits this burn would be executed in such a manner as to return

```
TARGET Ti BURN   13A

CRT    √SV SEL correct
       GNC 34  ORBIT TGT
       TGT NO - ITEM 1 + 10 EXEC
       √TGT Set data:
           T1 TIG = BASE TIME
           EL      + 0
           ΔT      + 76.9
           ΔX      - 0.9
           ΔY      + 0
           ΔZ      + 1.8
       COMPUTE T1 - ITEM 28 EXEC
       Record solution in PAD
```

Instructions for Ti burn (from STS-135 Rendezvous Checklist).

---

[16] In a filtered solution, the calculated burn obtained by the ground initialization state vector plus IMU-sensed delta velocity was further refined by means of additional relative navigation data. A propagated solution refined an on board filtered state vector containing relative navigation data from a previous data acquisition.

the Orbiter to the Ti point with the same desired lighting conditions. For a longer delay the burn would create an opening rate in order to eliminate the need for the crew to continuously monitor the target.

As long delays would usually involve at least one crew sleep cycle, if there was sufficient propellant then the delay maneuver would be targeted at a point in the following morning equivalent to that originally planned. This eliminated the prospect of having to awaken the crew to execute a trajectory control maneuver.

The flight rules mandated delaying Ti if a "GO for Ti" had not been provided by Mission Control with 5 minutes remaining to the burn. This call informed the crew that both the Orbiter and the target were ready for the final phase and the subsequent docking/grapple. If this confirmation was not given, the crew did not have any means of determining whether the target was ready. Delaying Ti was considered the most reasonable course of action. If the Ti delay resulted in a loss of timeline of at least one orbit, this provided time to regain communication with Mission Control without committing to proximity operations. In the event of losing communications after the "GO for Ti" confirmation, the burn would be executed because the target was known to be ready. The PEG 7 parameters for the Ti delay would be provided by the ground but owing to the time required to load the data into the MNVR EXEC display and then maneuver to the appropriate attitude this had to be provided at least 5 minutes before the Ti burn. If a Ti delay was invoked and the data for the maneuver was not forthcoming due to a loss of communications, then the flight rules simply called for

```
TARGET MC 2 BURN  18B  (Final)

CRT      √SV SEL correct
         GNC 34  ORBIT TGT
         TGT NO - ITEM 1 + 12 EXEC
         COMPUTE T1 - ITEM 28 EXEC
         √TIG change

         ┌─────────────────────────────────────────────────────────────┐
         │  IF TIG CHANGE < –3 OR > +7 MIN                              │
         │  Set BASE TIME to (Nominal MC 2 TIG –3 or +7 min as appropriate) │
         │  LOAD      - ITEM 26 EXEC                                    │
         │  TGT NO - ITEM 1 + 19 EXEC                                   │
         │                                                              │
         │  √TGT Set data:                                              │
         │       T1 TIG  = BASE TIME                                    │
         │       EL      + 0                                            │
         │       ΔT      + 27.0                                         │
         │       ΔX      – 0.9                                          │
         │       ΔY      + 0                                            │
         │       ΔZ      + 1.8                                          │
         │  COMPUTE T1 - ITEM 28 EXEC                                   │
         └─────────────────────────────────────────────────────────────┘

         Set EVENT TIMER counting to MC 2 TIG
         Record solution in PAD
         GNC 33  REL NAV
CRT      FLTR TO PROP - ITEM 8 EXEC
```

Instructions for MC2 burn (from STS-135 Rendezvous Checklist).

adding 3 feet per second to the X axis component of the delta-V of the final Ti burn solution. This would add more than sufficient energy to the orbit to null the closing rate for all reasonable Ti positions and would give the crew an opportunity to execute an NCC maneuver to refine the trajectory.

After the Ti burn, the flight plan included a series of four midcourse correction maneuvers (MC) to refine the trajectory of the Orbiter by taking advantage of more sensor data to trim out dispersions from the Ti burn. Each burn would be targeted to reach the point to make the next burn. As per procedure, these maneuvers would be executed by computing three solutions and using the final one. Generally speaking, all the MC burns were the same but MC-2 was based upon the elevation angle to the target rather than a calculated burn time. The rationale for this derived from the fact that analyses established that executing the burn with an elevation angle constraint would produce significantly smaller dispersions at the point where the pilots were to take manual control.

Since the elevation angle was set up as a constraint for this burn, to calculate the final solution the pilot had first to recall the data set (*TGT NO – ITEM 1 + 12 EXEC*) and compute the solution (*COMPUTE T1 – ITEM 28 EXEC*). The TIG for the burn calculated along with the delta velocities was compared with the nominal TIG in the Rendezvous Checklist. If it was either 3 minutes earlier than or more than 7 minutes after the nominal TIG, then a TIG CHANGE was performed. This involved setting up the BASE TIME as nominal MC-2 TIG –3 or +7 minutes as appropriate. The data set consisting of the input for the MC-2 burn, the BASE TIME, and a null elevation angle was then loaded (*LOAD – ITEM 26 EXEC*) and computed (*COMPUTE T1 – ITEM 28 EXEC*) to obtain the final definitive solution.

It might have been necessary to perform an out-of-plane or planar null maneuver between MC-1 and MC-2 to place the Orbiter in the same orbital plane as the target. To do this, the pilot would monitor the REL NAV display for when the out-of-plane

---

MANUAL OUT-OF-PLANE NULL [19A]

```
         GNC 33  REL NAV
CRT      When Y = 0:
F7           FLT CNTLR PWR - ON
             DAP: A/AUTO/PRI
             DAP TRANS: as reqd

             THC: Null YDOT

             If –Z AXIS TRACK,
                 +YDOT = FWD THC left
                         AFT THC right
             If –Y S TRK TRACK,
                 +YDOT = FWD THC down
                         AFT THC out

F7           FLT CNTLR PWR - OFF
             DAP: A/AUTO/ALT
             When rates nulled:
                 DAP: VERN(ALT)
```

Instruction for an out-of-plane maneuver (from STS-135 Rendezvous Checklist).

```
TARGET MC 3   19B
CRT    √SV SEL correct
       GNC 34  ORBIT TGT
       TGT NO - ITEM 1 + 13 EXEC
       √TGT Set data:
           T1 TIG = BASE TIME + 0/00:17:00
           EL       + 0
           ΔT       + 10.0
           ΔX       − 0.9
           ΔY       + 0
           ΔZ       + 1.8
       COMPUTE T1 - ITEM 28 EXEC
       Record solution in PAD
```

Instructions for MC3 burn (from STS-135 Rendezvous Checklist).

distance between the two orbits (Y) became zero. To assist in determining the exact moment of this nodal crossing, the NODE time parameter on the REL NAV display provided an estimate of when it would occur. At that point the crew would null the rate (YDOT) by means of RCS inputs from the THC. Depending on the axis used for tracking the target (–Z or –Y) flight procedures gave indications of the directions in which to input the RCS firings to null the YDOT. For this maneuver the DAP had to be in AUTO with the primary RCS jets selected (PRI). After the maneuver, either the vernier (VERN) or (ALT) DAP submode would be chosen.

MC-3 and MC-4 were executed in the same manner. MC-3 targeted the Orbiter to the MC-4 point, which in turn targeted it to coast up to a point 600 feet below the ISS on the +R-Bar. In the nominal case, MC-4 would be mostly a non-zero +X burn. It is worth noting that the NCC, Ti, and MC-1 burns where all scheduled relative to the initial baseline (Ti TIG), which was usually several minutes before orbital noon. This was because the on board software was unable to calculate these times. However for MC-2 to MC-4 the software automatically redefined the baseline (MC-2 TIG). This was evident in the different computations of the T1 TIG for MC-3 and MC-4 which were calculated as the BASE TIME set up for MC-2 plus a time specified in minutes depending upon which burn was to be performed.

For each burn from NCC onwards the flight rules required that it be finished by T1 TIG plus 90 seconds. This constraint was necessary since Lambert solution errors grew as the time between T1 TIG and burn completion increased. In the worst case, after 90 seconds the error would be slightly larger than the trim errors (0.2 feet per second) and would then increase rapidly. Experience showed that rendezvous burns usually had a duration of 30 to 40 seconds. If it looked like a burn would take longer than 90 seconds it would be terminated and retargeted for a new T1 TIG.

After getting the preliminary burn solution for MC-4 the flight plan called for the commander to configure the Orbiter for the imminent R-Bar approach by using the UNIV PTG display to point the +Z axis of the Orbiter towards the center of Earth. In this way, after MC-4 the R-Bar tracking could begin with the +Z axis of the Orbiter maintaining LVLH attitude-hold without regard for the target's inertial attitude. At this point, manual control would be initiated in order to maintain the target within the reticle of the COAS.

430 Orbital dancing

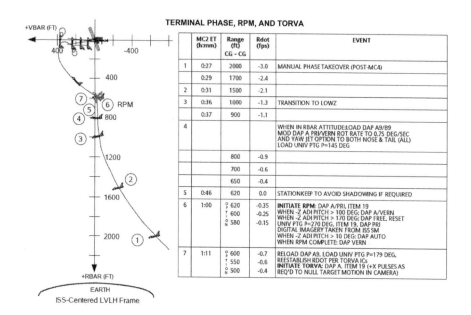

Terminal phase to approach ISS.

Soon after the MC-4 burn the commander would take full manual control of the Orbiter through to docking. Although this burn was designed to place the Orbiter at a point 600 feet below the target on the +R-Bar, dispersions in the sensor data and the navigation software usually meant that some braking burns were required to arrive at the designated point with a relative vertical velocity of a magnitude under 0.25 feet per second. During this coasting phase, the Low Z thruster mode would be activated to avoid plume impingement. On arrival at the desired point, and after confirmation from Mission Control, the Orbiter could initiate the R-Bar pitch maneuver for TPS photography by astronauts on the ISS. This maneuver was started by setting up in the UNIV PTG display a body vector which was rotated +145 degrees in pitch from the +X axis of the Orbiter and executing the tracking option targeting the body vector to point at the center of Earth. In this way, the Orbiter would automatically start to turn around the Y axis in order to point the body vector towards the selected target. About half way through the rotation, with the belly of the Orbiter facing the target, the crew would set up a new body vector, this time equal to the +Z axis of the Orbiter, which was to be pointed towards the center of Earth. With this new body vector set up, the Orbiter could continue the rotation until the +Z axis pointed at the center of Earth, thereby reinstating the attitude it had at the start of the maneuver. It is worth noting that due to the dynamics of the maneuver and orbital mechanics, the maneuver did not follow a circular path but rather an ellipsoidal profile which terminated with the Orbiter in the same attitude, but slightly behind and below the position it held when the maneuver started.

With ground authorization to proceed, the approach to the ISS could be initiated.

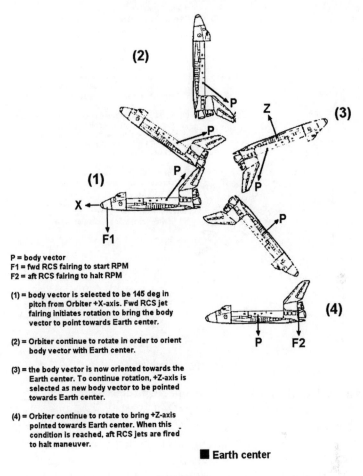

RPM logic.

Firstly, a new body vector had to be set, this time equal to the −X axis of the Orbiter. Again maintaining as a target the center of Earth, this made the Orbiter rotate about the Y axis. During this rotation the commander would apply +X translational inputs so the Orbiter could climb to the V-Bar of the ISS. The combination of rotation and translations would drive the Orbiter in a curve which had a radius of at least 250 feet relative to the center of gravity of the ISS.[17] Generally, the commander had to use heuristic knowledge of orbital mechanics to "bend" the curve for a smooth arrival at the V-Bar, about 400 feet ahead of the front docking port of the ISS. This maneuver was fully manual in terms of translation but automatic in

---

[17] It is important to understand that while the translational commands were manually given by the commander, the rotational motion (which resulted in RCS jet firings) was accomplished automatically in accordance with the DAP settings and UNIV PTG display.

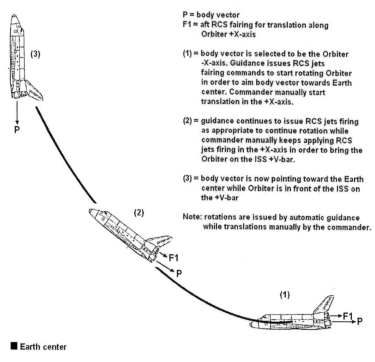

V-Bar approach logic.

rotation as a result of the DAP and UNIV PGT display settings. The precise point of arrival on the +V-Bar depended upon the skill of the commander. On crossing the V-Bar, the commander would execute a –X translation to null the upward motion. This initiated the natural station-keeping resulting from the fact that the centers of gravity of the two vehicles were at essentially the same altitude. To resume the approach, the commander would perform a –Z translation to set up the desired closing rate. As the Orbiter was closing towards the ISS during this phase its motion was retrograde relative to the ISS and indeed its own orbital motion. This caused the Orbiter to slow down, lose the energy required to maintain that orbital altitude, and descend with respect to the ISS. To counteract this tendency the commander had to execute +X translations, producing a series of small "hops" on the V-Bar whose amplitudes were designed to maintain the docking target mounted on the module on the front of the ISS on the centerline of the Orbiter's docking camera and within an approach corridor of 5 degrees. At a range of 30 feet, the commander had to report to the ISS and to Mission Control that the final approach was underway at a rate of no more one-tenth of a foot per second.

Orbiter docking system

Although one of the earliest Orbiter configurations studied sported a docking device, the chosen configuration lacked a means of physically docking with a space station.

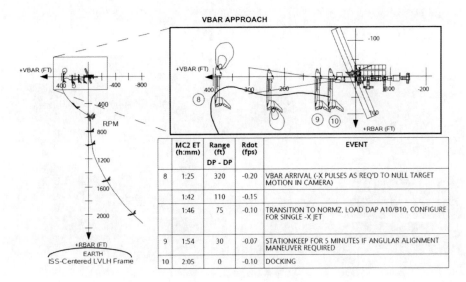

V-Bar approach to the ISS.

When the joint program with the Russian Mir space station started as a prelude to the assembly of the International Space Station, the Orbiters which flew these missions were fitted with the so-called Orbiter docking system.[18] This consisted of an external airlock, a supporting truss structure, a docking base, the avionics required to operate the system, and an androgynous peripheral docking system (APDS) developed by the Russians to achieve the capture, dynamic attenuation, alignment and hard docking by using essentially identical docking mechanisms on each vehicle. Whilst the docking mechanism on the Orbiter was intended to be active, that on the target was typically passive. The active part of the system consisted of a structural base ring that housed a dozen pairs of structural hooks (one active and one passive), an extendable guide (active) ring with three petals, a motor-driven capture latch within each guide petal, three ball screw/nut pairs connected using a common linkage, six electromagnetic brakes (dampers), and five fixer mechanisms that permitted motion of the active ring only on the Z axis.

During launch the active docking ring was in the retracted position, the structural hooks were open, and the capture latches on the petal were closed. In preparation for docking, on the second flight day the ring was extended to its ready-to-dock position. As the ring was being driven by two direct current electrical motors at a rate of about 4.3 inches per minute, the fixers were activated to keep the ring aligned. One second after the ring achieved its ready-to-dock position, the ring motors and the fixers were automatically commanded off and the system was powered down.

---

[18] Because *Columbia* never flew missions to either Mir or the ISS it was not retrofitted in this manner.

434 Orbital dancing

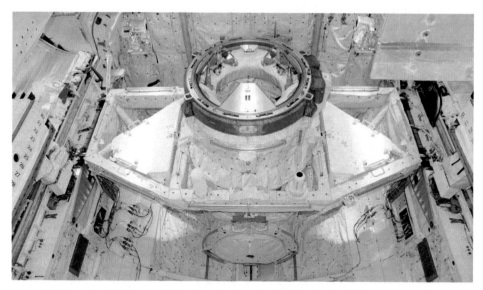

Orbiter docking system in its installed configuration in the Orbiter cargo bay.

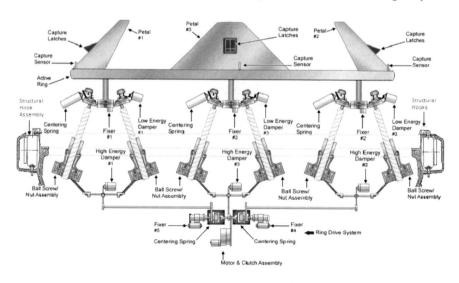

Orbiter docking system schematics.

Once the Orbiter was stationed on the V-Bar (or R-Bar) for final approach to the space station the docking system was powered back on. During this time it was also necessary to close the inner airlock hatch of the Orbiter, verify the function of the airlock fan, and switch on the docking lights. A television camera mounted inside the docking system that viewed a target at the center of the station's system provided the crew with a visual indication of the progress of the approach. All the commander had to do was to maintain the docking camera centered on the target.

Docking ring in retracted position.

Docking ring in extended position.

Docking started when the maneuvering Orbiter brought the petals of the active mechanism into contact with those of the passive mechanism. Minor misalignments of the two mechanisms were automatically corrected as the interface on the Orbiter was displaced and rotated in response to the relative velocities of the vehicles so that the capture ring could latch onto the opposing androgynous interface ring. Docking

436 **Orbital dancing**

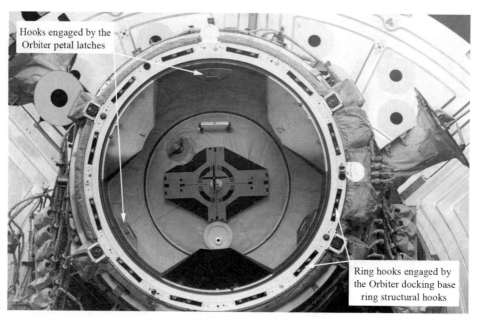

Docking target on the PMA-2 module.

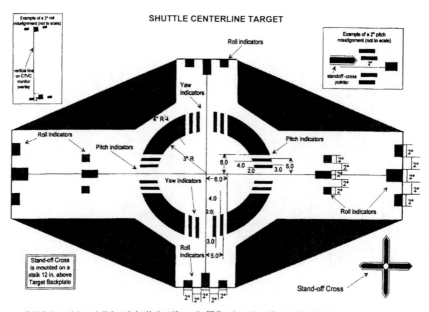

Detail of docking target.

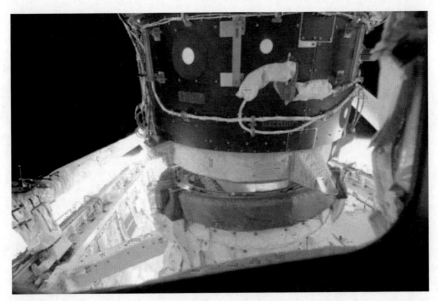

After capture, hard dock was delayed until relative motion between the Orbiter and the ISS was damped and alignment achieved. The picture shows an example of the large angular misalignment between the two docking mechanism after capture.

of two vehicles as massive as the Orbiter and Mir (later ISS) was complicated by the significant offset of the Orbiter's center of gravity from the longitudinal axis of the docking mechanism. This made capture more difficult by reducing the effective mass of the active vehicle at the docking interface. To remedy this problem, at contact the crew would use the preset post-contact thrusting (PCT) sequence to deliver the force required to achieve capture without exceeding dynamic load limits. To enhance the likelihood of success this sequence had to be initiated within two seconds of initial contact but not once capture had been achieved, in order to prevent placing excessive loads on the mechanism. If PCT was successful, capture would occur with the latches on the three petals of the active ring engaging those on the passive ring. Five seconds later the dampers were automatically energized to dampen out any relative motions. As both vehicles were in a state of free drift, this would effectively attenuate relative loads, rigidize the platform, and draw the vehicles into a fully docked configuration. At this point the dynamics of the linked vehicles were such that a pendulum motion occurred about the overall center of gravity. This motion was due to two causes. One was that the docking interface was not on the line joining the centers of gravity of the two vehicles. Also, the gravity gradient tended to rotate the asymmetrical stack into a stable attitude. Thirty minutes were allocated to allow this dampening to complete. It is important to understand the difference between dampening the relative motions of the two vehicles soon after capture and this pendulum-like dampening in the gravity gradient. In the first case the translational motion of the Orbiter towards the station had to be dampened in order to prevent the slowly moving spacecraft from "crashing" against the station. Then the relative motion between the vehicles

438 Orbital dancing

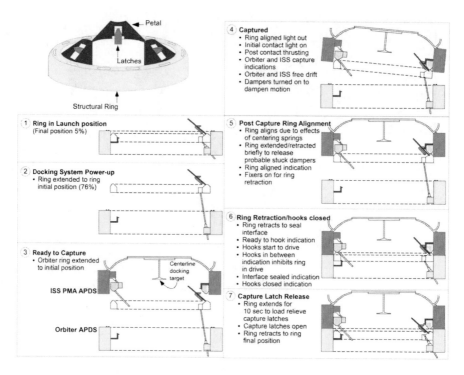

Orbiter docking sequence. (Courtesy of www.nasaspaceflight.com)

continued as a result of the pendulum effect, causing a periodic misalignment of the docking mechanisms. The two systems had to be fully aligned with each other for a successful hard docking, which is why it was necessary to wait for the pendulum motion to end.

When the two docking rings were aligned, the docking sequence could resume by retracting the active ring in order to draw the Orbiter in. During this phase the five fixers made the capture ring rigid and prevented relative vehicle misalignments from accumulating during retraction. Structural attachment occurred on completion of the capture ring retraction, with the 12 active hooks engaging their passive counterparts. As the hooks drove closed, the mating surfaces would compress the pressure seals to create an airtight vestibule through which the crew could pass between vehicles. As a backup procedure the active hooks on the docking mechanism of the station could be used to latch onto the passive hooks of the Orbiter. Once the hooks were closed, the docking sequence called for relieving the loads in the capture ring. This involved re-extending the ring for ten seconds, opening the capture latches and finally retracting the ring. With the docking process complete, the system was powered down.

Prior to undocking, the external and internal airlock hatches were closed and the station verified ready for undocking. This included releasing any hooks located in the docking ring of the station present to provide a structural margin. The docking lights and the television camera would then be activated, the airlock vestibule depressurized and a leak check performed. Then the docking system would be powered up. Undocking occurred by commanding the hooks to open. As they released, four spring plungers compressed between the mating surfaces with a combined force of about 700 pounds would impart a small opening rate. The Orbiter would then perform a separation burn and power down its docking system.

Trajectory control sensors

The successful execution of a rendezvous profile and ensuing proximity operations depended not only on careful pre-mission planning but also on an array of trajectory control sensors which enabled the crew to maintain situational awareness through to either grappling a satellite or docking with a station.

The primary control sensors available on board were the rendezvous radar and the star trackers, assisted by the COAS and payload bay CCTV cameras as backup. Due to budgetary constraints the rendezvous radar was merged with the Ku-band antenna of the communications system, hence during rendezvous and proximity operations only one function at a time could be used. The radar was also built with the capability of tracking a target which had a transponder out to a range of 300 nautical miles and skin tracking a passive target to a specified maximum of 22 nautical miles. It would have been capable of starting skin tracking at 27 nautical miles but the 22 mile limit was imposed in order to prevent a recurrence of the "eclipsing" problem encountered by STS-32 during its rendezvous with the LDEF satellite. This anomaly was caused by the Doppler effect which, owing to the distance, changed the pulse frequency and resulted in erroneous range readings. When functioning correctly, the radar measured the azimuth, elevation, range, and range rate of the target.

Star trackers were typically used to determine the attitude of the Orbiter in inertial attitude so as to realign the IMU platforms, but during rendezvous operations they provided the relative angle line-of-sight between the Orbiter and its target. They were activated in "target track mode", which was enabled using the S TRK/COAS CNTL (SPEC 22) display. Before commanding the selected star tracker to begin tracking a target, the Orbiter had to be pointed towards the target. Once in the correct attitude the tracking mode would be selected and the star tracker would start to search for the target in its field of view. Once acquired, it would continue to follow the target and provide accurate horizontal and vertical angles.

The data provided by the radar and the star tracker were used by the rendezvous navigation software to maintain an accurate estimate of the state vectors of both the Orbiter and its target. This was enabled very early in the rendezvous timeline using the REL NAV (SPEC 33) display, and used the Super-G algorithm to propagate the state vectors. To propagate the target state vector, Mission Control had to uplink an updated state vector no later than 15 hours prior to enabling the software. As soon as the radar and star trackers began to provide useful data, it was made available to the rendezvous navigation software to generate a more accurate (filtered) solution of the propagated Orbiter state vector. This milestone also marked the moment in the Rendezvous Checklist at which to start executing on board targeted burns because the state vector of the Orbiter was more accurate than that computed by the ground.

Radar data could not be used at ranges closer than 80 feet, since signals reflected off different locations on the target could give rise to noisy data and therefore highly inaccurate range and range rate measurements. For the final phase of proximity operations the crew had to resort to visual ranging methods such as subtended angles in the COAS or tilt angles using pairs of CCTV cameras. Not only did this impose an increased workload on the crew during an already busy and delicate phase of the mission, these techniques were imprecise and did not provide a direct range rate measurement. Furthermore, on-orbit use of CCTV was often impaired by lighting conditions that "bloomed" the images.

Ironically, the fact that the highly skilled Gemini and Apollo crews successfully conducted each rendezvous mission in the first decade of the space program had the effect of delaying the search for a backup means of measuring range and range rate. The baseline sensors of the Orbiter were successfully used to null the final approach velocity to zero in order to facilitate grappling, but no sensors existed for achieving a specific *non-zero* approach velocity with precision. This issue was far from trivial, since studies for docking with Space Station Freedom started as early as 1987 and docking with a station was altogether different from grappling a satellite. Whereas it was sufficient for the Orbiter to station-keep somewhere in the vicinity of a target for grappling, to perform a docking it was necessary to achieve a specific velocity at a specific time. If Freedom had been built, then it would have renewed the requirement for an on board method of measuring range and range rate close in, both as backup to the radar and to ease the workload on the crew. In order to save money, off-the-shelf commercially available sensors were evaluated by several missions and then flown as standard equipment for rendezvous and proximity operations.

Starting with the first docking mission to Mir, the crews on rendezvous missions have been able to make use of two light detection and ranging (LiDAR) devices: a manual hand-held LiDAR (HHL) and a trajectory control system (TCS). The HHL was tested for the first time by STS-49 and was similar to a hand-held police radar. On being manually aimed out the window of the Orbiter it provided a line-of-sight range by measuring the round trip light time to bounce a laser pulse off the target. It was able to provide range readings accurate to 6 inches from as near as 12 feet out to 1,500 feet, or to 5,000 feet for a target equipped with a retro-reflector. The TCS was tested by STS-51, STS-63 and STS-64. It was mounted on the payload bay sill and could automatically locate and track a target that was equipped with an optical retro-reflector to provide line-of-sight range, range rate, azimuth and elevation data. It was particularly useful in a docking operation because it could provide accurate relative state information in the final approach phase. For a target as large as a space station, the rendezvous radar became highly degraded at ranges less than 600 feet because its beam tended to wander across the structure. In contrast the TCS provided accurate measurements in close. To use the data provided by the HHL and TCS effectively, a software package called the rendezvous and proximity operations program (RPOP) was developed. It took data from the rendezvous navigation software and from the LiDAR sensors and plotted on a standalone laptop the current motion of the Orbiter relative to the station, providing cues to enable the crew to execute the approach and docking precisely. The HHL, TCS and RPOP jointly became known as "rendezvous tools". The crew had available both the rendezvous navigation software with its state vector filtered using the data supplied by the radar and star trackers, and the RPOP software with its own state vector. Although the rendezvous tools could be used even prior to performing the MC-3 burn they were nominally utilized from the start of the manual approach. The radar data often remained available to the crew as the Orbiter arrived on the +V-Bar, but at some point during proximity operations the Ku-band antenna was switched from radar mode to communications mode in order to transmit video to Mission Control.

To give the crew additional visual cues in the very last part of the approach, they could apply transparent overlay rulers on at least two CCTV screens on the aft flight deck to indicate the alignment of the approach corridor and the distance to docking system contact.

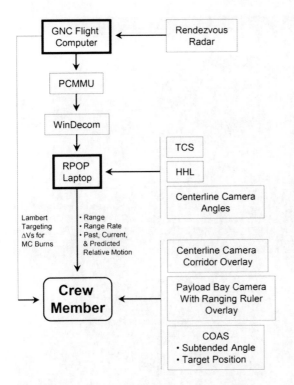

Simplified GNC and rendezvous tools architecture: RPOP, TCS, HHL and cameras provided piloting cues after the last targeted burn.

In the event of a radar failure (as occurred on STS-92 and STS-131) TCS, HHL and COAS subtended angles could be used earlier in the profile than was usual. It is worth noting that in order to save money the rendezvous tools were not certified to the same criticality level as the GNC software in the 1970s. Crews were trained to fly proximity operations, dockings and undockings without the use of rendezvous tools.

Undocking and separation

To complete a docked mission, the hatches between the vehicles would be closed and leak checks carried out. After powering up and checking the docking mechanism and loading the DAP with the proper parameters, the command would be sent to release the docking mechanism. At physical separation, the commander of the Orbiter would verbally communicate this fact to Mission Control and the station. In order to prevent inadvertent firings during the initial separation, the pre-undocking procedures called for placing the DAP in its free mode. It would be placed in LVLH mode after the docking devices had disengaged. On the aft flight deck the pilot would control the

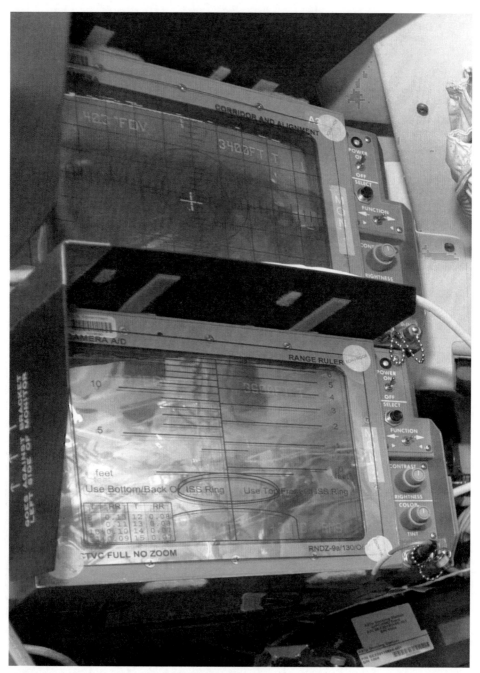

Camera ranging ruler overlay (bottom) and corridor alignment overlay (top) on two aft flight-deck CRT monitors during rendezvous.

FLYAROUND 9A

√Flyaround terminate criteria per 4A

```
*      If Breakout required during flyaround           *
*      Go to SHUTTLE NOSE IN-PLANE BREAKOUT            *
*      (CONTINGENCY OPS), 5-16 >>                      *
```

1. √DAP: A/AUTO/VERN(PRI)

2. Flyaround start from +Vbar
   GNC UNIV PTG
   TGT ID   √+ 2
   BODY VECT   √+ 5
   P   + 90 (-RBAR)
   Y   √+ 0
   OM   √+ 0
   √ERR TOT – ITEM 23 (*)
   TRK – ITEM 19 EXEC (CUR - *)

   THC: Maintain flyaround range of 650 ± 50 ft (CG-CG)
   Maintain ISS CG inside ± 15 degree vertical and
   ± 20 degrees horizontal on C/L camera

3. Prior to –Rbar crossing (Aft ADI P = 270):
   GNC UNIV PTG
   P   + 0   (–VBAR)
   TRK - ITEM 19 EXEC (CUR - *)

4. Prior to –Vbar crossing (Aft ADI P = 0):
   GNC UNIV PTG
   P   + 282 (+RBAR)
   TRK - ITEM 19 EXEC (CUR - *)

5. At flyaround completion – 10 minutes:
   If radar not tracking target:
   INITIAL RADAR ACQ 10A

6. When flyaround complete (in -Vbar attitude),
   Go to SEP BURNS 8B

Fly-around procedure after undocking (from Rendezvous Checklist).

withdrawal of the Orbiter, remaining within a tight corridor of 8 degrees. One minute after physical separation the VERN or ALT DAP RCS mode would be selected and +Z pulses fired at 10 second intervals to develop an initial opening rate of 0.15 feet per second. When the separation was 30 feet this would be increased to 0.2 feet per second. The Orbiter would withdraw to a distance of about 600 feet, always holding the docking mechanism centered on the docking target of the station. At that point a fly-around could begin. This involved setting up different body vectors to track the center of Earth as the maneuver progressed.

Let us assume that the Orbiter withdrew to a point on the +V-Bar of the station. The first body vector to set up would be the –Z axis, which in the undocking attitude was pointing at the station. UNIV PTG was used to realign the –Z axis towards the center of Earth. The maneuver would start with an RCS firing on the +X axis that would cause the Orbiter to move vertically relative to the station. With the rotation rate

444 Orbital dancing

UNDOCKING, STATIONKEEPING, TORF, AND FINAL SEPARATION

	UNDOCK ET (h:mm)	RANGE (ft) DP-DP	EVENT
	-0:03	0	ORBITER AND ISS IN FREE DRIFT TO BEGIN UNHOOKING (ISS LVLH PYR 0, 0, 0, ATTITUDE)
1	0:00	0	UNDOCKING; DAP B/ALT; MODE TO LVLH; MAINTAIN CORRIDOR
	0:01	2	SELECT VERNS; PERFORM DAP B +Z NORMZ BURNS AT 10 SEC INTERVALS TO BUILD OPENING RATE TO 0.15 FT/S
	>0:03	>30	DAP B +Z NORMZ BURNS AT 10 SEC INTERVALS TO BUILD OPENING RATE TO 0.3 FT/S
2	0:05	50	RE-SELECT -X JETS
		75	TRANSITION TO LOWZ
3	0:32	>600 (CG-CG)	ISS BEGINS 90 DEG YAW TO +/-YVV ORBITER MODES TO AUTO AND BEGINS STATIONKEEPING BETWEEN 600 FT AND 700 FT
4	0:59	~650 (CG-CG)	BEGIN 1/2 LAP TORF BETWEEN 600 FT AND 700 FT
5	1:22		SEP 1: 1.5 FT/S +X RADIAL DOWN BURN
6	1:50	>6000 (CG-CG)	SEP 2: 7 FT/S -X RETROGRADE BURN
7	NEXT DAY		SEP 3: 7 FT/S SINGLE OMS RETROGRADE BURN

Planned separation profile and events (STS-135).

set up by the tracking mode, the pilot would perform as many translational burns as necessary to hold the center of gravity of the station inside +/–15 degrees vertical and +/–20 degrees horizontal of the docking camera centerline. The combination of rotation and translations would yield a curved trajectory. With the Orbiter ascending, its center of gravity would be higher than that of the station causing it to slow down relative to the station. On crossing the –R-Bar, the body vector was pointed towards the center of Earth. To continue the fly-around, the body vector was changed to the +X body axis of the Orbiter, which would reach the desired orientation towards Earth at the –V-Bar crossing. Following the same conceptual scheme, depending upon the mission profile, the Orbiter would continue to loop around the station and then either depart after crossing +R-Bar or continue the fly-around for an additional 90 degrees and depart after crossing the +V-Bar. In either case, a series of burns would thereafter be performed to produce an increasing opening rate from the station.

ORBITAL FLIGHT RULES

On-orbit, the Shuttle was not only subjected to the rules of orbital mechanics but also to literally thousands of flight rules designed to produce a successful mission and, if a failure situation developed, a safe re-entry. It would be impossible to explain all these rules, many of which require in-depth knowledge of the systems. Here we will limit the discussion to the rules that describe the duration of a mission and the attitudes to adopt in flight.

As regards mission duration, the flight rules dictated that each mission had to last at least five days to enable the crew to perform their primary objectives and to have sufficient time to recover from any SAS arising from being weightless.[19] The rules also specified that there must be sufficient on board consumables to assure a two day extension if the primary landing site was unavailable due to adverse weather or if an Orbiter system developed a fault that would compromise a safe re-entry. Although an alternative landing site would always be available, landing at the primary site (i.e. the Kennedy Space Center) was the preferred solution for optimum post-landing ground processing. In the event of a problem with one of the on board systems, the additional time in orbit would allow the crew and Mission Control to investigate the issue and determine the best re-entry configuration. As occurred many times, management of the consumables on board the Orbiter enabled a mission to be extended for one or two days to address additional objectives. Even in these cases, the two day extension had to be protected. This meant that if the mission was extended for one additional day of activities, there had to be sufficient resources for three days: one day plus two days of extension.

For missions involving unstowing/stowing of payloads, the flight rules allowed a payload to be operated until the day prior to that of the nominal re-entry. If a

---

[19] Space Adaptation Syndrome (SAS), popularly known as "space sickness", is a form a motion sickness experienced by astronauts as their vestibular systems attempt to adapt to the onset of weightlessness.

problem was encountered in stowing and/or securing the payload, the crew could conduct a contingency EVA on the nominal day of re-entry and that would be postponed to the first extension day. Since the Orbiter was provided with a high level of redundancy the flight rules permitted a mission to continue after a single failure in any system, as that system would still have single-fault tolerance. In the case of failures resulting in less than fail-safe redundancy, the flight rules called for a real-time assessment of the failure mode and the remaining capability, health, and history of the system in order to eliminate generic failures, common ground processing errors, etc. Based on this assessment, actions would be taken either for a safe mission continuation or for a safe early re-entry. If a system that was critical for re-entry failed and put the Orbiter in a single-fault tolerant state, the flight rules allowed the mission to continue for at least 72 hours in order to accomplish deployment of a primary payload and to ensure crew recovery from SAS. For an Orbiter placed in single-fault tolerant condition as a result of multiple failures of several systems that were critical for re-entry, the flight rules instead urged an early termination because multiple failures were indicative of a generic failure. In the case of failures that placed the Orbiter in a zero-fault condition, termination of the mission would result in a deorbit maneuver at the earliest practical time for a landing on US territory. The flight rules also had the option of remaining in space despite violation of GO/NO-GO criteria for fight continuation in order to allow time to better understand the failure, provide adequate time for the crew to rest, and provide time to accomplish activities which would enhance the re-entry/landing conditions.

Despite the vastness of space, on several occasions debris avoidance maneuvers were performed in order not to get too close to a passing item of "space junk". In this scenario the rules allowed Mission Control to trade off the estimated risk of collision against the operational consequences of such a maneuver. If the risk of collision was deemed to be very high the rules allowed for curtailing an EVA, for awakening the crew to take action, and for eroding the margin of consumables. Normally, however, such maneuvers were minor (typically 30 to 60 feet per second) and could readily be added to the timeline without any impact on the mission. Any avoidance maneuver had to be executed in such a manner as to ensure that another maneuver for the same object would not become necessary later in the mission.

The vast majority of "orbital debris" consists of tiny objects that are impossible to track from the ground. For this reason, it is not unusual for a spacecraft, especially in low Earth orbit, to be hit by these small but highly energetic bullets. To limit as much as possible any damage to an Orbiter, and in particular to its radiators and windows, the flight rules recognized the worst-case attitudes to be either facing the payload bay in the direction of travel or nose forward with the payload bay facing up or out of plane. The best attitude to minimize debris damage was tail forward and payload bay facing down. The flight rules recommended keeping the Orbiter in an intermediate attitude in order to minimize as much as possible the time spent in the worst attitudes. And to further minimize the risk of damage the rules stated that any given attitude not be held for a period exceeding 48 hours. However, this rule was interpreted loosely because some mission objectives required the Orbiter to maintain a specific attitude for longer than that duration.

13

Returning home

FALLING FROM ORBIT

On-orbit checkout

Sad moments are rare during space missions, but if one must be found, probably this would the time to return home. Although after a few weeks in space an astronaut is eager to see their family again, leaving the surreal and magical environment of space has saddened each man and woman to have had the privilege of flying in orbit.

For astronauts on board the Shuttle, preparations to return home started with the on-orbit checkout operations during their last full day in space. The flight software was placed in major mode 801 in order to perform a two-part system checkout lasting 15 minutes to verify the readiness of the Orbiter for re-entry. The first part required the use of one of the auxiliary power unit/hydraulic systems with the particular unit being specified by Mission Control, and involved gimbaling the left and right main engines into their re-entry positions and cycling the aerosurfaces, hydraulic motors and hydraulic switching valves. On finishing these actions, the auxiliary power unit was deactivated. For the second part of the checkout the Orbiter was placed in free drift and the crew started to test the accelerometer assemblies, the microwave landing system, the TACAN, the radar altimeter, the rate gyro assemblies, and the air data transducer assemblies. Any sensor that failed the test would be deselected for re-entry. Next, the crew would carry out procedures intended to verify all flight control system switch contacts and all switches critical to re-entry. Then there was a test of the nose wheel steering system and procedures to verify proper operation of the controllers and their switches. Finally, the head-up displays would be tested and adjusted for re-entry. On finishing the checkout, the flight software would be returned to OPS 2 for nominal orbital flight.

The flight plan for the last full day in orbit also called for stowage and cleaning of the crew cabin because during re-entry no one wanted to be hit by an item that was not properly fixed in place! A hot-fire test was undertaken to verify the operation of seldom used RCS jets, thus shortening the ground turnaround time. The Orbiter was placed in free drift and the RCS jets fired one by one. It was then put in inertial hold to dampen the rates prior to adopting a specific attitude. The crew would spend the remainder of the day leisurely enjoying the magic of microgravity and the view of the planet below.

Deorbit burn

In order to re-enter the Earth's atmosphere a satellite must lose some of the energy which made its orbit stable. Generally speaking, an object in low orbit travels at an average speed of 17,500 miles per hour, generating the centrifugal force required to counteract the pull of gravity and enable it to fall around the world endlessly.[1] To re-enter, it is necessary to lose some of this energy so that gravity can prevail and draw the object down until the ever increasing drag of the upper atmosphere captures it.

On a Shuttle, the final preparations for re-entry started around four hours prior to the nominal time for the deorbit burn by the crew starting the Entry Checklist. The first two and a half hours were spent transitioning the Orbiter from a spaceship to an aircraft. For example, the Ku-band antenna used for communication and rendezvous activities was stowed inside the payload bay. Any radiators that had been deployed during the mission were stowed. The hydraulics lines of the thrust vectoring system of the main engines were repressurized to enable the engine bells to be stowed to prevent thermal damage. The doors of the payload bay were closed and its lighting turned off. GPCs 1 to 4 were placed in PASS OPS 3 (the operational sequence for re-entry) with the fifth GPC running BFS OPS 3. The star trackers were deactivated and their doors closed. And finally, all members of the crew participated in configuring and verifying switches for re-entry.

With one and half hours remaining to the deorbit burn, the commander and pilot would be in their seats to receive a voice read-up from Mission Control of the data required for that burn, along with data to be used in case of engine failures during the maneuver. This data was initially hand written into the deorbit entry landing preliminary advisory data (DEL PAD), which was a page in the Entry Checklist used by the pilots later when the "targeting" was entered into the MNVR display. This was for a PEG 4 navigation scheme because the deorbit burn was intended to bring the Orbiter to a specific altitude called the entry interface (EI) at a given up-range distance from the selected landing site and with a specific relationship between horizontal velocity and vertical velocity. Mission Control also gave supplemental data on the remaining OMS propellant quantities. This was written into the OMS PLPRT PAD of the Entry Checklist, which would serve as a quick reference to understand what action to take based on engine failures occurring during the burn. And Mission Control updated the state vectors of both the PASS and BFS by direct uplink without any participation of the crew.

---

[1] This is true only with the assumption that there are not effects such as atmospheric drag or gravitational perturbations which, if not countered in a timely manner, can reduce the perigee altitude over time and cause the object to re-enter the atmosphere.

In terms of execution, the deorbit burn was similar to other OMS burns during the mission but in this case more attention was devoted to preparations to deal with any failures that might occur. Unlike other orbital maneuvers, the deorbit burn included a point of no return that committed the Orbiter to re-enter the atmosphere even if it was stricken by a serious failure. An engine failure could cause the Orbiter to arrive at the entry interface at a time or in a condition that would jeopardize its safe return home, so the crew had to be able to rapidly reconfigure the OMS and attempt to achieve a safe re-entry. Forty minutes before the deorbit burn the Entry Checklist called for an OMS engine gimbal test to verify that the thrust vectoring system was working. Afterwards the pilots chose a configuration for the OMS burn that took into account any failures to the OMS system earlier in the mission. Ideally the deorbit burn was made in such a manner that equal amounts of the propellant were burned from each OMS pod, using both engines. This was to avoid shifting the center of gravity of the Orbiter along the Y axis past the permitted limit, and to guarantee that the remaining propellant was within the structural limits of the OMS tanks at touchdown. For off-nominal situations, three other configurations were available. One foresaw the use of just one OMS engine fed by propellants drawn from both pods. Another involved all of the aft RCS jets fed exclusively by OMS propellant in order to preserve the RCS propellant to carry out maneuvers during re-entry. The third option would have been implemented if a failure of either the oxidizer or fuel system of one of the OMS pods caused that engine to be classified failed. In this case the burn would be performed with the other engine by drawing oxidizer from one pod and fuel from the other. If necessary, the system could be manually reconfigured later in the burn to supply the working engine from its own pod in order to control the center of gravity on the Y axis. This configuration was referred to as "mixed cross-feed". OMS burns during a mission were normally calculated to guarantee a balancing of the propellant loads and prevent an excursion of the center of gravity on the Y axis. If for some reasons this balancing was not achieved, it would be possible to remedy it during a nominal deorbit burn by starting to feed both engines from one pod and then feeding each engine from its own pod after the imbalance had been eliminated. This configuration was referred to as "unbalanced propellant". Generally speaking, the rationale for these OMS and RCS plumbing configurations was to arrange the propellant lines in a way that would require the least number of changes in valves/switches if a problem were to develop during the burn.

With 25 minutes remaining to the deorbit burn the target data that was written on the DEL PAD was entered into the MNVR display using the DPS keyboard. Then the flight software was switched from major mode 301 to major mode 302, the only one in which the burn could take place. The GO/NO-GO authorization from Mission Control would be issued 5 minutes later. If the decision was to proceed, the Orbiter would adopt the burn attitude either automatically by AUTO DAP or manually by the RHC. As the burn was a retrograde one, the vehicle adopted an inertial, tail-first, belly-up attitude.

With the tension growing on the flight deck, with just 5 minutes remaining one of the three APU was started and kept in low pressure. In the event of the failure of an

DEORBIT BURN FLIGHT RULES
ONE-ORBIT LATE AVAILABLE ENT-1a/135/D/A

	FAILURE	PRE TIG Delay (max) One Orbit	PRE TIG Delay (max) One Day	POST TIG Stop Burn, > Safe HP	
	APU/HYD				
1	No APU operating	...X...			
	DPS				
2	RDNT fail, Split		...X...	X	
3	1 GPC	...X...			
4	BFS		...X...		
5	GPC BITE (Multiple GPCs)		...X...		
	ECLS				
6	2 Av Bay Fans in Bay 3	...X...			
7	2 Av Bay Fans in Bay 1 or 2		...X...		
	ELEC				
8	H2 Manf or TK leak (not in depleted tk(s) _____)	...X...			
9	2 MN Buses		...X...	X	
10	CNTL CA1 (No BFS Engage in GPC 3/5)		...X...		
11	Multi Φ AC BUS (unshorted)	...X...			
	GNC				
12	1 MLS (if reqd), IMU, or TACAN (C-band not avail)	...X...			
13	IMU Dilemma		...X...	X	
14	RHC Dilemma	...X...			
15	2 IMUs (Not targeting EDW/NOR)		...X...		
16	2 ADTAs (Winds>80kts or Dynamic Wx)	...X...			
	HOOK VELCRO			**HOOK VELCRO**	
17	2 FCS Channels (same surface) (Not targeting NOR/EDW)	...X...			
18	2 AAs, RGAs (Not targeting EDW/NOR)	...X...			
	OMS				
19	Prplt Tank	...X...		X	①
20	Ignition (neither eng ignites)	...X...			
21	Both OMS Eng fail			X	②
22	Prplt Lk after LAST LOS	perigee adjust			
	AFT RCS				
23	2 jets, same direction, same pod	...X...			
24	Prplt Lk after LAST LOS	...X...			
25	1 AFT RCS PROP TK fail	...X...			
	COMM				
26	MCC GO for DEORBIT not rcvd		...X...		

① Stop Burn > OMS PRPLT FAIL HP (Ref: DEL PAD/BURN Card)
② Stop Burn > OMS ENG FAIL HP (Ref: DEL PAD/BURN Card)

Deorbit burn flight rules.

essential on board system, the deorbit burn could be postponed. Specific cue cards in the Entry Checklist described the procedures for each type of failure and Mission Control would provide more insight. The EXEC button on the DPS keyboard would be pushed 15 seconds prior to the burn to allow the maneuver to proceed. During the preparations, Mission Control would have advised the crew how long ignition could be delayed beyond the nominal time for the burn and retain the downmode capability to achieve the target conditions at the entry interface. If the OMS quantities were less than 11 per cent, the RCS jets would make an "ullage burn" to settle this propellant in the aft section of the tanks in preparation for engine ignition. In the unlikely event of an OMS ignition failure both engines would be shut down, as would the APU that had been started, and the deorbit would be canceled in order to analyze the problem. However if all went well, all the crew had to do during the two and a half minutes of burn was to monitor the AUTO DAP. In particular, they would monitor delta-VTOT, TGO, the decreasing weight of the vehicle as it burned propellant, and good attitude control on the ADI. As the burn neared completion they would monitor the reducing perigee (HP) approaching the targeted value, and TGO and delta-VTOT approaching zero. If when TGO was zero the engines were not automatically shut down then the switches would be turned off manually to prevent an overburn. Then the residuals in the velocity components would be manually trimmed using the THC until they were within 2 feet per second of the nominal post-burn values. A delta of 2 feet per second equated to an energy error at the entry interface of 30 nautical miles, which was well within the capability of the guidance system.

As stated above, a deorbit burn had the peculiarity of having a point of no return. To understand why, it is necessary to appreciate that its task was to lower the perigee of the orbit into the atmosphere so that aerodynamic capture could be guaranteed. As the burn progressed and the Orbiter was gradually slowed, the crew could monitor on the MNVR display the *current* perigee of the orbit; that is to say, the perigee height that would apply if the burn were to be halted at that moment. In the case of a failure that cut short the burn, the big question was whether the Orbiter would be captured anyway, or remain in a stable orbit. It would have been desirable to have at least one day to allow the crew and Mission Control to work out a solution to the problem. For this reason the "safe HP" limit (the point of no return) was added to the deorbit burn procedures. This was defined as the minimum perigee height that would allow the Orbiter to remain in orbit for 24 hours before drag threatened imminent orbital decay. If the burn failed after the current perigee was below safe HP, then the crew had to do everything possible to complete the maneuver because the Orbiter would not have sufficient propellant to regain a stable orbit in order to obtain the time to troubleshoot the issue and schedule a second deorbit attempt. It is important to note that safe HP was not a fixed value as it depended upon the apogee (HA) of the orbit, which varied from mission to mission and even during a mission as orbital maneuvers were carried out. For this reason the bulk of the deorbit procedures were instructions advising how to respond to possible failures in the various OMS configurations (two engine burn, one engine burn, RCS burn, unbalanced propellant, and mixed cross-feed). Generally speaking, the response was to arrange an OMS cross-feed configuration to balance propellant consumption and maintain the optimum center of gravity in terms of both the X axis and the Y axis.

The RCS jets could be used to perform a nominal deorbit burn or to finish a burn in which multiple failures forced both OMS engines to shut down after having passed the safe HP point and before the targeted HP was achieved. As translations using the RCS were not guided, the maneuver had to be flown manually by using the THC to control the four +X aft jets and the RHC to control the attitude of the vehicle. In this case, the RCS jets would draw propellant from the OMS to preserve their own for maneuvering in the early part of the re-entry as the Orbiter was transformed from spacecraft to hypersonic glider. If an RCS burn using OMS propellant was unable to achieve a deorbit with the right conditions RCS propellant could be used, providing that enough aft RCS propellant was left for re-entry. And if this was insufficient, the next option was to fire to complete depletion the –X RCS forward jets. In this case there was no limitation on the amount of propellant that could be consumed since the forward RCS was not required for maneuvering during re-entry.[2] And if this measure was not enough there were two other options, one designed for the primary landing site and the other for an alternative site. But this last possibility was only available for landings intended for Edwards Air Force Base in California, allowing a landing at a backup site further down-range in New Mexico.[3]

If all went well and the Orbiter was slowed sufficiently for a nominal re-entry, the

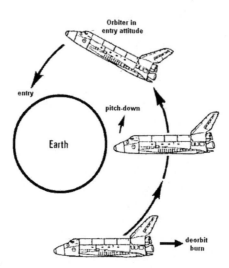

Maneuvering to entry interface attitude.

---

[2] A forward firing RCS would have disrupted the smoothness of the airflow around the nose with severe consequences for maneuverability and stability during re-entry.

[3] Both of these remedies will be explained in greater detail below.

crew had several tasks to perform prior to reaching the entry interface half an hour later. On completing the burn, the flight software was switched to major mode 303 and the OMS engines were stowed to protect their bells from the heat of re-entering the atmosphere. The Orbiter was then maneuvered, either automatically or manually, to the attitude for encountering the atmosphere and starting the long glide home. This required making a pitch down rotation. To understand why, recall that in the attitude for the deorbit maneuver the Orbiter was tail-first and belly-up in an inertial frame of reference. Maintaining this attitude along the not yet path, half an orbit later, its belly would be facing down but with the nose not yet elevated at the proper angle of attack for re-entry.

By pitching immediately after the deorbit burn and holding attitude through to the entry interface the Orbiter had the desired angle of attack when it made contact with the atmosphere. By this expedient, maneuvering to the re-entry attitude used less aft RCS propellant. Once in attitude, the commander verified the aerosurface actuators. About 18 minutes prior to the entry interface all the propellant in the forward pod of the RCS system was dumped by burning it, directing all the firing along the Y axis. This dump was done not only to manage the position of the center of gravity but also to minimize the amount of propellant remaining in the forward pod that might pose a hazard in the event of a hard landing. Five minutes later the two remaining APUs were started and all three placed at nominal pressure. With about ten minutes to go, the crew could relax in expectation of riding inside the most astonishing light show.

The atmospheric re-entry of a Shuttle required a great deal of planning. From the point of view of guidance it began about 5 minutes in advance of the entry interface, arbitrarily defined as an altitude of 400,000 feet. At EI-5 minutes the Orbiter was at an altitude of about 557,000 feet, some 4,400 nautical miles from the planned landing site and traveling at 25,400 feet per second. From this point through to an altitude of

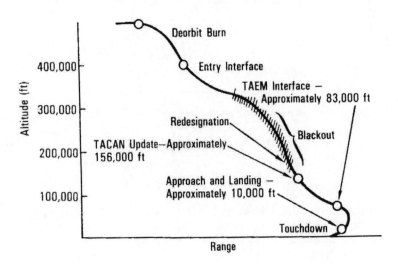

Entry flight profile.

about 83,000 feet and a speed of 2,500 feet per second the Orbiter flew the so-called re-entry phase designed to allow it to survive the intense thermal and structural loads resulting from penetrating the atmosphere at very high speed. Then the vehicle was guided through the so-called terminal area energy management (TAEM) designed to place it on the correct heading with the correct energy to carry out the approach and landing phase.

ATMOSPHERIC ENTRY

Entry corridor

To appreciate the complexity of re-entry, it helps to visualize the game of throwing stones onto water. The seemingly easy task is to make a stone skip on the surface at least several times before it plunges to the bottom. To do this, it is necessary to throw the stone (preferably a flat one) at such an angle that on coming into contact with the surface it bounces into the air again, and so on until it loses energy and simply drops into the water and sinks. This bears a conceptual similarity to a spacecraft re-entering the atmosphere in that the vehicle can use this "skipping" effect to lose energy.[4] The vital issues are the angle of the trajectory at the entry interface and the

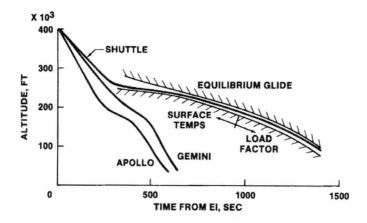

Entry corridor comparison.

---

[4] Apollo missions returning from the Moon used this "skipping" effect, with the first "dip" slowing the capsule enough for it to be captured for the actual re-entry. In this case the motivation was to ameliorate the thermal loads. As aerobraking, repeated dipping into a planetary atmosphere provides a propellant-free means of lowering the high point of an initial orbit. This was demonstrated by Mars Global Surveyor in 1997.

attitude of the vehicle. The guidance system of the Orbiter would ensure a specific attitude for the entry interface, and then steer the vehicle to the landing site whilst remaining within trajectory, thermal, and structural limits.

Once the Orbiter was in the upper atmosphere its large wings began to generate a degree of lift. This combined with the centrifugal force due to the relative velocity of the vehicle over the surface. These forces tended to drive the vehicle away from the atmosphere while gravity was drawing it down. A safe re-entry required balancing these opposing forces. If the angle at which the vehicle penetrated the atmosphere were too steep then it would burn up. If the angle were too shallow then the lift and the centrifugal force would be sufficient to cause the vehicle to climb back out of the atmosphere and be inserted into a low orbit from which it could not attempt another controlled re-entry.

After the entry interface the guidance system had to continue to control the flight path angle of the Orbiter as it penetrated deeper into the atmosphere, in order to balance the forces acting upon the vehicle. On a graph of altitude versus time since the entry interface, this equilibrium glide boundary would be a curve with each point being an exact balance between gravity and the vertical component of the combined lift and centrifugal force. If the Orbiter were to stray above this curve the equilibrium would have been broken due to excessive lift, and it would climb back into space. Entry guidance therefore had to keep it below this curve to guarantee that gravity would prevail. This was a "soft" boundary because the Orbiter would not break up in the event of a violation.

The entire surface of the Orbiter was plastered with tiles and blankets that served as a protection system against the immense heat generated during the early portion of re-entry. Although the TPS could withstand very high temperatures it was limited in terms of the heat load that it could accommodate. In addition to ensuring the Orbiter remained below the curve on the equilibrium glide boundary graph, the guidance had to maintain an attitude and a flight path angle that would ensure that temperatures did not exceed the TPS limits. In one sense the high temperatures were the result of the drag generated by penetrating the atmosphere but it was not simply a case of friction of the air against the vehicle. As the air was compressed it formed a shockwave. The shape of the vehicle was designed to ensure that this intensely hot shockwave did not come into contact with its surface. Nevertheless if the drag increased, the temperatures would also increase. If the drag decreased, the Orbiter would spend a longer period at a lower temperature, allowing heat to flow through the tiles and reach the "backface" that was in contact with the aluminum structure.

Since the aerodynamic flow field over the Orbiter's surface was quite complex, mathematical models representing heat rates on specific vehicle locations or control points were used for trajectory design. Depending on the particular flight condition and the maximum allowable temperature at each point, any of these locations could represent the limiting constraint on the trajectory. Given the flight-specific high drag thermal boundary and backface temperature constraints, the nominal drag profile that would be flown through the high Mach portion of re-entry could be calculated. This trade-off between high surface and backface temperatures was called the "thermal tradeline". On a graph its curve lies below the equilibrium glide boundary, with each point being the surface temperature at exactly the TPS limit. Guidance had therefore to hold the Orbiter below the equilibrium glide boundary and above this tradeline.

Apart from generating heat loads, drag also imposed accelerations on a structure. In comparison to Gemini and Apollo, the Orbiter was greatly limited in its maximum allowable load factor. For example, while the Gemini and Apollo capsules routinely endured 4 g to 7 g and Apollo had design limit of 12 g, the Orbiter was designed with a maximum normal load factor of only 2.5 g. This was therefore a third constraint to be managed by the guidance system. The atmospheric flight path profile not only had to prevent the Orbiter from burning up or climbing back out of the atmosphere, it had also to limit the loads imposed on the structure. In fact, excessive drag could cause a structural failure even before the limits of the TPS were reached.

A similar constraint was imposed by the aerosurfaces, since the flight path profile in the thicker air at lower altitude had to ensure that the dynamic pressure would not overcome the maximum hinge moment generated by the hydraulic actuators on the wing aerosurfaces.

The thermal tradeline, load factor limit, and aerosurfaces constraint were "hard" boundaries because violation of any of them would cause severe structural damage, and in the worst case complete loss of the vehicle.

When plotting all these constraints, it is easy to see how the region left between all the curves came to be called the "entry corridor". The task of the guidance system was to fly in the corridor, never crossing any of these curves. Due to the conditions imposed in terms of maximum temperatures, load factors, and dynamic pressure, the safety margin left little scope for errors.

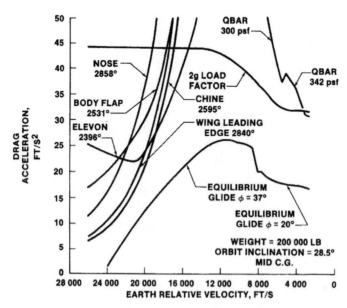

Representation of constraint boundaries for the Orbiter re-entry.

All these constraints could be plotted in a dynamic pressure/velocity state space but the use of dynamic pressure as a guidance control parameter would require that it be determined from sensed vehicle accelerations and attitudes for all valid angles of attack and Mach numbers, with similar data for lift and drag coefficients. Owing to considerable uncertainties in the model predictions for these quantities the constraint boundaries were reformulated into a drag acceleration/Earth-relative velocity (D-V) state space with drag acceleration being the rate of change of velocity measured by the IMUs. This required only an estimate of the lift-to-drag ratio (L/D) of the Orbiter, which was probably the most predictable aerodynamic parameter and had the benefit of being able to be directly measured in flight. In addition, the available range was uniquely defined by the drag acceleration profile. For a safe re-entry, the Orbiter had to fly the corridor between these constraint boundaries.

This difficult task was further complicated by the fact that the corridor width (and hence the safety margins) depended upon specific characteristics of the mission. The Shuttle had to operate over a large range of orbital inclinations with different vehicle weight and center of gravity positions. Inclination affected the relationship between inertial and Earth-relative velocity, and shifted the equilibrium glide constraint in the D-V plane to narrow the corridor as the inclination increased. Vehicle weight was the driver for the location of all the surface temperature boundaries, and an increase in the weight narrowed the corridor because the Orbiter had to fly a trajectory in which larger aerodynamic forces were required to achieve equivalent accelerations. And as the center of gravity shifted, the body flap deflection to trim the vehicle altered the airflow and increased the temperatures of the associated control points. However, in practice the elevon and predicted body flap positions were balanced to ensure that neither became excessively hot.

Entry guidance

Due to the complexity of the boundaries defining the entry corridor, from a guidance point of view re-entry was split into five major phases each depending on the current relative velocity of the Orbiter and designed not to violate the constraints of the entry corridor.

Between entry interface and when the sensed dynamic pressure reached a given threshold, the Orbiter was in the "pre-entry phase". This was the simplest in terms of guidance. The vehicle required only to maintain an LVLH attitude of +40 degrees pitch and 0 degrees in both roll and yaw. The start of closed-loop guidance marked the onset of the "temperature control major phase" which maintained the temperature well within the surface and backface limits. At a relative velocity of 19,000 feet per second the vehicle had been slowed sufficiently that the temperature was no longer a constraint. Flying through ever thicker air and still at a very high relative velocity, the next constraints to be considered were the structural limits. After the temperature control phase the Orbiter therefore entered the "pseudo-equilibrium glide phase", so named because it was based on a drag-velocity profile for equilibrium glide flight, defined as flight in which the flight path angle remained constant. This flight regime provided the maximum down-range capability. This phase allowed guidance to bring the Orbiter to the "constant drag phase" designed for flying a constant drag profile of (nominally) 33 feet per second-squared and for gradually lowering the angle of attack from the initial 40 degrees to 36 degrees. Upon terminating the constant drag phase, guidance started to compute the so-called "transition phase" that maintained a linear drag profile (as a function of energy) in order to null range errors and set up for the TAEM interface at an altitude of 83,000 feet, a speed of 2,500 feet per second (Mach 2.5) and a range from the landing site of 52 nautical miles. Throughout this transition phase the angle of attack was gradually lowered to achieve 14 degrees at the TAEM interface.

Entry flight control was accomplished using the aerojet DAP which generated effector and RCS jet commands to control and stabilize the Orbiter during its return from orbit. As a function of q-bar, the aerojet DAP was "flight adaptive". Its closed loop compared feedback against commands in order to generate an error signal and rate commands, with the jets and aerosurfaces being commanded as appropriate to achieve the desired rate. It gradually transitioned from the jets to the aerosurfaces as the Orbiter changed from a spacecraft to a hypersonic glider to subsonic glider to a high-speed 100-ton tricycle.

The aerojet DAP was activated by the transition to major mode 304, typically at EI-5 minutes, and operated right through to wheel stop on the runway. In the early portion of re-entry, attitude was maintained by the aft RCS jets to control both roll and pitch. At a sensed dynamic pressure of 10 pounds per square foot the ailerons became effective in controlling roll and the roll RCS jets were superseded. At 20 pounds per square foot the elevons became effective in controlling pitch attitude and the pitch RCS jets were superseded. At this point, the only RCS jets still operating were those for yaw. These were superseded when the speed fell below Mach 3.5 and enabled the rudder to function. This was due to the high angle of attack maintained for most of re-entry, which placed the rudder in the shadow of the body of the vehicle. Only after the angle of attack was considerably reduced on slowing below Mach 3.5 could the rudder operate and the RCS yaw jets be superseded. However, before this threshold the aerosurfaces of the tail functioned as speedbrakes past Mach 10 and assisted the elevons operate as elevators and ailerons in imposing lateral directional stability. And a predetermined profile of body flap adjustments increased the effectiveness and reduced the thermal stress on the elevons.

Flying during entry

Atmospheric re-entry can also be thought of in terms of energy management. When in orbit, the energy of a spacecraft is the sum of its kinetic and potential energies. To come back home it must lose all of this energy in a controlled manner. The only way to do so on a planet that possesses an atmosphere is to generate drag. Drag is not only necessary for losing energy, it also defines the range that can be flown. For example, if a vehicle is 4,000 miles from the runway and traveling at Mach 25 it is essential to cover at least that distance whilst slowing down to landing speed! Thus drag must be managed to ensure that there remains sufficient energy to achieve the destination.

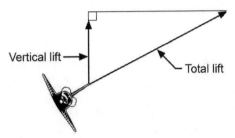

Bank to control vertical lift. (Courtesy of www.nasaspaceflight.com)

For this reason, during re-entry the guidance system would work to maintain the Orbiter on a predetermined drag/acceleration profile and guide it through a corridor that was limited on one side by altitude and velocity requirements for ranging (i.e. to reach the runway) and on the other by the thermal, q-bar, and Nz loading constraints. Ranging was achieved by adjusting the drag and velocity to maintain the vehicle in the corridor.

Based on the distance from the runway, the drag acceleration required varied in that the more distant the runway the less drag was necessary. Flying in the extreme upper atmosphere and in a nose-up attitude with 40 degrees of pitch, the Orbiter not only generated lift but also drag that made it lose ever more energy. With the force of gravity pulling downwards and the Orbiter losing energy, little by little it penetrated deeper into the atmosphere. The rate at which its altitude decreased was related to the drag generated. Because drag is a function of velocity, the faster the "sink rate" or loss of altitude (Hdot), the higher is the velocity and the drag.[5] As a consequence of their close relationship, drag acceleration and Hdot are interchangeable terms. For a conventional aircraft, variations in altitude are commanded by changing the pitch or the angle of attack (alpha) so that to descend you push on the stick for reduced lift. However, for the Orbiter running at a high Mach number (M > 2.5) in order to maintain the thermal constraints of the entry corridor, the angle of attack had to be held within a very narrow range, rendering it impractical to control lift by varying the pitch. At such high speeds the only way to control Hdot (and therefore drag) was by banking commands. Banking the Orbiter had the effect of changing the vertical component of lift so that the imbalance with weight caused the vehicle to sink at a rate determined by the bank angle, in this manner generating the desired drag. Normally closed-loop guidance was initiated at an Hdot of about –400 feet per second. Then at a rate of –240 feet per second guidance commanded a bank at about 3 degrees per second in the direction of the landing side. If for any reason the Orbiter did not start to bank at that point, Hdot would become positive and the lift vector would cause the vehicle to skip back out of the atmosphere. For this reason the crew would monitor the initial bank command on activation by guidance to determine whether the actual drag measured was converging to the calculated drag. A failure of guidance to achieve the requisite drag would prompt them to take manual control.

In a bank the total lift produced a horizontal component that caused the vehicle to slowly veer off course, so the first roll command would cause the Orbiter to turn

---

[5] It is worth remembering that drag is proportional to the square of velocity.

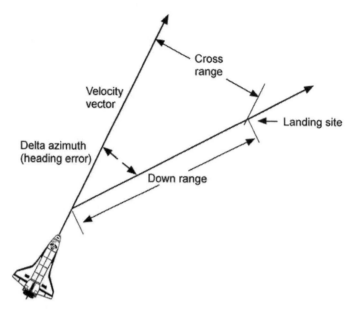

Downrange and cross-range concepts.

in the direction of the landing site, thus reducing the angle between the velocity vector and the landing site known as the heading error or delta azimuth (Delaz). As the turn continued, Delaz passed through zero and increased on the other side. By giving the Orbiter a cross-range capability of 750 nautical miles, hypersonic banking enabled it to use landing sites that were not directly on the orbital ground track. At a Delaz of 10.5 degrees, guidance commanded the Orbiter to bank back toward the landing site as the first of several so-called "roll reversals". Clearly, in making such a reversal the vehicle had to come all the way back to a roll angle of zero. In doing so, the vertical component of lift started to increase, slowing the sink rate and forcing the drag below the desired value. Therefore shortly after the zero roll angle, guidance overbanked by 20 degrees, meaning that if guidance was satisfied with 60 degrees of left bank prior to the reversal it would then bank 80 degrees right. Due to the slow response in drag change during an overbank, the overbank was held for about ten seconds because this was deemed more than sufficient to make up the lost drag and therefore come back to the required value. A more rapid means of removing the drag error was called alpha modulation, in which the angle of attack could be changed $+/-3$ degrees to produce a change in drag of about 10 per cent. Guidance used alpha modulation during a roll reversal just as the bank angle reached zero. It also ordered a rate of 5 degrees per second for the roll reversals in order to minimize the drag and Hdot errors. In this way the time spent at a small bank angle was minimized. After the first roll reversal, guidance continued to command bank in order to achieve the desired altitude rate. Subsequent reversals were made at a Delaz of 17.5 degrees. Past Mach 4 they were made at a Delaz of just 10 degrees. It is important to note that roll reversals were not used for dumping energy but because the very small "alpha envelope" of the Orbiter did not allow pitch to be used to control the sink rate.

In the event of a delta-V deficiency at OMS shutdown during the deorbit burn, the Orbiter would find itself in a high energy state. One way of dumping energy was to pre-bank the vehicle so that on reaching the entry interface it would have already begun to slice downwards through the atmosphere at a steeper angle in order to attain the proper trajectory for the planned landing site. In the opposite case of a low energy state the Orbiter needed to be held at the lowest drag possible. The solution was to bank at a small constant angle which would cause it to ascend above the equilibrium glide boundary, lose lift, and descend again. Once in the thicker air (below the equilibrium glide boundary) more "excess lift" would be generated to push the vehicle up again. This pattern would be repeated as often as necessary to create an oscillation about the equilibrium glide boundary called a phugoid. With this phugoidal motion the Orbiter could skip across the top of the atmosphere in a controlled manner and increase its down-range capability sufficiently to make the landing site.

Monitoring entry

Between re-entry and TAEM, a sequence of five ENTRY TRAJ displays became available to the crew for monitoring guidance and the trajectory against the planned re-entry profile and, if necessary, fly it manually. While the ENTRY TRAJ 1 display appeared on switching to MM304 the others kicked in at predetermined velocities, each being a continuation of its predecessor. In the center of each display there was a dynamic plot of either velocity or energy on the vertical scale versus range on the horizontal scale. A small vehicle "bug" was placed on the display as a function of velocity and range (or energy and range). Each display showed two sets of lines. The dashed lines were for a specific reference drag value. But they were not calculated in real-time and were therefore merely illustrative. The solid lines were for two major categories of trajectory lines. The outer left and right ones always displayed the drag limits, respectively the high drag line and the low drag line (equilibrium glide). The middle lines were almost all guided trajectory lines, meaning that if the bug was on one of these lines and the reference drag was achieved, the Orbiter would "fly down" the line.[6] Of course, guidance would never fly the bug down a drag limit line. In real-time the algorithm calculated the reference drag and showed it on the drag scale on the left side of the display, with an arrow pointing to the calculated value desired and a triangle showing the actual drag sensed by the IMUs. Directly correlated with this was the reference Hdot, it being the Hdot value needed to achieve the desired drag. If the actual drag was less than the reference drag, guidance would steer the vehicle to a more negative Hdot. The reference Hdot was shown on the lower right corner of the

---

[6] The reference drag was the value calculated by guidance as optimal for achieving a landing.

display. At the base of the display, directly below the bug, there was a rough estimate of the reference Hdot. As a bug flew down a trajectory line, it left a triangular trailer every 15 guidance cycles (28.8 seconds) for trend analysis. For ENTRY TRAJ 3 to 5 the trailers were displayed every 8 guidance cycles (15.63 seconds).

On initiating closed-loop guidance, a guidance box appeared that moved together with the Orbiter bug to reflect drag errors; that is to say, the difference between the actual sensed drag and the calculated reference drag. This guidance box left a trail of dots for trend analysis. Ideally, the nose of the bug would lie inside the box and there would be a dot inside every triangle.

The astronaut experience

Charles F. Bolden, who flew four Orbiter re-entries from the front row of the flight deck, describes it as follows, "The most spectacular period of time during re-entry, especially if it occurs at night, is the initial re-entry heating when you come into the atmosphere and the vehicle just – it heats up and it glows. The way I describe it to schoolkids, it's like being inside a light bulb and you're the filament in the light bulb. You see this sort of a dull red glow start at the base of the windshield and then what we call St. Elmo's Fire." But the show was far from being static. "Before you know it, it gets orangeish and then pinkish, and then the whole windscreens in the front are obscured, like you are in a cloud, or like you're inside a light bulb. You can't see out of them because of the bright light that's coming off the tile." But back towards the tail the show was completely different. "If you look out the overhead window, it's a different world. Up there you see the plasma ... charging and discharging and it's

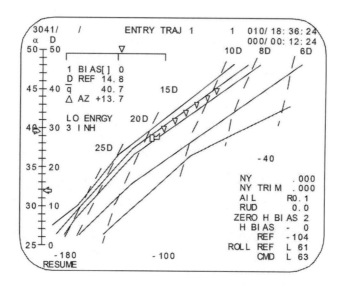

Example of re-entry trajectory display.

Glowing plasma beyond the window of a Shuttle commander wearing an "overall" flight suit. Note the ENTRY TRAJ display on his monitor.

like there's a monster up there, and this thing forms and then collapses." His conclusion, "It's an absolutely fascinating light show."

Vance D. Brand, who commanded three Shuttle flights, offers further insight into the light show, "Around Mach 20 you could see white beams of shockwaves coming off the nose. If you had a mirror ... you could look up through the top window and see ... a pattern and the fire going over the top of the vehicle, vortices and that sort of thing." But sometimes the spectacle of re-entry could provide a surprise. Like during STS-5 when Brand had to take manual control from the autopilot to do a flight test maneuver. "First, I pushed down from 40 degrees to 35 degrees angle of attack, then up to 45 [degrees] and then back [to 40 degrees]. That's just a few degrees, but when I did that Joe Allen, who was sitting in the center seat as flight engineer, said that a shockwave came from the nose and it came up and attached to the window right in front of us. That was a little worrisome, because he knew it was hot. The shockwave was very hot. But about as soon as it got there I was on my way to a higher angle of attack, so it walked back to the nose. So there were interesting things like that, that happened."

It is easy to agree with Terry J. Hart, who was a mission specialist on STS-41C, "Watching the fireball around the vehicle was really breathtaking." But re-entry was not simply a spectacular light show, it produced other interesting effects on objects and people. As Hart explains, "I had a camera that I was holding during re-entry. I remember letting it go and it would just kind of sit there, of course, when we were weightless. And as we started to hit the upper parts of the atmosphere I watched the camera accelerate forward as you let go, because the vehicle was decelerating. Even before your body felt it, you could see objects start to go forward."

Steven A. Hawley flew five Shuttle missions, and says, "The whole time it's like you're going down the side of the mountain in a pickup truck. You feel like you're kind of falling forward in your seat, because you are decelerating. And that lasts an hour. It's kind of fun when the onset of Gs happens and you think, 'Oh, jeez, it must be two or three Gs by now.' But the one little mechanical gauge we've got in the whole Orbiter is this little G-meter that sits up by the commander, and it's showing about two-tenths or something. Your [body's] calibration for what Gs feel like is different." This alteration in sensations provided scope for some fun, "On subsequent flights, it's kind of fun, having had that experience, to tape over [the G-meter] and have people guess when we've got half a G, say, and everybody guesses way too early."

Through the Shuttle program, all re-entries were flown with automatic guidance because no anomalies ever developed.[7] From this point of view, STS-2 represent an exception because the flight plan called for making twenty-nine manual maneuvers to investigate the stability margin and performance of the Orbiter. Manually flying a machine such as the Orbiter during re-entry was a dream come true for Joe Engle, the commander of that mission. He explains why this was so important, "We were very anxious to see how much margin the Shuttle had in the way of stability and control authority, how much muscle the surfaces had at different Mach numbers, hypersonic Mach numbers and angles of attack." Another interest was what to do in the event a deorbit burn required excessive cross-range to reach the landing site. As explained above, cross-range was controlled by the bank angle and roll reversals because it was not permissible to do so by varying the angle of attack. But as Engle says, if it was absolutely necessary, "You *could* increase the cross-range ability by decreasing the angle of attack. It allowed the leading edge of the wing to heat up a bit more and would cut down on the total number of missions that a Shuttle could fly but it would allow you to get that extra performance, that extra range, to make it to the landing site." Although this had been tested in a wind tunnel, it was necessary to obtain real field data. "How much the leading edge would heat up and just how much more lift-to-drag, turning ability and cross-range ability that would give you was theoretically known but the wind tunnels are very susceptible to a lot of variables ... so you really want to know for sure what you have in the way of capabilities if you ever have to use them." As to when this might be necessary, Engle notes, "You don't know when you may have a payload you weren't able to deploy, so you have maybe the center of gravity not in the optimum place and you can't do anything about it. And just how much maneuvering will you be able to do with the vehicle in that condition? How much control authority is really out there on the elevons? And how much cross-range do you really have if you need to come down on an orbit that isn't the one that you really intended to come down on?" Engle describes what he did during that flight to obtain data to explore these issues, "I pulsed the vehicle in all three axes – in pitch, a step input, roll inputs and rudder kicks for yaw – to see what the effectiveness of the surfaces were during re-entry, the

---

[7] The fatal re-entry of *Columbia* on mission STS-107 is obviously excluded.

effectiveness of the flight control system, and how quickly the vehicle would damp out after being disturbed. [And] at various Mach numbers we swept the angle of attack, I believe it was plus or minus 5 degrees, to see how much more cross-range a 35 degree angle of attack gave us than 40 degrees did, and above 40 degrees to 45 degrees to see if we could pull up to a higher angle of attack should for some reason we desire to lower the heat on the leading edge of the wings. That would cost us down-range and cross-range ... so if that was necessary the flight planners could program where the deorbit burn should be. If you didn't have as much range, you could make the burn a little bit later so that you weren't as far from the landing site as nominally planned." The tests were successful. As Engle recalls, "It was something that, like in anything, there was good healthy discussions on, and ultimately the data showed that it was really worthwhile to get. Those maneuvers we did on STS-2 were programmed into the automatic entry flight control system so that subsequent to that, those maneuvers continued to be made and data continued to be gotten, but it was done automatically by the computer."

TERMINAL AREA ENERGY MANAGEMENT

At 60 nautical miles from the runway threshold and at an altitude of approximately 82,000 feet above ground level (AGL) traveling at an Earth-relative velocity of 2,500 feet per second, the guidance system automatically switched to major mode 305 to initiate the second leg of the journey home, the terminal area energy management (TAEM). By now the Orbiter had completed its metamorphosis from a spacecraft to an aircraft, thereby enabling guidance to govern it using more traditional piloting techniques, such as varying the angle of attack to control altitude. The task of guidance in TAEM was to align the Orbiter with the runway at just the right amount of energy for the approach and landing.

As with re-entry, TAEM guidance had to fly within a specific corridor, this time defined only by the dynamic pressure, the main requirement being to limit normal accelerations (Nz). To understand why, it is necessary to recall that the greater the dynamic pressure the more hinge moment the hydraulic actuators needed to impart to the aerosurfaces to achieve a given effect on the aircraft. As TAEM guidance had to control the energy of the vehicle in the presence of unknown winds by using altitude as its control parameter, if the dynamic pressure were excessive then the actuators would not be able to be operate the elevons properly and the result would be loss of flight control stability. Imposing a limit on the dynamic pressure placed a limitation on the normal acceleration (Nz) since a high Nz would mean a high rate of descent, which is to say a higher relative wind velocity and a higher dynamic pressure.

Like the re-entry phase, TAEM guidance was split into subphases: the acquisition phase, the heading alignment phase, and the pre-final phase.

The acquisition phase started at the TAEM interface, and guidance simply steered towards an imaginary inverted cone known as the heading alignment cone (HAC). This was tangent to the extended runway centerline, and nominally located 7 nautical miles from the runway threshold. Although Shuttle pilots were required to

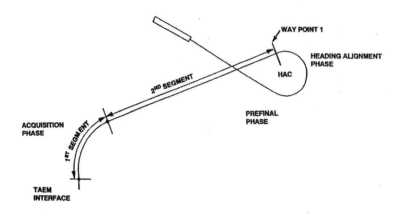

TAEM flight profile.

have served as military test pilots with thousands of flight hours of experience, on a nominal mission they were able to fly the Orbiter manually for no more than a few minutes. As during the ascent, the re-entry was automatic and the pilots were supervisory until the eager commander took manual control some 15 seconds prior to HAC interception. As per the flight rules, at the commander's discretion the pilot could get a very quick taste of how to fly the vehicle in the atmosphere by flying the turn around the HAC. Nominal entrance to the HAC used the "overhead approach" because the HAC was on the *opposite* side of the extended runway centerline from the Orbiter's initial position upon its transition to re-entry guidance. When the Orbiter reached the tangent point (Waypoint 1) between its trajectory and the HAC, the heading alignment phase could start. A roll command angle ensured the Orbiter executed a turn to follow the HAC in a spiral arc of decreasing radius. At the HAC exit point, known as the nominal entry point (NEP), the Orbiter ought to be lined up with the runway (+/–20 degrees) at the proper flight path angle and airspeed to begin the steep glide to the runway. From this point on, the commander would fly the remaining portion of the journey. The exit point marked the start of the pre-final phase in which the Orbiter pitched down in order to acquire the steep glide slope, increased airspeed, and banked to line up with the runway. When on the centerline, on the outer glide slope, and on airspeed, the approach and landing phase could start.

As explained earlier, if the Orbiter did not initially have the right energy, the re-entry guidance would do everything that it could to deliver the Orbiter to the TAEM interface with the right energy. If it was still in an incorrect energy state on reaching the TAEM interface, the TAEM guidance had several options available to resolve the situation. For example, if the Orbiter had excess energy, the HAC turn angle was less than 200 degrees and the range to the runway was more than 25 nautical miles, guidance would command a constant 50 degree bank[8] to steer the vehicle away from

---

[8] Since the S-turn was most likely to occur in the supersonic regime, this bank was usually limited to 30 degrees in order to protect the ground from a large sonic boom.

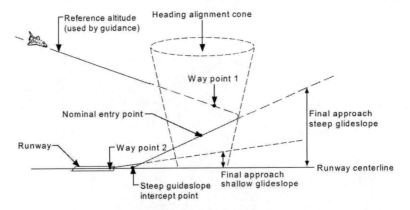

HAC flight path geometry; side view.

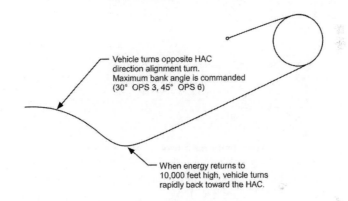

TAEM S-turn profile.

the HAC until the excess energy was dissipated. Once the Orbiter had shed sufficient energy, guidance would rapidly steer towards the HAC in order to resume a normal acquisition phase. This S-turn should not be confused with the re-entry roll reversals. The S-turn was a feature of TAEM guidance and, if required, was used to dissipate excessive energy. Although the ground track of a roll reversal resembled that of an S-turn, it did not dissipate energy. The Orbiter dissipated energy in the re-entry phase by banking the lift vector for increased drag. A secondary effect of banking was the turning motion that steered the Orbiter away from the landing site. The roll reversals were to prevent the vehicle from flying too away from the landing site. Rather than being a means of controlling energy, the roll reversals controlled cross-range. To use roll reversals to dissipate energy, a much larger Delaz angle would have been needed than the small angles used during re-entry.

Other techniques available to an Orbiter which had excess energy involved fully opening of the speedbrake below Mach 0.95 or diverting to a different runway. This decision would be made manually by the crew in conjunction with Mission Control

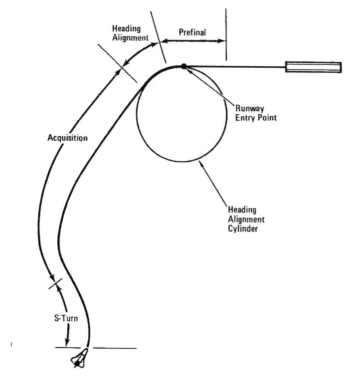

TAEM flight path profile; bird's view.

(there were no on board alerts or messages) and while the geometry allowed a turn of sufficient magnitude to acquire the new HAC. There were many runways at Edwards Air Force Base to choose from, but there was only one strip at the Kennedy Space Center and the only choice available was from which end to approach it. The final means of shedding energy was to fly manually outside the HAC, taking care not to stray too far lest the vehicle then end up low on energy! There were several methods available to an Orbiter in this situation. For example, if it was below Mach 0.95 the speedbrake could be completely closed. Guidance could also pitch the nose up so as to generate more lift, increasing the L/D ratio to reduce the rate at which the vehicle lost energy. And, of course, there might be the option of choosing a different runway.

TAEM guidance would nominally fly an overhead HAC but as an option to save energy it could be set to reach the HAC by a straight-in approach, meaning it would not have to cross the extended runway centerline in order to turn on the HAC back towards the runway. In this case the HAC would be placed on the *same* side of the extended runway centerline as the Orbiter. For the straight-in approach the vehicle would turn on the HAC for less than 180 degrees in order to use less energy for flying the HAC and have just enough for landing. Depending on the HAC turn angle in the overhead approach, the range to fly to the runway could be shortened by several miles, thereby saving some of the already low energy.

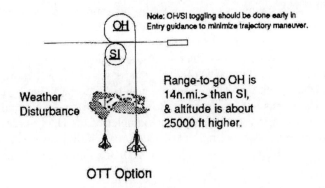

HAC intercept profiles for overhead (OH) and straight-in (SI).

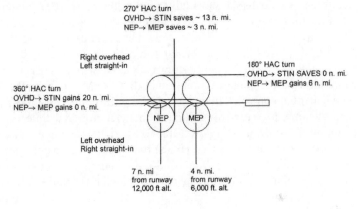

HAC turns.

In fact, for the four orbital flight tests of the program *Columbia* landed using a straight-in approach to the HAC, with the last one being a cylinder rather than a cone. Starting with STS-5, the TAEM guidance was set to fly an overhead conical HAC approach with the option of manually switching to a straight-in approach. The main reason for introducing the overhead approach was that it would offer a better means of avoiding a thunderstorm when landing at the Kennedy Space Center. The altitude difference between the straight-in and overhead approaches increased the likelihood of successfully flying either under or over a weather disturbance along the approach path. And in an overhead approach, guidance would fly the Orbiter with a higher energy state to provide additional margin in making the runway. Any excess energy could then be dissipated in subsonic flight, in the vicinity of the runway. The flight rules required that the decision to switch to a straight-in approach be made no closer than 45 nautical miles to the runway, as otherwise the heading errors caused by that change could nullify any potential gain in energy.

If switching to a straight-in approach was still insufficient, guidance could fly the Orbiter around a HAC located closer to the runway, at the so-called minimum entry point (MEP). The saving of several miles in the approach could make the difference between a landing and a crash. The drawback was that a MEP would greatly decrease the time on final approach, with less time for controlling the Orbiter prior to landing. The conditions for switching to either a HAC straight-in approach or a MEP would be announced to the crew by means of display messages and a system management alert. The pilots would then manually select the new HAC to be flown by guidance.

If the Orbiter was low on energy at low altitude in the heading alignment phase, guidance could automatically shrink the diameter of the HAC to decrease the range to the runway. HAC "shrinking" was based on the altitude error and the remaining turn angle of the HAC, and was inhibited if the remaining HAC turn was less than 90 degrees.

TAEM displays

The crew had three specific displays available to monitor the TAEM guidance. These provided all of the information needed to determine whether the guidance had issues in steering to dissipate energy at the desired rate. The first two, named VERT SIT 1 and VERT SIT 2, displayed the same format and data. In particular, each presented a plot of altitude (y axis) versus range to runway (x axis), with an energy-over-weight (E/W) scale monitoring the guidance function in the TAEM region as the basis for a decision to take manual control. Three lines on these altitude-range plots represented the maximum dynamic pressure (upper line to the left), nominal trajectory (center), and the maximum lift-to-drag ratio (lower line to the right). A numerical value along each line stated the equivalent airspeed (EAS) that the Orbiter would need in order to maintain that line. Ideally, the bug would stay on the nominal trajectory. A vertical scale located on the right of the display provided information on the vehicle's current total energy (kinetic and potential) expressed in feet and represented by a triangle on the scale. In the center of the scale there were three overbright tick marks with the abbreviation NOM.[9] The middle (shorter) tick mark represented the nominal energy reference while the overbright ticks above and below it represented E/W values of +8,000 feet and –4,000 feet respectively, together defining the "E/W corridor". The TAEM guidance sought to maintain the Orbiter at an energy within this corridor to protect against a degraded navigation altitude which could drive the E/W off the reference profile. If the value strayed out of this corridor, guidance would command a pitch angle to drive the Orbiter back inside the E/W limits. Other important marks were present on this scale. In particular, above the nominal energy marks there was a mark labeled STN (S-turn). A triangle above this mark would represent a situation in which guidance had to command an S-turn to shed excess energy so as to return the triangle to the E/W corridor. On the other end of the scale, a mark labeled MEP was the low energy state in which the Orbiter would be able to reach the runway only by using a MEP HAC. Finally, an arrow on the bottom part

---

[9] Note that the NOM label might not have lined up with the nominal tick marks, since the label was static and the tick marks were dynamic and could change position along the scale.

of the scale provided an indication of the energy level at which it would be necessary to switch to a straight-in approach to the HAC.

The difference between the two displays was that VERT SIT 1 was available for altitudes between 100,000 feet and 30,000 feet, and VERT SIT 2 applied from there to 8,000 feet. At the TAEM interface on a nominal mission, the Orbiter bug would appear on the VERT SIT 1 display approximately one-quarter of the way down from its top edge.

Since the VERT SIT 1 display monitored the majority of the supersonic portion of the TAEM guidance with the Orbiter's kinetic energy being much greater than its potential energy, the crew would monitor the E/W scale rather than the altitude-range lines because these lines were determined to maintain the Orbiter inside its dynamic pressure corridor and did not reflect the current energy state of the vehicle. The crew would monitor the altitude-range lines on the VERT SIT 2 display after the Orbiter was subsonic because at subsonic speeds the energy management problem required managing the potential energy rather than the kinetic energy. In other word, the E/W scale monitored primarily the kinetic energy while the altitude-range plot monitored primarily the potential energy, the parameter that controls the range that a glider can fly. The energy pointer was therefore not very useful in deciding whether the Orbiter could make the runway. In fact, below 20,000 feet the E/W scale was automatically deleted from the display. To determine whether the Orbiter could reach the runway, the pilots used the airspeed and the bug on the altitude-range plot. If the bug was on the low line with the Orbiter flying at maximum L/D, or on the high line with the Orbiter flying at its maximum airspeed, then the runway was just in range on a no-wind day. It is worth noting that both the low and high lines on the altitude-range

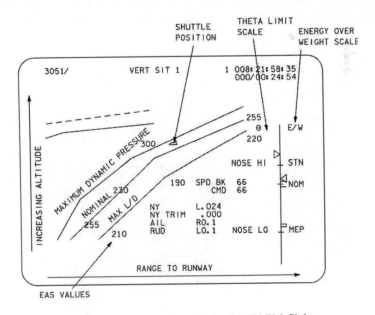

Example of navigation display for TAEM flight.

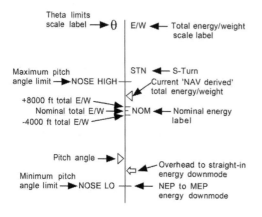

E/W scale on the TAEM display.

plot presumed there was no downmoding. An S-turn, switching to a straight-in approach, selecting the MEP or a different runway would have significantly affected the range that could be flown.

The horizontal situation display (HORIZ SIT – SPEC 50) showed a "God's-eye view" of the Orbiter and the runway approach. The center of the display contained a circle representing the HAC, with its diameter being either the diameter of the HAC at the time of HAC intercept in the acquisition phase or at the current altitude during the heading alignment phase. The HAC on this display always agreed with the HAC that guidance was flying. The runway was also represented on the HAC circle. At the base of the display was a fixed bug pointing straight up that represented the Orbiter, with the "world" rotating around it. In other words, the view offered by this display was a ground plane projection containing the vehicle and landing site in a frame of reference that was fixed to the vehicle. The scaling automatically changed so that the ground track resolution increased as the vehicle neared its target. Its future path was shown by three circular heading predictors for positions at respectively 20, 40 and 60 seconds ahead based on current flight conditions. A horizontal bar located to the right of the HAC circle was the remaining time to HAC intercept. Twenty seconds prior to

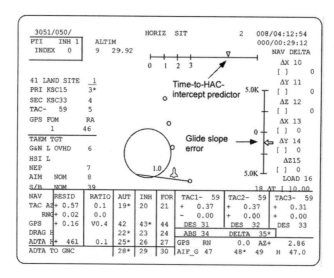

HSI display.

intercept, the triangular cursor would start to flash, and then at ten seconds it would start to move. At intercept, the scale automatically changed to depict the HAC radial error as an indication of how well the turn on the HAC was going on. A vertical scale placed on the right side of the display depicted the glide scope error, that is to say the altitude error from the TAEM guidance.

At the base and right and left sides of the display, various other navigation and guidance information was provided. It was always initialized for an overhead HAC approach but it permitted the crew to switch to a straight-in approach. It also allowed manual switching of the HAC intercept point from the NEP to the MEP.

APPROACH AND LANDING

With the Orbiter now less than two minutes from touchdown and with the runway in sight, the astronauts knew they would soon be able to hug their loved ones, but they had to remain focused. In the main, the objective of the approach and landing (A/L) was to land on the runway within specific constraints that included the distance to the runway, groundspeed, airspeed, and altitude rate. The maximum groundspeed was 225 knots, which was the maximum tire speed. The targeted airspeed was 195 to 205 KEAS depending on vehicle weight. The altitude rate at touchdown was to be less than 4 feet per second.

A two-glide slope, two-flare trajectory was employed to approach the runway and make an unpowered landing with a low L/D. As with the previous phases of re-entry,

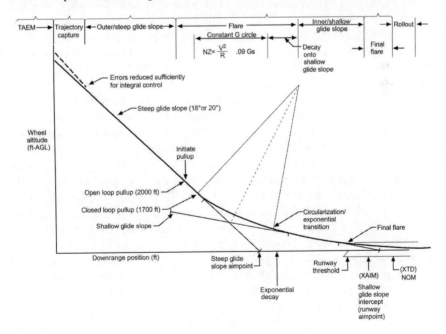

Approach and landing trajectory.

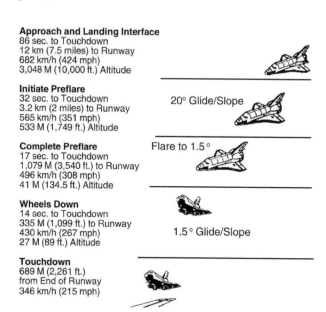

Approach and Landing Interface
86 sec. to Touchdown
12 km (7.5 miles) to Runway
682 km/h (424 mph)
3,048 M (10,000 ft.) Altitude

Initiate Preflare
32 sec. to Touchdown
3.2 km (2 miles) to Runway
565 km/h (351 mph)
533 M (1,749 ft.) Altitude

20° Glide/Slope

Complete Preflare
17 sec. to Touchdown
1,079 M (3,540 ft.) to Runway
496 km/h (308 mph)
41 M (134.5 ft.) Altitude

Flare to 1.5°

Wheels Down
14 sec. to Touchdown
335 M (1,099 ft.) to Runway
430 km/h (267 mph)
27 M (89 ft.) Altitude

1.5° Glide/Slope

Touchdown
689 M (2,261 ft.)
from End of Runway
346 km/h (215 mph)

Approach and landing interface to touchdown events.

A/L guidance was divided into phases: trajectory capture, outer glide slope, preflare and shallow glide slope, final flare, and rollout.

The trajectory capture phase was designed to fly from the TAEM interface up to acquisition of the outer glide slope. Simply put, its task was to allow the commander to perform, if necessary, some last minute adjustments in the position of the vehicle relative to the runway centerline. Large errors in position were best removed during the TAEM pre-final phase, and the transition to trajectory capture would be delayed to eliminate such errors. The transition would occur anyway at an altitude of 5,000 feet above ground level to provide some time to bring the Orbiter through the outer glide slope. During this phase, the body flap was automatically retracted to the trail position, stowing this aerosurface for the remainder of the flight. If this did not occur, or if the flap was not within 5 per cent of the trail position, an alert would flash on the HUD and the crew would stow it manually.

Confirmation that the flight software had automatically switched to A/L guidance was shown by a flashing "A/L" symbol on the VERT SIT 2 display and by the HUD with the display of the CAPT mode. The switch from TAEM pre-final guidance to A/L guidance was based on a logical check of altitude error, lateral position error, dynamic pressure error, and flight path angle error, and would nominally occur at an altitude of 10,000 feet but in some cases it could occur at the exit from the HAC at 12,000 feet.

The outer glide slope (OGS) consisted of a sloped line segment intercepting the ground at a specified distance from the runway threshold, also referred to as the "aim point". The task of the OGS was to accelerate to a high airspeed and ensure that the Orbiter was at the proper position for the maneuvering of the subsequent phases. For this reason, it was designed to be as shallow as possible, providing the lowest descent rate and the least demanding transition to a 1.5 degree shallow glide slope, yet steep

Approach and landing

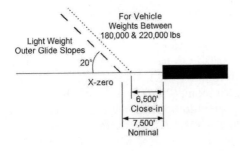

OGS profile.

enough to retain sufficient speedbrake reserves to accommodate varying winds and trajectory dispersions. Hence the OGS angle depended on the weight of the vehicle. Specifically, for a vehicle weighing less than 222,000 pounds the angle was set at 20 degrees whilst for heavier vehicles it was set at 18 degrees. In either case the slope was much steeper that the 3 degree approach of a commercial airliner. On the OGS, the commander would fly pitch and roll/ yaw in CSS mode, while leaving guidance to automatically open/close the speedbrake to maintain the reference airspeed value of 300 KEAS. In fact from the start of A/L guidance, speedbrake and pitch commands were related in such a way that if a pitch command resulted in an airspeed above or below the reference value, the speedbrake would close or open in proportion to that error in order to return the airspeed to the reference value. This logic continued until the altitude was 3,000 feet, when the "smart speedbrake" logic took over. Introduced with STS-51D in April 1985 this simply retracted the speedbrake to an angle which would result in a touchdown at a speed of 195/205 KEAS at a point 2,500 feet from runway threshold. This angle of retraction was maintained to an altitude of 500 feet, when there was an adjustment to account for any change in winds and atmospheric conditions since passing through 3,000 feet. No further adjustments were made. The adjustment altitude of 500 feet was chosen because the time to fly from 3,000 to 500 feet was approximately the same as to fly from 500 feet to touchdown. This allowed time for the adjustment to influence the energy state of the vehicle. It should be noted that owing to degradation of the handling qualities of the Orbiter near touchdown, the maximum permissible nominal speedbrake command at this point was 50 per cent.

As stated, the extension of the outer glide slope segment intercepted the ground at the aim point, which for nominal approach was 7,500 feet from the runway threshold. But some atmospheric conditions could force the aim point nearer the threshold. For this reason, there was a "close-in aim point" set 6,500 feet from the threshold. If this was chosen then guidance automatically added 1,000 feet to the total "down-range", relocating the touchdown point 1,000 feet further forward to ensure that the Orbiter would have the same touchdown energy target despite the geometry having changed. In this case a larger speedbrake command would be needed in order to dissipate the additional energy.[10] Guidance also offered an option

---

[10] In comparison to a nominal aim point, in a close-in aim point scenario the Orbiter was delivered closer to the runway. At the moment of transition to the preflare and inner glide slope it would have more energy than in the nominal scenario. Hence the speedbrake had to be opened wider to provide the Orbiter with the same energy as the nominal scenario at touchdown.

known as the "short-field speedbrake" for use in the case of limit violations on either rollout margin or brake energy. When this was selected, guidance would open the speedbrake an additional 12 to 13 degrees at an altitude of 3,000 feet and hold this through to touchdown so as to dissipate an additional 1,000 feet of touchdown energy. As a consequence the Orbiter would land 10 KEAS slower than planned whilst maintaining the nominal touchdown point 2,500 feet from the runway threshold. In energy terms this converted the additional 1,000 feet of dissipated down-range distance into a touchdown velocity 10 KEAS slower than nominal. The short-field speedbrake could be utilized on any occasion where the touchdown groundspeed was predicted to exceed the limit of 214 knots of the flight rules. This would be decided by the flight director and manually selected by the crew using the SPEC 50 display. The objective would be to obtain additional margin from the 225 knots certified for the tires. This option could also be selected in the event of a landing with a failed or leaking main gear tire because it would diminish the stress on the tire upon coming into contact with the ground. It could also be employed for TAL/ECAL runways to gain an adequate margin for the rollout. A third speedbrake option that the crew could set in SPEC 50 was the so-called "emergency landing site speedbrake" for a heavyweight landing on a runway with a usable length of less than 8,500 feet. In this situation, the speedbrake would be set to dissipate an *additional* 1,500 feet of energy over the short field option; 2,500 feet in total.

To assist the commander in orientating the Orbiter towards the selected aim point during the OGS visual aids were placed on the terrain. In particular, the nominal aim point was indicated by a rectangle and the close-in aim point by a triangle, both of which were colored for high visibility. On lakebeds and light-colored terrain the aim points were usually black, whereas for vegetation they were usually white. At both aim points PAPI lights were installed to assist the crew in identifying and flying the OGS. Such lights are commonly used by pilots on final approach to verify they are approaching the runway along the correct path and at the right slope. They are highly directional and their color appears to change with viewing angle. If the Orbiter was higher than the nominal OGS its pilots would observe three white and one red light, or four white lights. If the Orbiter was lower then the pattern would be

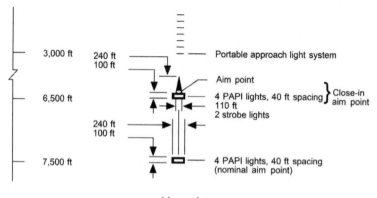

Aim points.

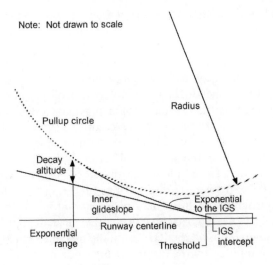

Preflare geometry.

one white and three red lights, or four red lights. Two white and two red lights would be observed if it the approach was correct.

At 2,000 feet of altitude the commander started to execute the preflare maneuver, a 1.3 g pull-up to smoothly transition from the OGS to the shallow/inner glide slope (IGS). The maneuver trajectory was a circular arc tangent to the OGS followed by an exponential portion tangent to the IGS. As part of this transition, the altitude rate was reduced from 190 feet per second to just 12 feet per second. To protect the main gear struts against unacceptable loadings, the rate at touchdown had to be between 6 and 9 feet per second depending on vehicle weight and crosswind. The task of the IGS was to further dissipate the energy of the vehicle and deliver it to the runway at a suitable altitude rate. It was a 1.5 degree approach[11] in order to maintain a reasonable ground clearance on emerging from a preflare that would require only minor maneuvers for the actual landing.

IGS was initiated at an altitude of 300 feet and coincided with the deployment of the landing gear. For the first four missions, velocity served as the cue for when to deploy the gear; specifically 270 knots, which corresponded to an altitude of 200 feet on the nominal energy profile. But this put the deployment very late in the approach and too near the runway. In the case of STS-1 a higher than expected energy meant deployment occurred at just 85 feet. Had the gear been deployed earlier, some of this energy would have been dissipated by the increased drag. For STS-2 the opposite situation developed. The Orbiter was low in energy and deployment at an altitude of 400 feet further reduced its energy state. On STS-3 the Orbiter was once again high in energy and was actually accelerating past 270 knots as the gear deployed at just 87

---

[11] This is why the IGS was also called the shallow glide slope.

feet, a few seconds before touchdown. Some of the people watching the landing were shocked that the gear came down so late. Starting with STS-5 altitude served as the cue for when to deploy the gear. The nominal procedure was for the pilot to arm the gear at the start of the preflare (2,000 feet) and command deployment at 300 feet $+/-$ 100 feet to satisfy the design requirement of having the gear down and locked at least five seconds prior to touchdown. Certainly, gear deployment must occur by 200 feet. Although altitude was the main cue for deployment there was the additional rule that it must not occur at an airspeed exceeding 321 knots, because the high aerodynamic loads on the doors would interfere with the normal opening sequence.

To fly the preflare and IGS the pilots would refer to the "ball and bar lights" on the left side of the runway. The ball light was placed 1,700 feet beyond the threshold and the bar light 500 feet further on. Beginning at 1,000 feet AGL in the preflare, the commander would progressively bring the ball and bar into his field of view and note the rate at which the ball was approaching the bar. As the ball approached the bar he would relax the nose-up input on the RHC until the ball intercepted the bar. As long as the ball was superimposed on the bar the Orbiter was precisely on the 1.5 degree glide slope. An additional visual aid were rectangles on the runway itself. Aiming the vehicle straight towards these runway marks resulted in good IGS control and a safe threshold crossing height.

It was essential to keep the Orbiter lined up with the runway centerline through to preflare, with deviations being corrected as soon as they developed in order to enable the commander to keep his attention focused on the pitch preflare tasks. This would avoid the pitch axis and final shallow glide slope departing the commander's field of view, possible resulting in an early, excessively fast and potentially hard landing.

It is important not to confuse the approach on the IGS with a conventional ILS. In making an ILS approach a conventional aircraft flies a stabilized path leading to the runway. The IGS was a reference path for flying the Orbiter from preflare through to final flare. In other words the IGS represented a path which, if flown, would give the Orbiter the best possible speed and distance from the runway

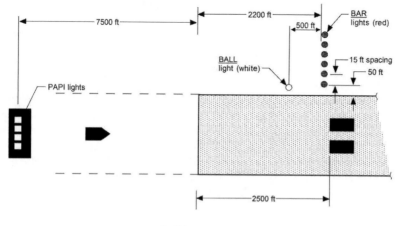

Ball bar geometry.

threshold for initiating the final flare and touchdown. On an ILS approach the pilot of an aircraft is required to precisely follow the electronic path for a considerable time. The Orbiter spent only several seconds on the IGS prior to starting the final flare. In flying the IGS, Shuttle pilots found the "bang-bang" technique useful for making pitch attitude corrections. This involved making small, rapid inputs with the RHC and returning it to the detent, then waiting until the effect was apparent prior to making a further correction. It was less effective at greater landing weights because the center of gravity moved forward, but was helpful in accommodating the peculiar handling qualities of the Orbiter in landing.

Just prior to crossing the runway threshold the commander would initiate the final flare at an altitude in the range 30 to 80 feet. This was a smooth increase in pitch to reduce the sink rate to an optimum of 4 feet per second for touchdown. In performing this maneuver the commander had to keep in mind two important factors. Firstly, the pronounced "ground effect" created by the wing on attaining 50 feet which eased the task of landing. Secondly, there was a rather annoying handling quality of the Orbiter during the early reaction to a pitch input. Due to the relative locations of the center of gravity, the aerodynamic center, the control surfaces, and the flight deck, an upward deflection of the elevons in response to a pitch-up command would actually cause a momentarily *decrease* in lift; the desired *increase* in lift would occur only after the Orbiter had rotated to a higher angle of attack. If the commander wished to reduce the sink rate, it was necessary to time the pitch-up to avoid a premature touchdown. To further complicate the situation, any little effects sensed by the pilots following a pitch-up input risked having them apply a greater input than required. Good practice was therefore to minimize the number of inputs as the Orbiter neared the ground in order to avoid an early touch down or a balloon effect.

Re-entry navigation hardware

In order to accomplish a pinpoint landing, the guidance system required accurate and reliable measurements of the state vector of the Orbiter from entry interface through to touchdown. The Super-G model which constantly processed the data provided by the IMUs was complemented during re-entry by several other navigation sensors.

One of these was the TACAN, a Cold War global tactical air navigation system for military and civilian aircraft operating at L-band frequencies (1 GHz). The three TACAN receivers installed on board the Orbiter were used to determine slant range and magnetic bearing from a TACAN or a VHF omni-range tactical air navigation (VORTAC) ground station. Each unit was provided with two antennas, one on the lower fuselage and one on the upper forward fuselage, both covered by thermal tiles. With a maximum range of 400 nautical miles, during re-entry TACAN was available from an altitude of 160,000 feet through to 1,500 feet AGL. Acquisition started when the distance to the landing site was approximately 120 nautical miles, at which point the system began to interrogate the navigation stations the Orbiter would meet during its descent, locking onto the first one that became available. To provide the

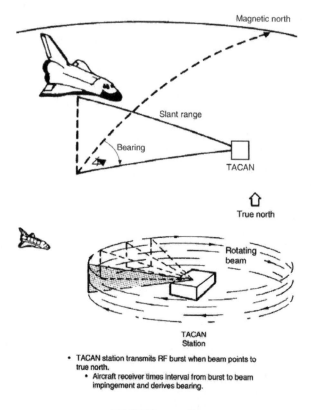

TACAN operation.

range and bearing data needed to update the positional components of the state vector at least two of the three TACAN units had to be locked onto the same station. As the descent continued, the distance to the remaining stations was computed and the next-nearest was automatically selected when its range was less than that to a previous station. When the distance to the landing site was down to approximately 20 miles only one station was interrogated. Carrying three TACAN units provided a redundant management scheme. That is, the three units were compared to detect any significant difference. If all units were good, the middle values were selected for range and bearing. If a fault was detected a fault message was generated.

Since the data provided by the TACAN could not be used to calculate altitude, at Mach 5 the pilot deployed the air data system, a barometric array of sensors designed to provide data on the movement of the Orbiter in the air mass. It consisted of two air data probes, one on each side of the forward lower fuselage to sense the air pressure caused by the vehicle passing through the atmosphere. This was used to update the altitude component of the state vector, provide guidance in calculating steering and speedbrake commands, update the flight control laws, and provide display data for the various GNC displays and the commander's and pilot's flight instruments. Each air data probe sensed the static pressure and the upper, center and

Approach and landing 481

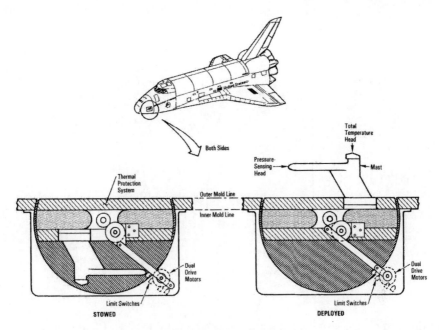

Air data probe assembly.

lower angle-of-attack pressures. These four pressure-port sensors were joined by two temperature sensors. All pressure and temperature sensors on a given probe were connected to a pair of air data transducer assemblies (ADTA) which elaborated the pressures and temperatures for transmission in digital form to the GPCs and flight instruments. Once again, a redundant management scheme guaranteed purity and correctness in the data used by the navigation software. This began with the GPCs comparing the pressure readings from the four ADTAs. If all the pressure readings agreed within a specified range, one set of readings from each probe was averaged and processed by the software. If one or more pressure signals of a set of pressure readings failed, that set's data flow would be halted and the software would proceed using data from the other ADTA of that probe. If both sets failed for one probe, only data from the other probe would be used. However, temperature data from all four ADTAs was sent to the software.

After the Department of Defense decommissioned its TACAN stations the fleet of Orbiters were equipped with three GPS receivers. These were used in the same fashion, but rather than updating the existing state vector as was done by processing the TACAN and air data probe data they determined an entirely new state vector.

Since in the final stage of its descent the Orbiter was an aircraft, it was natural to use a conventional microwave landing system (MLS) to enable the Orbiter to obtain accurate three-dimensional position information to compute state vector components in steering to the runway. The ground station had an elevation shelter located near the projected touchdown point and an azimuth/distance-measuring equipment shelter at the far end of the runway. The Orbiter would acquire the signal when fairly

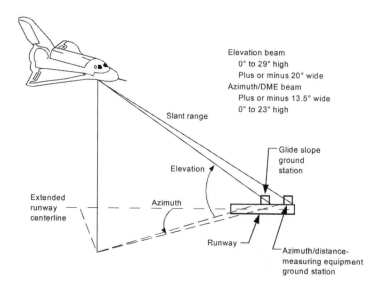

MLS operation.

near the landing site, usually at an altitude of about 18,000 feet. Final tracking would start at the approach and landing interface some 8 nautical miles from the azimuth/distance-measuring equipment and at an altitude of about 10,000 feet. The three MLS units on board would feed the primary flight instruments with useful data on elevation angle, azimuth angle, and range relative to the MLS station to enable the navigation system to compute accurate state vector components and the commander to fly precisely to the touchdown point.

Head-up display

A head-up display (HUD) is a common feature of modern aircraft that presents flight data on the pilot's line of sight through the windscreen. It is particularly useful during approach and landing because it means the pilot never needs to divert his attention from the approaching runway.

On the Orbiter the HUD was an optical miniprocessor that cued the commander and pilot during the final approach to the runway. It allowed out-of-window viewing by superimposing flight commands and information on a transparent combiner in the line of sight to the window, and the transmissivity of the combiner enabled a pilot to look through it and see actual targets like the runway. For example, at 9,000 feet on final approach in an overcast with its base at 8,000 feet, the lighted outline of the runway would be displayed on the combiner. Once below the overcast, the lighted outline of the runway would be superimposed on the real runway.

A three-position MODE switch located below the HUD allowed for an automatic sequencing of formats and symbology (NORM position), or for a manual selection to "declutter" the display (DCLT position). A succession of DCLT would serially

Approach and landing

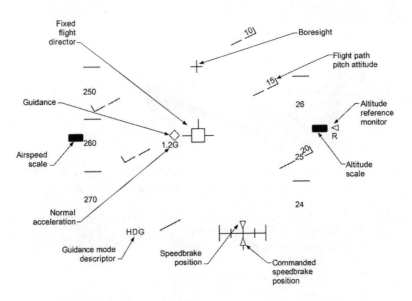

HUD display during approach and landing; no declutter.

eliminate display elements in the following order: the first activation would remove the runway symbology; the second would replace the airspeed and altitude tapes with digital values and delete the horizon/pitch attitude scales but retain the horizon line when it was in the field of view; the third would remove all symbology except the boresight; and the fourth would return the HUD to its original form with all symbols displayed. The third and final MODE switch setting ran a self-test of the display for a period of five seconds.

Introducing the HUD to the Orbiter was a task assigned to Jack Lousma, veteran of the Skylab program, "Somebody discovered that a head-up display was being used in fighter airplanes and might be of some use in the Shuttle for landing ... The idea of the head-up display is that if you put all airspeed and altitude information on the windscreen, then you don't have to take your eyes off the runway." Lousma started a thoughtful investigation to explore how advantageous it would really be, "I went to various companies that were making the airplanes that had them ... I would fly in the airplane or sometimes just fly the simulator just to get an idea of what the head-up display was all about." The military were using the HUD for tasks such as weapons delivery, air-to-air combat, and low-level navigation, but *not* as a landing aid. But he saw its potential, "I decided that we could use this for approach and landing because the Shuttle is a glider, and you only have one chance to do it right. Any kind of help is probably good. I developed an engineering plan [as a preliminary to] a management plan." Management eventually accepted the recommendation, but since *Columbia* had already been built and it would have been too expensive to add a new system, the first five missions were landed without the HUD. The other Orbiters were built with the HUD installed. The system was added to *Columbia* during its first period of programmed maintenance.

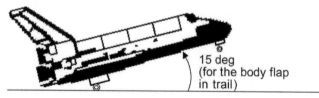

$$\text{ENERGY RESERVE} = \frac{V_{\text{Tailscrape}} - V_{\text{Touchdown}}}{\text{Deceleration}_{\text{Avg}}}$$

TAILSCRAPE — The maximum pitch attitude (deg) the vehicle can land at and not scrape the tail (body flap).

ENERGY RESERVE — The time (sec) the vehicle could have continued flying after touchdown until the tailscrape condition.

Energy reserve and trail scrape.

Touchdown

The A/L trajectory was designed to result in a touchdown about 2,500 feet past the runway threshold at a safe speed and sink rate in order to provide an energy reserve and Orbiter pitch attitude at touchdown.[12] Energy reserve was defined as the time in seconds that the Orbiter could fly until the pitch attitude reached the tail-scrape angle of 15 degrees. This time was a function of weight, since heavier vehicles had to fly faster to produce aerodynamic lift equal to their weight. Consistent energy reserve and consistent pitch attitude were both key factors that helped pilots to land with sink rates within vehicle limits. Other parameters influencing the energy reserve were winds and air density, vehicle velocity errors, and off-nominal factors such as the altitude at which the landing gear was deployed, GNC air data errors, tile damage, a stuck speedbrake, and NAV errors.

At touchdown, which was when the main landing gear wheels first made contact with the runway, a sensor sent a weight-on-wheels (WOW) discrete signal to inform the flight control system that the Orbiter did not need to be flown anymore. A series of events took place very rapidly, and once again the pilots had to be supremely alert. The rudder and aileron trim were frozen, and the speedbrake opened to 100 per cent to help to slow the vehicle on the runway. Meanwhile, an automatic sequence of load balancing started to move the ailerons in a manner designed to balance the loads on the tires. These commands were a function of true airspeed, lateral acceleration, and the yaw rate of the vehicle. At the same time, the brake isolation

---

[15] The Orbiter landed at a speed in the range 195 to 205 KEAS depending on the vehicle weight, which was fast for an aircraft.

valves opened to allow the main wheel brakes to operate later on, during the final part of the rollout. The Orbiter maintained an angle of attack of about 8 degrees in order to generate lift and reduce the maximum loads on the main landing gear.

Derotation

With the main gear already spinning on the runway, a pilot might easily presume that completing the landing would be child's play. But the Orbiter was an aircraft like no other and bringing the nose gear wheels down required a different technique from a conventional aircraft. As Joe Engle explains, "When you touch down, you have to keep coming back on the stick a little bit and ease the nose down, and keep coming back on the stick, because as you slow down, the dynamic pressure's getting lower, so the force on the surfaces is less; they have less force, less muscles, so you have to keep deflecting the surface more and more in order to control the rate of derotation on nose-gear slapdown."

But the Orbiter's flight software had been designed with a peculiar behavior, as Engle says, "The flight control system in the Orbiter is a rate command system and that means that ... if you don't ask for anything in the stick, if you don't come out of detent on the rotation hand controller in pitch and ask it to do a pitch rate one way or the other, it tries very hard not to. It will do everything it can not to, and that's true even after it touches down." The direct consequence of this was that when it came to land an Orbiter, pilots had to think in the opposite direction. As Engle puts it, "In the Orbiter you don't do that. As a matter of fact, if you don't deflect the stick, and the nose starts down, the flight control system senses a pitch rate with no request from the pilot, so it brings the elevons back up to try to hold that attitude. It's kind of an unnatural thing. Plus it happens after you've come through the re-

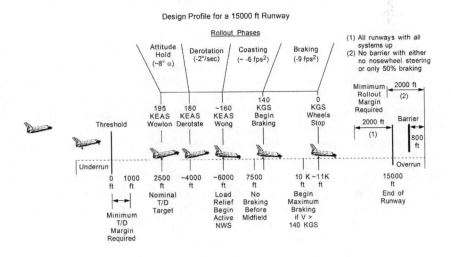

Rollout phases.

entry and made the approach, a steep approach, and gotten the bird on the ground and you're kind of in a relaxed mode, 'I'm back home safe,' and so you're not ready for something new like that. And before you know it, the bird has slowed down, the [elevons are] all the way up, saturated in deflection, trying to keep the nose up, and you really don't want the nose up because it's going to come down and slap down real hard then." Therefore, as soon as the main gear was on the ground, a Shuttle pilot had to "make a conscious effort ... to go forward on the stick. You must initiate the derotation and get the nose started down so that by the time the [elevon] surfaces get almost to full up, why your nose is on the ground or nearly on the ground. It's a different technique, is all." Once a pilot understood the *technique* for a correct derotation, the next step was to know the *correct time* to push the nose gear down. Shuttle commander Thomas Kenneth Mattingly provides some useful insight, "You really don't want to land too slow, because if you land below about 165 knots you could hold the nose up. So you don't put the nose down above 160, [you do it] not much below that."

A nominal derotation would be started at around 185 knots, a speed selected as a compromise between the maximum load on the main gear, the maximum load on the nose gear, the time between touching down and starting derotation, and the need to get the nose gear on the runway to begin braking. Derotation was originally achieved by applying a forward motion to the RHC but this was revised to pushing the RHC trim switch forward. This change was made after it was noticed that pilots tended to push forward the RHC very rapidly, almost as a step input, causing the nose landing gear to make a hard contact with the runway. The trim switch allowed for a smoother derotation. Obviously, in the event of trim switch failure the RHC could still be used for derotation. Although it lasted only a few seconds, derotation required good crew coordination. The pilot had to recite the equivalent airspeed and make a "derotate" announcement when it reached 185 knots. At this point, the commander would input the nose down command in order to generate a pitch rate of 1 to 2 degrees per second and the pilot would start to call out the pitch rate displayed by the needles of the ADI.

Velocity was not the only important parameter in starting the derotation, the onset rate of the derotation was equally important, as Mattingly explains, "Because of that big fat fuselage, the nose will stay up in the air. In the simulator it will stay up there to maybe 70 knots. But there's no way to let it come down without overloading the structure and breaking the fuselage. So you really need to get there and get the nose down before you lose elevator control, because otherwise this thing will just fall." This is what happened on STS-9 as the result of a fault in the flight control system. Brewster H. Shaw was the pilot, and before describing what happened that day he offers some useful background, "There are gains in the flight control system, and the gains change depending upon what phase of flight you're in. When you're flying a final [approach] there are certain gains that make the vehicle respond in a certain way to the inputs the pilot makes on the stick. When the main gear touches down, the gains change so you can derotate the vehicle and get the nose gear on the ground in an appropriate way." Shaw along with his commander, the veteran John Young, were ready to land the Orbiter safely in the way they had been instructed. As

Shaw explains, "The way we trained, it was loose control, so you'd start it down and then you'd let the RHC ... come back into detent, and it would just fall nicely down." However, without the crew knowing anything about it, something had been changed in the flight control system. "Well, they tightened up the gains. When John got the rate going, he let [the RHC] come back to detent as always. And it stopped! The nose stopped right there ... He had to get back in the loop and command it on down. So that delay meant we were decelerating all the time. We were going pretty slow when the nose gear hit the ground, and the ailerons [which were] basically saturated couldn't control the rate, so it fell through." This was *not* a textbook nose gear touch down. "So ... John's flying the vehicle I'm giving him all the altitude and airspeed calls and everything, and you feel this nice main gear gently settling onto the lakebed. From downstairs where the rest of the guys were ... you hear 'Yay!' and clapping ... Then John gets the thing derotated and we're down to about 150 knots or so when the nose gear hits the ground with a smash! So they change from 'Yay! Yay!' to 'Jesus Christ! What was that?' ... That was just really funny ... And poor John, he was embarrassed because of the way the nose gear hit down. But it wasn't his fault. They had changed this thing without him being able to practice using the new flight control system."

On the other hand, as Mattingly points out, "If you put your nose down too soon, then the negative angle of attack on the wings adds to the load on the tires and you can blow the tires."

On 30 March 1982 *Columbia* was about to make its third landing and everything seemed to be nominal, then the unforeseen happened. Jack Lousma, who was in the left seat, has a vivid memory of that day, "Nominally, the nose gear is held up in the landing position by the attitude-hold control function for aerodynamic braking until slowing to 185 knots, then it is manually lowered to the runway. In the STS-3 case, the nose began to lower to the runway immediately after touchdown. To hold it up I executed a quick pitch-up input with the rotational hand controller, with no apparent effect. I repeated the same input and the nose began an unexpected and rapid pitch-up, whereupon I quickly lowered it to the runway." Watching video footage of the landing, it is impossible not feel apprehensive upon seeing the Orbiter rapidly pitch up and almost scrape its tail on the runway. An investigation established that it was not Lousma's fault but a flight control system problem. In his words, "a divergence or instability in the longitudinal control software for the Shuttle landing configuration caused the unexpected pitch-up behavior".

Most Shuttle landings were very soft, gentler in fact than is usual for an airliner. Speaking of landing STS-4 with Hartsfield in the right seat, Mattingly says, "So he's calling off airspeed and altitude, and I've no idea what it's going to feel like to land. When I would shoot touch-and-go's in the KC-135 there was never any doubt when we landed; you could always tell. I was expecting bang, crash, squeak, something. Then nothing, and nothing. Finally, Hank says, 'You'd better put the nose down.' 'Oh,' I said, 'All right.' So I put it down and I was sure we were still in the air. I thought, 'Oh, God, he's right. We can't be very far off the ground.' Sure enough, we were on the ground and neither of us knew it. I've never been able to do that again in any airplane. Never did it before."

Drag chute

In the time between touchdown and the start of derotation the pilot deployed a drag parachute housed in the tail unit to assist the deceleration system and safely halt the vehicle on the runway for both nominal end of mission and aborts. The requirements for the chute included the ability to stop a 248,000 pound Orbiter making a TAL abort in a distance of 8,000 feet with a tail wind of 10 knots on a hot (103°F) day and maximum braking at 140 knots groundspeed.

The deployment sequence involved blowing off the compartment door and firing a mortar to extract a 9-foot pilot chute, whose inflation would pull out the 40-foot drag chute. The main canopy did not immediately deploy to its fully open position. Instead, it was reefed to 40 per cent of its full capability to reduce the opening shock on the vehicle and crew. After 3.5 seconds two redundant reefing line cutters let it open fully. The pilot chute would separate at this point. The main chute would be jettisoned when the vehicle had slowed to 60 knots. It would not be attempted at a speed under 40 knots in case the riser lines draped over the central main engine and damaged its nozzle.

For a typical heavyweight landing the nominal deployment sequence would start at 195 KEAS but for a lightweight landing this could occur as soon as touchdown. In both cases the deployment had to be concluded prior to completion of the derotation maneuver. The flight rules also allowed for an early deployment between touchdown and the nominal deployment time for a high energy landing in order to get the vehicle stopped on the runway. To avoid the chute imposing excessive stress on the structure of the Orbiter in the case of deployment at too fast a speed there were pivot pins in the mechanism that were designed to shear if the load exceeded 125,000 pounds, thus jettisoning the chute.

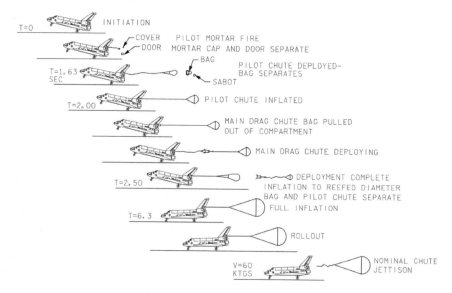

Drag chute deploy sequence.

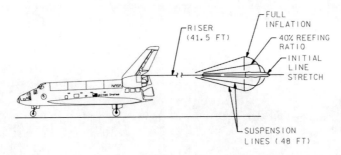

Drag chute configuration.

Pilot chute extraction

The introduction of the drag chute by STS-49 in May 1992 greatly improved the Orbiter's performance during the rollout. It not only saved the brakes, it also reduced wear on the tires. Because the chute was attached above the center of gravity of the vehicle, it create a nose-up moment that reduced the elevon control required during derotation and hence the load on the main landing gear. The deployment of the chute also imparted positive directional stability which minimized directional upsets caused by failed tires.

STS-95 was a mission noted for including amongst its crew John Glenn, the first American to orbit in space. It is less well known for being the first time that the drag chute was *not* deployed since its introduction. This was because the door of the chute compartment detached immediately after liftoff. Frances A. Ferris led a group from Rockwell tasked with conducting a rapid investigation. As she explains, "This failure was interesting because we had to figure out why the door fell off, and we had a vehicle on-orbit that didn't have a door. So we had the root cause investigation and another set of investigations about what do we do when we land? Is it safe to land?

Drag chute extraction and derotation.

We managed to obtain some video that showed the aft end of the vehicle. There was some browning and maybe a little charring of straps, but [otherwise] the drag chute compartment looked intact." This was very encouraging. Things looked even better when a thermal analysis was made. "The temperatures in that area are more severe during ascent than descent. The determination was made that we'd survive re-entry, that there wouldn't be any issue." Having cleared the Orbiter for a safe landing, a new doubt arose. As Ferris explains, "There was concern that if you … tried to deploy the chute, would it hang up and do more harm than good?" And it was not merely the landing. "There was a lot of discussion about if we deploy the chute are we going to destroy some evidence that would help us determine what happened and why … and what corrective action we'd need to take." In the end it was decided not to deploy the chute. On 7 November 1998 *Discovery* safely landed on runway 33 at the Kennedy Space Center. The ensuing investigation concluded that the sparkles used to burn the hydrogen leaking from the SSMEs just prior to liftoff must have left some pockets of unburned hydrogen which caused pressure waves at the moment of engine ignition. But this was insufficient to cause the door of the chute compartment to detach. That was traced, as Ferris explains, to "the pin that went in the hinge that held the door in place. If you got a pin that was on the low end of the material property strength, it could break with the engine ignition environment. We had never had that situation of the low margin pin and the high pressure environment at the same time." The fix was straightforward, as Ferris delightfully points out, "We put in stronger pins." No other failures occurred with the chute system.

Rollout

When the nose gear made contact with the runway it triggered the weight-on-nose-gear (WONG) discrete signal that told the flight control system to automatically move the elevons to 10 degrees down in order to reduce the loads on the main gear. The crew had to ensure that this load relief command took place, because its failure could impose unacceptably high loads on the main gear and result in tire failures and possible loss of control.

Approach and landing 491

Rollout. It is possible to see the downward elevon deflection to alleviate the loads on the main landing gear.

Nose gear touchdown also enabled nose wheel steering (NWS) via commands given by the rudder pedals to a steering actuator on the nose landing gear strut. NWS was a lateral acceleration command system, meaning that if the rudder pedals were not deflected no yaw command was given. Consequently, if the GPCs sensed an acceleration via the Nz accelerometers due to crosswind or differential braking, the nose wheel would be moved to remove this uncommanded acceleration. If either the commander or the pilot applied heel pressure to deflect his rudder pedals the GPCs would command a lateral acceleration to turn the Orbiter. For redundancy, a pair of hydraulic systems drove to the NWS actuator. If both systems failed, steering could still be achieved by differentially braking the main gear.

The start of the braking phase was a combination of groundspeed and position on the runway. Nominally, it began soon after nose wheel touchdown with the Orbiter at a groundspeed of less than 140 knots and having passed the midfield point on the runway. It would also start for a groundspeed in excess of 140 knots if the Orbiter was within 5,000 feet of the end of the runway. In this case maximum braking force would have to be applied. The selection of 140 knots for the start of braking was a compromise between the maximum possible braking speed, the need to limit the energy absorbed by the brakes, and the desire to halt the vehicle quickly.

Crews were trained to set WOW/WONG manually at nose gear touchdown to ensure the load relief measures and activate nose wheel steering. This was achieved using either the ET SEP or SRB SEP manual overrides whose original purpose had expired. The SRB SEP switch was favored since it was a toggle that was simple to operate, whereas the ET SEP switch was a more complicated over-center lever lock switch.

Landing site selection

The Orbiter came down unpowered and without any possibility of going around for a second approach, so a large and long runway was deemed necessary for a successful landing with a reasonable margin of safety. The perfect site was Edwards Air Force Base in California because it offered several dry lakebed runways as well as a hard runway. The lakebeds relieved the commander of any worries about not being able to stop the Orbiter before running out of strip. Another advantage was that the desert environment offered practically perfect weather conditions essentially all year round. However, a landing at Edwards meant that the Orbiter would need a ride back to the Kennedy Space Center atop a carrier aircraft.

To make turnarounds more efficient, in 1976 the Shuttle Landing Facility (SLF) was constructed in Florida, a few miles northwest of the Vehicle Assembly Building. This paved runway was 15,000 feet long with a 1,000-foot overrun on each end. Its width of 300 feet was similar to the length of a football field and there were 50 foot asphalt shoulders on each side. It was of 16 inches thick in the center but an inch less on the sides in order to allow water to drain away. To further assist in water drainage the strip sloped 24 inches away from the centerline to the edges. But the weather in Florida was much more variable.

The first six missions from April 1981 to April 1983 were all scheduled to land at Edwards where there was plenty of room for off-nominal approaches. Owing to wet conditions at Edwards, STS-3 used the Northrup desert strip at White Sands in New Mexico. STS-7 was meant to be the first mission to land in Florida, but marginal weather diverted it to California. The first landing at the SLF was mission STS-41B in February 1984. It is worth noting that Shuttle pilots preferred landing at Edwards because the SLF did not offer the margin of a dry lakebed. A misjudged approach to the SLF was likely to end up either in a ditch or in the adjacent swamps.

As any object which did not belong in the runway environment posed a hazard for a safe landing, workers checked up to 15 minutes prior to landing for the presence of any foreign object debris (FOD). One particular FOD were birds, since they could damage the thermal protection system of the Orbiter. This was a special concern for landings in Florida because the SLF was in the middle of a nature reserve. Special pyrotechnic and noise-making devices, and selective grass cutting, were employed to discourage birds from gathering around the runway.

The choice of the actual landing site was based on the weather conditions, which had to be within the limits defined for a safe Orbiter landing. The winds were usually the key factor in determining the final approach. In normal circumstances the Orbiter landed into the wind, meaning that if the wind direction was from north then it would approach from the south. Other factors were the Sun angle in the pilots' field of view. Of course a runway can be approached from either end, and therefore the SLF was available both as runway 15 and runway 33.

Post-landing operations

As occurs when a passenger airliner lands at its destination, the crew could not leave the Orbiter at wheel stop, they had to wait for almost an hour. When the vehicle came to a halt, 25 specially designed vehicles and 150 personnel closed in on it from where they had been waiting since at least two hours before the scheduled time of landing. Owing to the considerable hazard of having the Orbiter full of toxic fumes and gases, the convoy would first move to a staging position some 1,250 feet behind the vehicle. Safety assessment teams in protective attire and breathing apparatus used detectors to obtain vapor level readings around the Orbiter and test for possible explosive and/or toxic gases such as hydrogen, hydrazine, monomethyl hydrazine, nitrogen tetroxide, and ammonia. Once the area around the Orbiter was deemed safe for the convoy, this was moved close. In particular, two purge and coolant umbilical access vehicles took up position directly behind the Orbiter to access its umbilical area. But before starting purging and coolant operations an assessment of the hydrogen level in the aft fuselage was carried out. If hydrogen was detected, the procedure called for an emergency power down of the Orbiter and an immediate evacuation of the crew and the convoy personnel. Fortunately, this situation never arose and the purge and coolant vehicles were able to perform their tasks. On installing the carrier plates for the hydrogen and oxygen umbilicals the payload bay and other cavities were flushed out with cool and humidified air to clear any residual fumes. No more than one hour after wheel stop, cooling operations would start to enable the on board cooling systems to be shut off.

In the meantime, the crew prepared to exit the Orbiter. The crew hatch access vehicle was moved to the hatch side of the Orbiter and a "white room" positioned by the hatch. Once the hatch was open, a doctor would enter the Orbiter and perform a preliminary medical examination to clear the crew for egress. On emerging from the

CTV and "white room" linked to the Orbiter.

white room they followed a curtained ramp directly into a crew transport vehicle similar to the "people movers" of an airport. After doffing their bulky pressure suits, and putting on more comfortable clothes, those members of the crew who were fit to stand performed a walk around inspection of the Orbiter, then they were driven away in the van.

Responsibility for the Orbiter now passed from Mission Control in Houston to the Kennedy Space Center, whose personnel entered the vehicle to install switch guards and remove data packages from experiments. After a total safety downgrade, ground personnel prepared the vehicle for towing, including installing locking pins in the landing gear and attaching the tow bar. Meanwhile, engineering test teams issued commands to configure on board systems for the tow. Unless there were time-sensitive experiment removals on the runway, towing normally began within four hours of landing, and lasted two hours. On arrival at one of the three OPF hangars the Orbiter would start a new maintenance flow to prepare it for its next mission in space.

Or at least that was the routine while the Shuttle was operational. Now that the program has concluded, the surviving Orbiters are star attractions in museums.

Glossary

Ablative material: a material, especially a coating material designed to provide thermal protection to a body in a fluid stream through loss of mass.
Abort: to cut short or break off an action, operation or procedure with a vehicle, especially because of equipment failure.
Airlock: a compartment capable of being depressurized without depressurization of the Orbiter cabin, used to transfer crewmembers and equipment to and from the unpressurized payload cargo bay.
Bus: a main circuit, channel or path for the transfer of electrical power or, in case of computers, information.
Capture: the event of the remote manipulator system end effector making contact with and firmly attaching to a payload grappling fixture. A payload is captured at any time it is firmly attached to the remote manipulator system.
Cavitation: the formation of bubbles in a liquid, occurring whenever the static pressure at any point in the fluid flow becomes less than the fluid vapor pressure.
Coldsoak: the exposure of equipment to low temperature for a long period of time to ensure that the temperature of the equipment is lowered to that of the surrounding atmosphere.
Commander: the crewmember who has ultimate responsibility for the safety of embarked personnel and has authority throughout the flight to deviate from the flight plan, procedures and personnel assignments as necessary to preserve crew safety or vehicle integrity. The commander is also responsible for the overall execution of the flight plan in compliance with NASA policy, mission rules and Mission Control Center (See MCC) directives.
Damping: suppression of oscillation.
Deadband: that attitude and rate control region in which no Orbiter reaction control system (see RCS) forces are being generated.
Deorbit burn: a retrograde rocket engine firing by which vehicle velocity is reduced to less than that required to remain in orbit.
Docking: the act of joining two or more orbiting objects.
Drouge parachute: a small parachute used specifically to pull a larger parachute out of stowage.
Dynamic pressure: the pressure of a fluid resulting from its motion.
g: symbol representing the acceleration due to gravity.
Geysering: the accumulation of gaseous medium in a line and subsequent expulsion of liquid medium from the line by a bubble.
Gimbal: a device with two mutually perpendicular and intersecting axes of rotation, thus

Glossary

giving free angular movement in two directions, on which an engine or another object may be mounted; also, to move a reaction engine about on a gimbal so as to obtain pitching and yawing correction moments.

Heat exchanger: a device for transferring heat from one fluid to another without inter-mixing the fluids.

Inclination: the maximum angle between the plane of the orbit and the equatorial plane.

Longeron: main longitudinal member of a fuselage or nacelle.

Launch pad: the area at which the stacked Space Shuttle undergoes final prelaunch checkout and countdown and from which it is launched.

Mission: the performance of a coherent set of investigations or operations in space to achieve program goals.

Mission specialist: the crewmember who was responsible for coordination of overall payload/Space Shuttle interaction and that during payload operation phase, directed the allocation of the Space Shuttle and crew resources.

Multiplexing: simultaneous transmission of more than one signal through a single transmission path.

Orbiter: the manned orbital flight vehicle of the Space Shuttle system. Often it is simply referred to as the Space Shuttle or Shuttle.

PAD: Preliminary Advisory Data. Preadvisory data or information on spacecraft attitude, thrust values, event times, etc..., transmitted in advance of a maneuver.

Payload specialist: the crewmember who was responsible for the operation and management of the experiments or other payload elements that were assigned to him or her and for the achievement of their objectives. The payload specialist was an expert in experiment design and operation and might or might not be a career astronaut.

Pilot: the crewmember who was second in command of the flight and that assisted the commander as required in the conduct of all phases of Orbiter flight.

Pilot parachute: a small parachute used to retrieve a drogue or main parachute from stowage.

Pitch: an angular displacement about an axis parallel to the lateral axis (widthwise) of a vehicle.

Retrieval: the process of using the remote manipulator system and/or other handling aids to return a captured payload to a stowed or berthed position. No payload was considered retrieved until it was fully stowed for safe return or berthed for repair and maintenance task.

Rib: a fore-and-aft structural member of an airfoil used for maintaining the correct covering contour and also for stress bending.

Riser: one or more straps by which a parachute harness is attached to human or hardware harness.

Roll: rotational or oscillatory movement about the longitudinal (lengthwise) axis of a vehicle.

Space Shuttle system: Orbiter, external tank and two solid rocket boosters.

Spar: a main spanwise member of an airfoil or control surface.

STS: Space Transportation System. An integrated system consisting of the Space Shuttle, upper stages, Spacelab and any associated flight hardware and software.

Telemetry: the science of measuring a quantity or quantities, transmitting the results to a distant station and there interpreting, indicating or recording the quantities measured; also, the data so treated.

Tile: name used for rigidized ceramic fiber thermal insulation material employed on the Orbiter.

Translation: movement in a straight line without rotation.

Ullage: the amount that a container, such as a fuel tank, lacks of being full.

Bibliography

National Space Transportation System Reference Vol.1&2, NASA, www.ntrs.gov
Space Transportation System User Handbook, NASA TM-84765, www.ntrs.gov
Space Shuttle Vehicle Familiarization, USA, www.nasa.gov
Space Shuttle Technical Conference Vol.1 & 2, NASA-CP-2342, www.ntrs.gov
NSTS 1988 News Reference Manual, NASA, www.nasa.org
Shuttle Crew Operations Manual, USA, www.ntrs.gov
National Space Transportation System Overview, NASA-TM-102953, www.ntrs.gov
STS Investigator's Guide, NASA-TM-105573, www.ntrs.gov
Return to Flight Task Group, NASA, www.ntrs.gov
Columbia Accident Investigation Board Report, NASA, www.nasa.gov
Implementation of the Recommendation of the Presidential Commission on the Space Shuttle Challenger Accident, NASA, www.ntrs.gov
Space Shuttle Operational Flight Rules, NASA, www.nasaspaceflight.com
Crew Software Interface, USA006083, www.nasa.gov
Data Processing System Dictionary, NASA JSC-48017, www.nasa.gov
Caution and Warning Workbook, USA006019, www.nasaspaceflightnow.com
DPS Familiarization Workbook, USA005351, www.nasaspaceflight.com
Data Processing System (Hardware and System Software) Workbook, USA005350, www.nasaspaceflight.com
DPS Overview Workbook, USA006078, www.nasaspaceflight.com
System Management Training Manual, USA006033, www.nasaspaceflight.com
Space Shuttle Avionics System, NASA SP-504, www.ntrs.gov
Computers in Spaceflight, NASA-CR-182505, www.ntrs.gov
Space Shuttle Orbiter Payload Bay Door Mechanisms, NASA, www.ntrs.gov
Space Shuttle Orbiter Aft Heat Shield Seal, NASA, www.ntrs.gov
SRMS History, Evolution and Lessons Learned, NASA, www.ntrs.gov
Space Shuttle Orbiter Structure & Mechanisms, NASA, www.ntrs.gov
Mechanical Systems Training Manual, NASA, www.nasa.org
Androgynous Peripheral Docking System, USA008876, www.nasaspaceflight.com
Payload Bay Doors and Radiator Panels Familiarization Handbook, NASA, www.nasa.gov
SSME Evolution, NASA, www.nasa.org
Main Propulsion System Workbook, USA006162, www.nasaspaceflight.com
Space Shuttle Main Engine – Thirty Years of Innovation, NASA, www.nasa.gov
SSME Overview, NASA JSC-19041, www.nasaspaceflight.com

Bibliography

Space Shuttle Main Engine Orientation, Boeing BC98-04, www.nasaspaceflight.com
Space Shuttle Main Engine The First Ten Years – part 3 Start and Shutdown, Robert E. Biggs, www.enginehistory.org
SRB Overview, NASA JSC-19041, www.nasaspaceflight.com
Solid Rocket Booster (SRB) – Evolution and Lessons Learned During the Shuttle Program, NASA, www.ntrs.gov
Space Shuttle External Tank – System Definition Handbook SLWT Vol.1&2, Lockheed Martin, www.nasaspaceflight.com
External Tank Program – Legacy of Success, AIAA, www.ntrs.gov
Orbiting Maneuvering System Workbook, USA006500, www.nasaspaceflight.com
Reaction Control System Workbook, USA006163, www.nasaspaceflight.com
Environmental Control and Life Support System, USA006020, www.nasa.org
Electrical Power System Training Manual, USA005437, www.nasaspaceflight.com
Thermal Protection System of the Space Shuttle, NASA NASW-3841-72, www.ntrs.gov
RCC Plug Repair Thermal Tools for Shuttle Mission Support, NASA, www.ntrs.gov
Boundary Layer Transition Flight Experiment Overview, NASA, www.ntrs.gov
Inertial Measurement Unit Workbook, USA004488, www.nasaspaceflight.com
Star Tracker/Heads-up Display/ Crew Optical Alignment Sight Workbook, USA006082, www.nasaspaceflight.com
Guidance and Control Insertion/Orbit/Deorbit Workbook, USA006501, www.nasaspaceflight.com
Flight Procedure Handbook Ascent/Abort, NASA JSC-10559, www.nasaspaceflight.com
Intact Ascent Aborts Workbook, USA007151, www.nasa.gov
Contingency Aborts, USA005671, www.nasa.gov
Space Shuttle Day-of-Launch Trajectory Design Operations, NASA, www.nasa.gov
Ascent Guidance and Flight Control Workbook, USA, www.nasaspaceflight.com
Ascent Checklist STS-114, NASA JSC-48005-114, www.nasa.gov
Navigation Overview Workbook, USA006081, www.nasaspaceflight.com
Flight Dynamic Office (FDO) Console Handbook Vol.3, NASA DM-CH-07, www.nasaspaceflight.com
Shuttle OPS GNC, JSC-18863, www.nasaspaceflight.com
History of Space Shuttle Rendezvous, NASA JSC-63400, www.ntrs.gov
Rendezvous in orbit, NASA, www.ntrs.gov
Rendezvous and Proximity Operations of the Space Shuttle, NASA, www.ntrs.gov
Rendezvous and proximity operations workshop, NASA-TM-101895, www.ntrs.gov
Rendezvous STS-135, NASA JSC-48072-135, www.nasa.gov
Entry Digital Autopilot Workbook, USA006497, www.nasaspaceflight.com
Entry, TAEM and Approach/Landing Guidance Workbook, USA005512, www.nasaspaceflight.com
Entry Digital Autopilot Workbook, USA006497, www.nasaspaceflight.com
Flight Procedures Handbook Entry, NASA JSC 11542, www.nasaspaceflight.com
Flight Procedures Handbook Approach, Landing and Rollout, NASA JSC-23266, www.nasaspaceflight.com
Entry Checklist STS-135, NASA JSC-48020-135, www.nasa.gov
STS Missions Press Kit, NASA, www.jsc.nasa.gov
Several interviews from JSC Oral History Project, NASA, www.jsc.nasa.gov
Space Shuttle: The History of the National Space Transportation System, Dennis R. Jenkins, ISBN-10 0963397451

Index

Abort, xv, 10, 14, 24, 51, 63, 105, 128, 129, 180, 194, 217, 219, 292, 323, 329, 353, 357, 358, 360, 361, 362, 363, 364, 369, 371, 374, 375, 376, 380, 381, 382, 383, 386, 388, 488
Abort Once Around (AOA), xv, 11, 361, 373, 374, 380, 381, 382
Abort To Orbit (ATO), xv, 360, 361, 371, 372, 373, 374, 375, 380, 382, 388
   IY steering, 368, 371, 372
   Launch pad, 117, 339, 340, 341, 342
   Return to Launch Site (RTLS), xix, 10, 193, 194, 317, 361, 362, 365, 366, 367, 374, 376, 378, 380, 382, 384, 385, 386, 387
    region determinator, 374, 376
   Transatlantic Abort Landing (TAL), xix, 11, 361, 367, 368, 369, 370, 371, 374, 375, 376, 378, 382, 384, 386, 387, 476, 488
Advanced Flexible Reusable Surface Insulation (AFRSI), xv, 249, 250, 252
ASCENT TRAJ display, 353, 357, 358, 359, 382
Atlantis, 114, 178, 249, 255, 257, 265, 266, 271, 272, 324, 406, 416, 417, 418
Auxiliary Power Unit (APU), 49, 145, 146, 147, 291, 292, 293, 295, 296, 297, 298, 324, 340, 376, 381, 447, 449, 451, 453

Backup Flight Software (BFS), 20, 24, 25, 26, 301, 351, 345, 356, 357, 358, 448
Barauskas, Stan, 291, 292, 295, 296, 297

Behnken, Robert, 282
Bolden, Charles, 341, 462
Brand, Vance, 397, 463
Brandenstein, Daniel, 99
Brewster, Shaw, 23, 486

Chaput, Joe, 152, 153, 154
Challenger, 58, 79, 82, 99, 110, 129, 132, 133, 134, 249, 250, 262, 341, 353, 371, 400, 412
Collins, Eileen, 100, 270
Columbia, 13, 23, 64, 74, 78, 110, 164, 172, 174, 175, 177, 198, 234, 245, 247, 249, 250, 252, 254, 256, 259, 261, 262, 263, 264, 265, 266, 267, 268, 272, 273, 279, 281, 285, 289, 290, 297, 339, 341, 355, 408, 416, 433, 464, 464, 483, 487
   Accident Investigation Board (CAIB), 174, 273, 274, 281, 285, 408
Crippen, Robert, 263, 273, 342, 344, 345, 346, 349, 355, 397

Digital Autopilot (DAP), 332, 333, 334, 336, 338, 391, 392, 393, 396, 429, 431, 432, 441, 449, 451
   aerojet, 333, 334, 370, 458
   orbital, 333, 334, 389
   transition, 333, 370
   RCS, 389, 390, 394, 395, 443
   TVC, 389, 391
Direct Insertion (DI), 355, 356, 372, 373, 380
Discovery, 11, 111, 163, 224, 249, 255, 257, 267, 268, 269, 270, 272, 279, 290, 339, 340, 341, 371, 411, 412, 413, 490

Docking, 40, 41, 44, 68, 179, 267, 321, 324, 336, 397, 403, 404, 406, 407, 408, 427, 430, 431, 432, 433, 434, 435, 438, 439, 440, 445, 551
  orbiter system/mechanism, 432, 433, 434, 437, 438, 439, 441, 443
  sequence, 438

Endeavour, 111, 112, 249, 255, 257, 283, 290, 413, 414, 416, 418
Engle, Joe, 232, 464, 465, 485,
ENTRY TRAJ display, 461, 462
External Tank (ET), xvi, 115, 132, 147, 158, 159, 160, 161, 162, 163, 164, 165, 166, 167, 169, 170, 171, 172, 173, 174, 176, 177, 178, 276, 297, 333, 354, 355, 356, 364, 365, 366, 367, 370, 371, 374, 375, 383, 384, 385, 386, 492
  SRB attachment points, 130, 167, 168
  door, 51, 54, 252, 378
  Orbiter interface, 162, 163, 169, 170, 351
  separation, 356, 358, 365, 366, 370, 382, 384, 385, 386, 387

Felt Reusable Surface Insulation (FRSI), xvi, 249
Fibrous Refractory Composite Insulation (FRCI), xvi, 253
Flight deck, 5, 25, 26, 40, 41, 42, 54, 128, 148, 225, 234, 305, 311, 312, 320, 321, 322, 323, 326, 327, 330, 339, 349, 355, 359, 382, 416, 449, 462, 479
  aft station, 41, 43, 72, 74, 75, 320, 321, 324, 327, 333, 441
  analog cockpit, 324
  cockpit, 180, 270, 272, 322, 325, 326, 339, 340, 342
  crew compartment, 41, 305
  glass cockpit, 325, 329
  lower equipment bay, 40
  middeck, 40, 225, 232
Foreman, Mike, 282, 283
Fossum, Mike, 276, 288

Gardner, Dale, 99
General Purpose Computer (GPC), xvi, 1, 3, 5, 6, 7, 8, 9, 10, 11, 16, 18, 19, 20, 21, 22, 23, 25, 26, 28, 31, 34, 51, 115, 120, 121, 123, 124, 201, 230, 303, 306, 308, 311, 314, 315, 319, 325, 326, 332, 341, 342, 448, 481, 491
Gibson, Robert, 265, 266
Goetz, Otto, 107, 109, 110, 111
Guidance, 13, 20, 23, 25, 107, 123, 218, 304, 305, 313, 318, 320, 330, 342, 344, 345, 353, 354, 356, 357, 358, 359, 362, 363, 364, 367, 368, 370, 371, 374, 376, 382, 386, 387, 388, 391, 392, 424, 426, 451, 453, 455, 456, 457, 458, 459, 460, 461, 462, 464, 465, 466, 467, 468, 469, 470, 472, 473, 474, 475, 476, 479, 480
  adaptive, 345, 346, 347, 348
  first stage, 342, 350
  second stage, 349, 352
Guidance, Navigation & Control (GNC), 1, 5, 8, 11, 19, 20, 21, 22, 23, 24, 25, 299, 300, 302, 305, 317, 332, 333, 391, 394, 395, 402, 424, 441, 480, 484

Hart, Terry, 349, 350, 463
Hartsfield, Henry, 341, 487
Hawley, Steve, 78, 224, 288, 341, 463
Hopson, George, 105, 106, 112, 113
HORIZ SIT display, 472
High temperature Reusable Surface Insulation (HRSI), xvii, 239, 241, 250, 252, 253, 259
Hubble Space Telescope (HST), x, 46, 77, 78, 79, 411, 412, 413, 414, 415, 416
Husband, Rick, 266

International Space Station (ISS), x, 74, 75, 78, 165, 221, 234, 267, 274, 282, 285, 287, 336, 374, 380, 389, 397, 406, 407, 408, 409, 416, 418, 429, 430, 431, 432, 433, 437

Kramer-White, Julie, 38, 40, 79, 80

Launch pad, 23, 42, 59, 115, 116, 117, 125, 130, 158, 164, 170, 171, 173, 204, 205, 240, 342, 496
  abort, 339, 340, 341
Liftoff, 5, 72, 99, 101, 114, 115, 116, 118, 120, 121, 123, 125, 128, 129, 142, 145, 146, 158, 162, 171, 177, 219, 234, 248, 260, 295, 340, 342, 344, 345, 350, 358, 368, 399, 489, 490

Linnehan, Richard, 283
Low temperature Reusable Surface Insulation (LRSI), xvii, 239, 240, 252, 254, 256
Lousma, Jack, 371, 483, 487

Mattingly, Ken, 486, 487
Mir, 225, 397, 403, 406, 407, 408, 410, 418, 433, 437, 440
Mitchell, Royce, 127, 130, 132, 133, 134, 135
MNVR EXEC display, 356, 391, 420, 424, 426, 427, 448, 449, 451
Moser, Thomas, 37, 38, 40, 46, 48, 49, 53, 54, 59, 246, 273, 274
Mullane, Mike, 223, 224, 225, 265, 266, 339

Odom, Jim, 158
Orbital Maneuvering System (OMS), xviii, 179, 180, 181, 182, 183, 184, 185, 189, 191, 195, 196, 199, 201, 261, 315, 316, 330, 338, 353, 355, 356, 361, 362, 363, 367, 371, 376, 378, 379, 380, 391, 392, 448, 449, 451, 452
  engine, 179, 180, 181, 184, 189, 191, 197, 198, 200, 252, 364, 367, 371, 372, 373, 380, 381, 389, 392, 449, 452, 453, 461
  fuel dump, 362, 363, 364, 367, 370, 371, 375
  gimbaling system, 180, 181
  OMS 1 burn, 355, 356, 361, 373, 381
  OMS 2 burn, 355, 356, 361, 373, 380, 381, 419
  pod, 183, 184, 188, 189, 198, 199, 224, 252, 261, 262, 264, 271, 272, 449
ORBIT TGT display, 421, 423, 424

Payload, 3, 5, 7, 8, 9, 14, 20, 29, 32, 37, 40, 41, 46, 48, 53, 67, 68, 70, 72, 73, 74, 75, 77, 90, 110, 116, 126, 129, 157, 158, 165, 179, 212, 216, 221, 228, 234, 252, 316, 320, 321, 324, 331, 336, 338, 341, 349, 352, 353, 390, 396, 404, 405, 407, 412, 445, 446, 464, 495, 496
  bay (cargo), 23, 41, 42, 44, 45, 46, 48, 67, 74, 75, 78, 182, 210, 211, 226, 232, 234, 235, 276, 282, 285, 313, 324, 403, 405, 408, 412, 416, 417, 439, 440, 446, 448, 493
  bay doors, 40, 45, 51, 53, 54, 55, 56, 57, 59, 60, 61, 72, 182, 217, 219, 262, 264

retention, 46, 47, 72
  specialist, 324, 341, 496
Powered Explicit Guidance (PEG), 352, 353, 363, 364, 389, 391, 392, 424, 426, 427, 448
Primary Avionics Software System (PASS), 7, 8, 9, 10, 18, 20, 22, 23, 24, 25, 29, 33, 105, 301, 356, 357, 358, 448

Radiator, 60, 61, 216, 217, 219, 239, 252, 446, 448
Reaction Control System (RCS), xix, 179, 180, 182, 183, 184, 186, 187, 188, 189, 191, 192, 193, 194, 195, 199, 201, 202, 204, 252, 315, 330, 336, 337, 338, 355, 356, 362, 363, 365, 367, 371, 380, 381, 389, 390, 396, 397, 400, 403, 404, 405, 406, 408, 412, 415, 416, 426, 429, 431, 443, 447, 449, 451, 452, 453, 458,
  aft, 183, 189, 193
  DAP mode, 389, 390, 394, 443
  forward, 188, 193, 195, 452
  low-Z mode, 336, 405, 406, 407, 412, 413, 414, 415, 416, 430
  plume, 404, 411, 416
  primary, 201, 203, 336, 338, 362, 429
  wraparound, 180, 391
  vernier, 203
Reference frame, 301, 302, 303, 330, 334
  inertial (INRTL), xvii, 301, 302, 303, 330
  LVLH, xvii, 302, 303, 330, 334, 335, 336, 345, 357, 392, 395, 403, 405, 424, 429, 441
  M50, 300, 301, 302, 303, 306, 314
  body axes, 75, 302, 405, 314, 330, 332, 357, 392, 393, 405, 406
Reinforced Carbon-Carbon (RCC), xix, 250, 251, 256, 262, 263, 273–9 274, 275, 276, 277, 278, 279, 284, 289
Rendezvous, 40, 41, 62, 179, 300, 308, 310, 311, 317, 321, 324, 380, 389, 391, 394, 397, 398, 399, 403, 406, 408, 411, 415–22, 424, 426, 429, 439, 440, 441, 442, 448
  ORBT, xviii, 407, 408, 416
  Pitch Maneuver (RPM), xix, 267, 408, 411, 431
  Stable orbit approach, 400, 401, 403, 407, 408, 413, 419, 451

REL NAV display, 421, 422, 428, 429, 439
Ried, Robert, 244
Robinson, Stephen, 269, 270, 272, 276
Rominger, Kent, 282
Rotational Hand Controller (RHC), xix, 74, 314, 330, 331, 332, 336, 338, 356, 359, 391, 396, 397, 449, 452, 478, 479, 486, 487
    detent position, 331, 336, 338, 391, 396, 397, 497, 485, 487

Sellers, Piers, 276, 288
SM SYS SUM 1 display, 33
SM SYS SUM 2 display, 33
SM TABLE MAINT display, 33
Solid Rocket Booster (SRB), xix, 6, 116, 120, 127–34, 136–9, 141, 143–8, 150–6, 158, 161, 163, 169, 234, 265, 266, 314, 342, 347, 348, 359, 383, 385, 491
    attachment points, 161, 167, 168
    cold, 347
    crossbeam, 161
    factory joint, 133, 136
    field joint, 133, 134, 135, 136
    hot, 347
    ignition, 130, 142–4, 147, 342
    nozzle joint, 133
    o-ring, 134, 135, 136
    separation, 5, 138, 146–8, 218, 219, 348, 349, 352, 353, 357, 361
    separation motor, 148, 149
Spacelab, 23, 44, 234, 496
Space Shuttle Main Engine (SSME), xix, 52, 81, 83, 87, 88, 89, 90, 91, 99, 104, 107, 108, 109, 110, 111, 114, 116, 120, 121, 124, 158, 191, 198, 252, 314, 350, 355, 382, 490
    Block I, 111, 112
    Block IIA, 112, 114
    Block II, 114

    combustion chamber, 93, 96
    ignition system, 97
    net positive suction pressure (NPSP), 371
    nozzle, 100, 101, 102
    Phase I, 110
    Phase II, 110, 111
    start sequencing, 118
    thermal conditioning, 115
Standard Insertion (SI), xix, 355, 356, 373, 381
State vector, 11, 12, 20, 25, 299, 300, 304, 305, 306, 314, 315, 316, 317, 318, 319, 320, 357, 394, 420, 424, 426, 439, 441, 448, 479, 480, 481, 482
Station-keeping, 40, 400, 401, 402, 405, 412, 414, 432
Strain Isolator Pad (SIP), xix, 243–9, 259, 260, 261, 262

Thermal Protection System (TPS), xix, 241, 244, 248, 252, 254–8, 263, 267, 269, 272, 273, 274, 279, 280, 281, 285, 287, 288, 430, 455, 456
Thomas, Andy, 270, 282, 283, 284
Thomas, John, 135
Trajectory lofting, 347, 353, 363
Translational Hand Controller (THC), xix, 74, 75, 314, 331, 332, 333, 336, 338, 355, 356, 391, 426, 429, 451, 452

UNIV PTG display, 334, 393, 424, 429, 430, 431, 432, 443
    Tracking modes, 395

VERT SIT display, 470, 471, 474

White, Terrence, 241, 247

Young, John, 264, 346, 486, 487

About the author

Davide Sivolella was born in Pinerolo, Italy, in 1981. As a child, he developed a fascination with all kinds of flying machines, especially those which travel above the atmosphere. This passion for astronautics led to bachelor's and master's degrees in Aerospace Engineering from the Polytechnic of Turin (Italy). He then worked as a specialist in aircraft structural repairs of civil airliners in the United Kingdom. He thinks of aircraft as spacecraft that fly low and slow. *To Orbit and Back Again: How the Space Shuttle Flew in Space* is his first book, born from a life-long passion for the Space Shuttle program. Apart from his fondness for human space exploration, and writing about it, he also enjoys cooking, traveling, and landscape photography. He currently lives in England near London with his Spanish wife Monica.

Printed by Publishers' Graphics LLC